Sensors for Chemical and Biological Applications

Sensors for Chemical and Biological Applications

Edited by
Manoj Kumar Ram
Venkat R. Bhethanabotla

CRC Press
Taylor & Francis Group
Boca Raton London New York

CRC Press is an imprint of the
Taylor & Francis Group, an **informa** business

CRC Press
Taylor & Francis Group
6000 Broken Sound Parkway NW, Suite 300
Boca Raton, FL 33487-2742

First issued in paperback 2017

ISBN 13: 978-1-138-11792-1 (pbk)
ISBN 13: 978-0-8493-3366-8 (hbk)

Library of Congress Cataloging-in-Publication Data

Sensors for chemical and biological applications / [edited by] Manoj Kumar Ram and
  Venkat R. Bhethanabotla.
     p. ; cm.
  Includes bibliographical references and index.
  ISBN 978-0-8493-3366-8 (alk. paper)
  1. Biosensors. 2. Chemical detectors. I. Ram, Manoj Kumar. II. Bhethanabotla,
Venkat R.
  [DNLM: 1. Biosensing Techniques. 2. Environmental Monitoring. 3. Gases--analysis.
4. Nanotechnology. QT 36 S4783 2010]

R857.B54S456 2010
610.28'4--dc22   2009049510

Visit the Taylor & Francis Web site at
http://www.taylorandfrancis.com

and the CRC Press Web site at
http://www.crcpress.com

Contents

Preface

A chemical sensor is a device that is used for the qualitative or quantitative determination of the analyte through a chemical reaction. Thirteen contributions from authors who have embarked on research programs in the exciting areas of chemical and biological sensors are included in this book. Considerable effort has been made to increase the surface area of the sensing interface in sensors through the nanotechnological approach, and this effort has been discussed throughout the book. Chapter 1 outlines the nanostructured metal-oxide-based gas sensors, which have revealed higher sensitivity and selectivity of several harmful environmental hazard gases. Chapter 2 shows that sensing, aging, and mechanical characteristics of the conducting polymer can be improved by fabricating nanocomposite polymer membranes, which detect the toxic exhaust gases with better accuracy. Chapter 3 outlines the gas-sensing properties of ultrathin films of metal phthalocyanines and porphyrins, and their nanocomposite membranes for organic volatile gases. Chapter 4 focuses on the theoretical and experimental considerations of the piezoelectric gas sensor.

Recent advances in the design and fabrication of chemical and biological sensors for toxicity evaluation are summarized in Chapter 5. Chapter 6 discusses the applications of electronic noses and tongues in areas such as food, beverage, environmental, clinical, and pharmaceutical applications. Chapter 7 overviews the applications of sensors in food and environmental analysis. Chapter 8 focuses on the medical diagnosis, with particular emphasis on in-vivo measurement where either body or breath odor are collected and analyzed. Chapter 9 outlines the DNA biosensors that hold great promise for the task of environmental control and monitoring.

Detection of trace amounts of explosives, chemical warfare, and bioweapons in environmental samples is currently a labor-intensive laboratory technique requiring expensive, sophisticated instrumentation. Chapter 10 concentrates on sensors for explosives, chemical warfare agents, and fertilizers. Chapter 11 outlines the detection of chemical warfare agents, a potential source of serious environmental problems due to deliberate use, accidents, or improper disposal. Various chemical analytes are accomplished through optical detection technique that is discussed in Chapter 12. Chapter 13 outlines the comparison of the existing approaches toward individual and multispecific detection of biowarfare agents where a wide range of microorganisms, of both bacterial and viral origin, as well as purified protein toxins can be turned into bioweapons. In addition, Appendix A shows the few successful commercial companies and research institutions who are actively involved in the areas of sensor and biosensor applications.

We are fortunate to have assembled contributions from world-class authorities in this field, and sincerely thank all of them. In their enthusiasm for the field of chemical and biological sensors they have produced this book, which we believe, will be of unusual help to the increasing number of researchers in this field. We are indebted to Dr. Hanming Ding, Associate Professor, East China Normal University, Shanghai, China, for his careful attention in reviewing the chapters, his support, and his help in the organization of the book. I, Dr. Manoj K. Ram, would also like to take this opportunity to thank Dr. Hulya Demiryont, Chief Scientist; Mr. Tom Westfall, Technical Manager; Mr. David Moorehead, Senior Engineer; and Mr. Jay Wolfington, President, Eclipse Energy Systems, Inc., for their valuable suggestions throughout the preparation of this book. Last but not the least, I warmly acknowledge the gracious support of my wife Kumari R. Tarway and my children Natasha and Akash.

Contributors

Matt Aldissi
Fractal Systems Inc.
Safety Harbor, Florida

Graziana Bagni
Department of Chemistry
University of Florence
Via della Lastruccia
Sesto Fiorentino (Firenze), Italy

Andrew Baxter
Xenosense
Belfast, Northern Ireland

Venkat R. Bhethanabotla
Sensors Research Laboratory
Department of Chemical
Engineering
University of South Florida
Tampa, Florida

A. Bogomolova
Fractal Systems Inc.
Safety Harbor, Florida

Joanne Brennan
Applied Biochemistry Group
School of Biotechnology
Cambridge Antibody Technology
Cambridge, U.K.

Wenqing Cao
C-ACS
Los Alamos National Laboratory
Los Alamos, New Mexico

Arnaldo D'Amico
Department of Electronic Engineering
University of Rome "Tor Vergata"
Via del Politecnico
Roma, Italy

Alfredo Darmaninsheehan
Applied Biochemistry Group
School of Biotechnology
Cambridge Antibody Technology
Cambridge, U.K.

Anil K. Deisingh
Caribbean Industrial
Research Institute
University of the West Indies
St. Augustine
Trinidad and Tobago, West Indies

Corrado Di Natale
Department of Electronic Engineering
University of Rome "Tor Vergata"
Via del Politecnico
Roma, Italy

Hanming Ding
Department of Chemistry
East China Normal University
North Zhongshan Road
Shanghai, People's Republic of China

David Faguy
Department of Chemical
and Nuclear Engineering
University of New Mexico
Albuquerque, New Mexico

W. J. J. Finlay
Applied Biochemistry Group
School of Biotechnology
Biomedical Diagnostics Institute
National Centre for Sensor Research
Dublin City University
Dublin, Ireland

H. James Harmon
Physics Department
Oklahoma State University
Stillwater, Oklahoma

Stephen Hearty
Applied Biochemistry Group
School of Biotechnology
Biomedical Diagnostics Institute
National Centre for Sensor Research
Dublin City University
Dublin, Ireland

Ying Hu
Sensors and Actuators Lab
Microelectronics Center
School of EEE
Nanyang Technological University
Nanyang Avenue, Singapore

Dmitri M. Ivnitski
Department of Chemical
and Nuclear Engineering
University of New Mexico
Albuquerque, New Mexico

Brandy J. Johnson
Center for Bio/Molecular
Science and Engineering
Naval Research Laboratory
Washington, D.C.

Claire Jones
Xenosense
Belfast, Northern Ireland

Babu Joseph
Sensors Research Laboratory
Department of Chemical Engineering
University of South Florida
Tampa, Florida

Richard O'Kennedy
Applied Biochemistry Group
School of Biotechnology
Biomedical Diagnostics Institute
National Centre for Sensor Research
Dublin City University
Dublin, Ireland

Paul Leonard
Applied Biochemistry Group
School of Biotechnology
Biomedical Diagnostics Institute
National Centre for Sensor Research
Dublin City University
Dublin, Ireland

Eugenio Martinelli
Department of Electronic Engineering
University of Rome "Tor Vergata"
Via del Politecnico
Roma, Italy

Marco Mascini
Department of Chemistry
University of Florence
Via della Lastruccia
Sesto Fiorentino (Firenze), Italy

Amanda Oliver
Department of Math,
Science, and Engineering
Northern Oklahoma College
Stillwater, Oklahoma

Roberto Paolesse
Department of Chemical Science and
Technology
University of Rome "Tor Vergata"
Via della Ricerca Scientifica
Roma, Italy

Giorgio Pennazza
Faculty of Engineering
University "Campus Bio-Medico
di Roma"
Via Alvaro del Portillo, Italy

Manoj K. Ram
Eclipse Energy Systems Inc.
St. Petersburg, Florida

Claudio Roscioni
Azienda Ospedaliera S. Camillo-
Forlanini
Via Portuense
Roma, Italy

**Subramanian K. R. S.
Sankaranarayanan**
Sensors Research Laboratory
Department of Chemical
Engineering
University of South Florida
Tampa, Florida

Marco Santonico
Department of Electronic Engineering
University of Rome "Tor Vergata"
Via del Politecnico
Roma, Italy

Ihab Seoudi
Department of Chemical
and Nuclear Engineering
University of New Mexico
Albuquerque, New Mexico

Ravil A. Sitdikov
Department of Chemical
and Nuclear Engineering
University of New Mexico
Albuquerque, New Mexico

Sharon Stapleton
Applied Biochemistry Group
School of Biotechnology
Cambridge Antibody Technology
Cambridge, U.K.

Ooi Kiang Tan
Sensors and Actuators Lab
Microelectronics Center
School of EEE
Nanyang Technological University
Nanyang Avenue, Singapore

Ibtisam E. Tothill
Analytical Biochemistry
Cranfield Health
Cranfield University
Bedfordshire, U.K.

Susan Townsend
Applied Biochemistry Group
School of Biotechnology
Cambridge Antibody Technology
Cambridge, U.K.

Ebtisam S. Wilkins
Department of Chemical
and Nuclear Engineering
and Department of Biology
University of New Mexico
Albuquerque, New Mexico

Ozlem Yavuz
Fractal Systems Inc.
Safety Harbor, Florida

Sagar Yelleti
Department of Chemical
and Nuclear Engineering
and Department of Biology
University of New Mexico
Albuquerque, New Mexico

Weiguang Zhu
Sensors and Actuators Lab
Microelectronics Center
School of EEE
Nanyang Technological University
Nanyang Avenue, Singapore

1 Solid-State Gas Sensors

Ying Hu, Ooi Kiang Tan, Weiguang Zhu
Sensors and Actuators Lab, Microelectronics
Center, School of EEE
Nanyang Technological University
Nanyang Avenue, Singapore

Wenqing Cao
C-ACS, Los Alamos National Laboratory
Los Alamos, New Mexico

CONTENTS

1.1 INTRODUCTION

With the drastic growth in industrial development and population, the natural atmospheric environment has become polluted and is rapidly deteriorating. To prevent environmental disasters, before it is too late, it is imperative that such pollutants be monitored and controlled. A large variety of gases in the atmospheric environment, such as NO_X, CO_X, SO_2, H_2S, O_2, H_2, CH_4, and so on, need to be detected.

Conventional analytic instruments using optical spectroscopy, gas chromatography, and mass spectrometry are time consuming, expensive, and seldom used in real time in the real field [1]. The solid-state gas sensors, which directly utilize semiconductor/gas interactions, have the advantage of compactness, robustness, versatility, and are inexpensive to integrate as micro-array to monitor the various gas compositions in the environment. Since 1962, thin-film ZnO and porous SnO_2 ceramics were first demonstrated as gas-sensing devices. The solid-state gas sensors have gone through many developmental phases over the past 40 years [2]. Up to now, there are two main developmental trends: one is to develop portable instruments of hybrid-array gas sensors for a variety of gases; another is to study the new principles and manufacturing techniques of single sensor to explore different detection ranges of the different gases, including their application areas. Hybrid techniques are a combination of thick film, thin film, and integrated circuit technology and are advantagieous in terms of production scale, size of device, and encapsulation [3]. Whether it is hybrid-array gas sensors or single sensor, the need for higher performance with low-power, small-size, and relatively low-cost application in environments with numerous interferents and variable ambient conditions is the ultimate objective.

Besides, surface acoustic wave devices and optical gas sensors, the solid-state gas sensors including silicon-based chemical sensors, semiconducting metal oxide sensors, catalysis, solid electrolyte sensors, and membranes based on electrical parameters are widely studied [4]. Silicon-based chemical sensors include silicon, germanium, and compound semiconductor devices (e.g., GaAs) fabricated by planar technology for the field-effect transistor (FET). Semiconducting metal oxide sensors change electronic characteristics as a result of an interaction occurring at the solid-gas interface. Catalysis sensors are semiconducting oxide sensors responding to a temperature change and occur on the surface of the material itself. In solid electrolytes, the conductivity stems from mobile ions rather than electrons. Membranes, which are used as sensor elements themselves or as filters, are a key component in many types of sensing devices. Even with so many types that exist in solid-state gas sensor family, only the solid electrolyte, semiconducting metal oxide, and capacitor-type (one of FET) gas sensors are normally utilized as shown in Table 1.1. This chapter systematically illustrates these three representative sensors with different structures, chemical modifications, and sensing mechanisms for different gases in the atmospheric environment. Special attention is paid to the semiconducting metal oxide type over the others owing to its advantages of low cost, small size, simple structure, ease of integration, and no reference electrode [5].

The new concept of nanostructure as applied to the semiconducting metal oxide gas sensors does not only improve the sensing properties in terms of sensitivity and response time but also effectively reduces the operating temperature and improves integrated circuit density for hybrid-array gas sensors. It has also given rise to the accelerated development in semiconducting metal oxide gas sensors with low power, small size, and relatively low cost. The different synthesis techniques for the nanosized semiconducting metal oxide materials and varied gases sensors (NO_x, CO_2, CO, SO_2, O_2, etc.) in atmospheric environment are reviewed in this chapter. The extension of the operating temperature to the near-human temperature regimes and better sensing properties derived from the nanostructured $SrTiO_3$ oxygen gas sensors have expanded the applications to medical, environmental, and domestic fields that are not achievable with the conventional coarse materials.

TABLE 1.1
Solid-State Gas Sensors for Detecting Environmental Gases

Detected Gas	Types of Sensors	Sensing Materials
SO_2	Semiconducting type	Ceramic SnO_2 [7], CeO_2, doped-WO_3 [8–10]
	Solid-electrolyte type	Alkali metal sulfates [11,12], Ag-β''-alumina [13], Na-β-alumina [14], NASICON ($Na_2Zr_2Si_3PO_{12}$) [15], MgO-stabilized [16,17], sulfate-based solid electrolytes [15]
CO_2	Semiconducting type	Doped-SnO_2 and In_2O_3 [18–22]
	Solid-electrolyte type	K_2CO_3, NASICON/Na_2CO_3, NASICON/Na_2CO_3-$BaCO_3$, NASICON/Li_2CO_3-$CaCO_3$, NASICON/$NdCoO_3$ [23–28]
	Capacitor type	$BaTiO_3$-PbO [29], NiO-$BaTiO_3$, PbO-$BaTiO_3$, Y_2O_3-$BaTiO_3$, CuO-$BaTiO_3$, like CeO-$BaCO_3$/CuO [30–34]
CO	Semiconducting type	Doped-In_2O_3, SnO_2, and ZnO [35–38]
NO_2	Semiconducting type	Doped (Pd, Au, Cd)-SnO_2 [39,40], In_2O_3 [41], doped (Pd, Pt, Au)-WO_3 [42], TiO_2-WO_3 [43–45], SiO_2-WO_3 [46], Sn-W-O system [47,48], Al_2O_3-V_2O_5, NiO-CuO [49,50]
	Solid-electrolyte type	$Ba(NO_3)_2$ [23], Na-β/β''-alumina/$NaNO_3$ [51], Na-β/β''-alumina/$Ba(NO_3)_2$ [52], NASICON/$NaNO_2$+Li_2CO_3 [53], Y_2O_3-ZrO_2/$CdCr_2O_4$ [54], NASICON/pyrochlore oxide [55]
	Capacitor type	NiO/ZnO [56]
NO	Semiconducting type	In_2O_3-SnO_2 [57], WO_3 [58,59], Bi_2O_3-WO_3 [60], doped (Pd, Pt, Au)-WO_3 [42]
	Solid-electrolyte type	Y_2O_3-ZrO_2/$CdCr_2O_4$ [54], NASICON/$NaNO_2$ [61]
	Capacitor type	$SrSnO_3$-WO_3 [62]
O_3	Semiconducting type	SiO_2-CeO_2-In_2O_3 [63], Fe_2O_3-In_2O_3 [64], WO_3 [65], $Zn_2In_2O_5$-$MgIn_2O_4$ [66]

1.2 THE PRINCIPLES OF SOLID-STATE GAS SENSORS

Gas sensors for detecting the atmospheric environment must be able to operate stably under deleterious conditions, including chemical and/or thermal attack. Therefore, solid-state gas sensors would appear to be the most appropriate in terms of their practical robustness. Yamazoe describes a gas sensor as a device that basically performs two functions: the receptor and the transducer functions [6]. The interaction of gas with a solid metal oxide surface can be presented as a succession of two phenomena. The first is the exchange of molecular species at the solid-gas interface, such as adsorption, chemical, and electrochemical reactions, which gives the receptor function of recognizing a particular gas species. This is followed by a motion and redistribution of ions and electrons in the solid, which manifests as the transducer function of transforming the gas recognition into a sensing signal.

The sensing signal could be changed in the electrical resistance or capacitance of the sensor elements. To achieve gas sensors with good sensing characteristics, both the receptor and transducer functions must be sufficiently addressed. For example, to increase the gas sensitivity, it is important to promote the transducer function, whereas to increase the gas selectivity to a particular gas, the receptor function is especially important. Each type of solid-state gas sensors based on electrical parameters must satisfy certain criteria. Semiconducting metal oxide and capacitor-type gas sensors, while responding readily to modulations in gas composition, should be insensitive to temperature. Sensors operating on electrochemical principles cannot tolerate electronic conduction but must exhibit exceptionally high ionic conduction. It is often insufficient to predict sensor response from only its bulk electrical characteristics. Hence, the basic approaches for detecting the composition of gases with electrical ceramics include the following: some rely on the atmospheric dependence of electrical conductivity and dielectric constant exhibited by some semiconducting metal oxide and capacitor-type materials, respectively; others utilize ionic conductors as solid electrolyte membranes in electrochemical concentration cells. The following sections will introduce the principal characteristics of solid-electrolyte, capacitor-type, semiconducting metal oxide, and nanostructured gas sensors.

1.2.1 THE MAIN TYPES OF SOLID-STATE GAS SENSORS

1.2.1.1 Solid Electrolyte Gas Sensors

Since the doped ZrO_2 solid electrolyte (an oxygen ion conductor) oxygen gas sensors were successfully applied in the exhaust system of almost all modern automobiles in the mid-1990s, the other gases CO, NO_x, and short-chain hydrocarbons have been paid attention to closely [67]. Solid electrolyte gas sensors based on potentiometric have various configurations. Three types have been classified by Weppner, depending on whether the ionic species derived from the gas in question coincides with the mobile ion (Type I), the immobile ion (Type II), or neither of them (Type III) of the solid electrolyte used [68].

In Type III, there is an auxiliary phase attached on the surface of the solid electrolyte so as to be sensitive to the gas, and it is produced by a compound that contains the same ionic species as derived from the gas. The auxiliary phase can act as a sort of poor ion-conducting solid electrolyte, which forms a half cell of Type I or II as shown in Figure 1.1 [1]. Type III sensors can be divided into three subgroups depending on the types of the half cells combined [6]. Since a NASICON solid electrolyte potentiometric gas sensor using alkali metal carbonate as an auxiliary phase solid electrolyte is known to be sensitive to CO_2, Type III sensors have been of immense importance as sensors for oxygenic gases such as CO_2, NO_X, and SO_X [69]. A NASICON SO_2 sensor can be expressed as "Pt, SO_2, O_2/Na_2SO_4-$BaSO_4$/ NASICON/Na_2SiO_3, Pt" [15], where Na_2SiO_3 is the reference electrode in air and the binary composite of Na_2SO_4, and $BaSO_4$ is the auxiliary phase. The electromotive force of these solid electrolyte sensors can be expressed by a Nernstian equation as follows:

$$E = E_0 + (RT/2F) \ln PSO_x \tag{1.1}$$

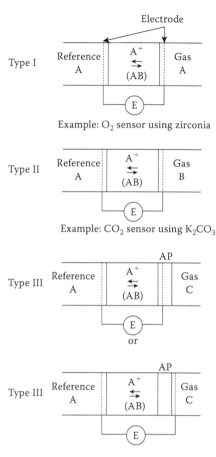

Example: O_2 sensor using zirconia

Example: CO_2 sensor using K_2CO_3

FIGURE 1.1 Three types of solid electrolyte gas sensors (AP: Auxiliary phase, Type III for CO_2 sensor using NASICON/Na_2CO_3). (Reprinted from *IEEE Sensors J.*, 1, Lee D.-D et al., Environmental gas sensors, 2001, with permission from Elsevier.)

where E_0 is constant, F is Faraday's constant, R is gas constant, and T is temperature.

1.2.1.2 Capacitor-Type Gas Sensors

The capacitor-type gas sensors are one type of the field-effect devices. The field-effect devices can be classified into two types: metal-oxide-semiconductor (MOS) capacitors and transistors (MOSFETs), as shown in Figures 1.2 and 1.3, respectively [70]. Since the first hydrogen sensor based on a Pd-gate MOS capacitor with Si substrate was reported by Lundström et al. [71,72], solid-state gas sensors based on these two structures have been intensely investigated [73].

In general, the semiconductor is silicon and the oxide is silicon dioxide. The potential on the gate, which is the metal, can be varied to control the charge distribution at the oxide-semiconductor interface. The catalytic gate MOSFET device differs from an ordinary MOSFET in that the gate is metalized with palladium, Pd,

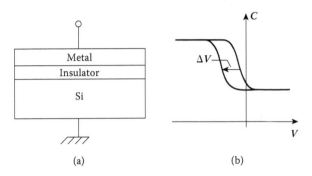

FIGURE 1.2 A Pd-gate MOS capacitor-type (a) structure (b) *C-V* curve shift.

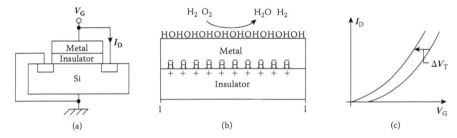

FIGURE 1.3 A Pd-gate MOSFET-type (a) Structure (b) Detection principle (c) I_D–V_G curve shift.

instead of polysilicon. The thin layer of Pd is highly permeable to hydrogen. When the device is exposed to hydrogen gas, the drain current versus gate voltage curve shifts toward lower voltage as shown in Figure 1.3. Hydrogen gas molecules are catalytically adsorbed on the surface of the Pd by dissociating into hydrogen atoms. These atoms diffuse rapidly through the Pd and get absorbed at the Pd-insulator interface, where they become polarized. The resulting interface dipole layer is in equilibrium with the outer layer of chemisorbed hydrogen. The dipole layer gives rise to an abrupt potential step through the structure, which is normally called the voltage drop, ΔV_T. The changes in the electrical properties of these devices upon gaseous absorption are reflected in I_D-V_G curve, shifts in flat-band voltage or changes in the threshold voltage, and the drain current of the devices. Utilizing selective gate structures or the combination of a catalytic metal gate and gas-sensitive oxide layer, the device can detect nonhydrogen-containing gases; MOS devices with a perforated metal gate or metal-oxide gate were used for the detection of carbon monoxide at high temperature (160°C) [74]. On the other hand, MOS capacitor-type gas sensors are often used for exploratory work since they are much easier to fabricate than MOS transistors [75–77]. The basic mechanism behind the gas response is identical to that of the Pd gate MOSFETs. However, the dipole layer is detected as a shift in the *C-V*

$$V_{FB} = V_{FBO} - \Delta V \tag{1.2}$$

where V_{FBO} denotes the flat-band voltage with no hydrogen in the ambient atmosphere. Their measurement can be directly in terms of the change of the dielectric constant of the films between the electrodes expressed as a function of gas concentrations. The capacitance changes for the sensors are typically in the range of pF and very dependent on the operating frequency and surrounding conditions, such as humidity and temperature. For example, an oxygen-sensitive Pd-SnO$_2$ capacitor-type sensor has been developed for the detection of O$_2$, CO, and H$_2$ at a relatively low temperature [78,79]. A spin-coated polyphenylacetylene conducting polymer film can respond to the various gases, such as CH$_4$, N$_2$, CO, and CO$_2$ [80]. CuO-BaTiO$_3$, AMO/PTMS, and CeO$_2$/BaCO$_3$/CuO capacitor-type gas sensors can detect CO$_2$ [30,81,82].

1.2.1.3 Semiconducting Metal Oxide Gas Sensors

Owing to the advantages of low cost, small size, simple structure, ease of integration, and no reference electrode for the semiconducting metal oxide gas sensors, special attention is paid to this type of gas sensors all over the world [5]. In 1962, thin-film ZnO [2] and porous SnO$_2$ [83] ceramics were first demonstrated to show that the semiconducting metal oxides are sensitive to the gas composition of a surrounding atmosphere, especially at elevated temperatures. After that, doping small amounts of noble metals (such as Pt and Pd) to these oxides were shown to be effective in improving their sensing properties [84]. In 1968, leakage monitors for town gas and liquefied petroleum gas (LPG) were available in the market [85]. Hence, the following three trends were developed in the semiconducting metal oxide gas sensors: First, to extensively study and understand their sensing properties; second, to research the key factors affecting the sensing properties of the sensors, such as the effects of grain size and the geometry of grain connection, the promoting of noble metals, and the effects of sensor configuration, and so on; third, to expand the practical applications of semiconducting metal oxide gas sensors [86–90].

1.2.1.3.1 Basic Mechanism

Semiconducting metal oxide sensors detect gases via the variations in their resistances (or conductances). The most widely accepted explanation for this mechanism is that negatively charged oxygen adsorbates play an important role in detecting inflammable gases such as H$_2$, CO, and certain toxic gases in the air. The variation in the surface coverage of the adsorbed oxygen dominates the sensor resistance. Actually, among the several kinds of oxygen adsorbate species O$_2^-$, O$^-$, and O^{2-}, the O$^-$ adsorbed oxygen in the surface of semiconducting metal oxides between 300°C and 500°C is found to be predominant [91]. The detail process for the n-type semiconducting metal oxides is that the adsorbed oxygen forms a spaces-charge region on the surfaces of the metal-oxide grains; it results in an electron transfer from the grain surfaces to the adsorbates as follows:

$$1/2O_2^{(g)} + 2e^- = 2O_{ads}^- \tag{1.3}$$

The depth of this space-charge layer (L) is a function of the surface coverage of oxygen adsorbates and intrinsic electron concentration in the bulk. Before the sensor is exposed to reducing gases, the resistance of an n-type semiconducting metal oxide

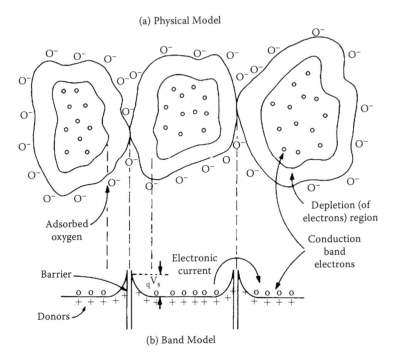

(a) Physical Model

(b) Band Model

FIGURE 1.4 Model for adsorbate-dominated n-type semiconducting powder in gas sensing: (a) Physical Model: three contacting powder grains are shown to illustrate how depletion region dominates the intergranular contact and (b) Band Model: corresponding to (a) indicating the Schottky barrier at the intergranular contact. (Reprinted from *Sens. Actuators*, 12, Morrison, S. R., Selectivity in Semiconductor gas sensors, 1987, with permission from Elsevier.)

gas sensor in air is high, due to the development of a potential barrier to electronic conduction at each grain boundary, as shown in Figure 1.4 [4]. When the sensor is exposed to an atmosphere containing reducing gases at elevated temperatures, the oxygen adsorbates are consumed by the subsequent reactions, so that a lower steady-state surface coverage of the adsorbates is established. During this process, the electrons trapped by the oxygen adsorbates are returned to the oxide grains, leading to a decrease in the potential barrier height and a drop in resistance.

This resistance change is normally used as the measurement parameter of a semiconductor gas sensor. The relative resistance (S) is defined as the ratio of the resistance in air to that in a sample gas containing a reducing component. The reactivity of the oxygen adsorbates is, of course, a function of both the type of reducing gas present and the sensor operating temperature. Therefore, the temperature T_M at which the maximum relative resistance S_M is observed is dependent upon the particular reducing gas present. Since semiconducting metal oxide gas sensors respond more or less to any reducing gas via this mechanism, the sensors usually suffer from cross-sensitivity, that is, the lack of selectivity to a specific gas.

Occasionally, the reducing gases interact directly with the sensor materials, for example, hydrogen is chemisorbed negatively on SnO_2 under certain conditions

[92,93]. Such phenomena are usually observed at lower temperatures. The reaction with the oxygen adsorbates and the electrical interactions with intermediate products of the reducing gases lead to more complicated sensor responses [94,95]. The response of n-type semiconducting metal oxide gas sensors to oxidizing gases such as O_3 or NO_2 is relatively simple. The sensor resistance increases upon exposure to these gases as a result of their negatively charged chemisorption on the grain surface. Therefore, the sensitivity is a function of the amount of chemisorption, provided that the surface coverage of oxygen adsorbates remains constant.

Since the charge carriers in p-type semiconducting metal oxides are positive holes, the resistance in air is low because of the formation of negatively charged oxygen adsorbates, and the extraction of electrons from the bulk eventually enhances the concentration of holes in the grain surface. The consumption of oxygen adsorbates by reaction with reducing gases leads to an increase in resistance, which is the reverse for the case of n-type metal oxides. Conversely, the adsorption of oxidizing gases on p-type metal oxides results in a decrease in the resistance.

In general, the measurement of a higher resistance in the presence of a sample gas than that in air is accompanied by some problems associated with signal handling (the higher the resistance, the worse the signal-to-noise ratio). This explains why the n-type semiconducting metal oxides are usually employed as sensor materials for reducing gas detection: because of their lower resistance in reducing gas.

The relationship between conductivity (σ) and gas partial pressure is as follows [96]:

$$\sigma \propto P_{gas}{}^{\beta} \tag{1.4}$$

where P_{gas} is the partial pressure of the gas and β is generally in the range 0.5–1.0 depending on the mechanism.

1.2.1.3.2 The Types and the State-of-the-Art

Since 1960s, the most common conventional types of semiconducting metal oxide sensors are the "Figaro sensors" designed by N. Taguchi (Figure 1.5a) and the planar format sensors, thick- or thin-film sensors (Figure 1.5b). The fabrication of the thick-film devices comprises (i) processing of the sensing materials powders, (ii) preparation of the paste, and (iii) fabrication of the thick-film device using the screen-printing technology. The sensing layer is typically 20–30 μm thick [97]. On the other hand, the thin-film sensors can be prepared by sputtering evaporation with subsequent oxidation, chemical vapor deposition, or spray pyrolysis. In the spray pyrolysis technique to provide thin film of SnO_2, one sprays, essentially using an atomizer, a metal organic form of tin or tin chloride dissolved in a suitable solvent. The solution is sprayed onto a heated substrate, and the tin compound reacts with oxygen or water vapor to form the oxide. It is one way to avoid vacuum systems. With vacuum deposition techniques a heater element is usually deposited first, followed by an insulator such as silica. The third layer is sensing material tin oxide.

A new slide-off transfer printing method of fabricating multilayer sensors is shown in Figure 1.6. Typically, the oxide films were separated from the mount paper by soaking them in water before they were transferred on the alumina substrates. After that, they were printed with interdigitated Pt electrodes (the gap between electrodes was

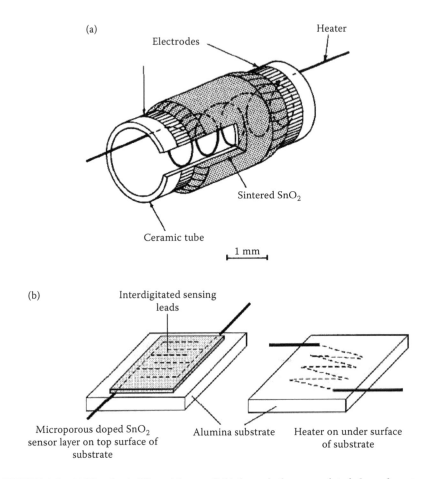

FIGURE 1.5   (a) The classic "Taguchi sensor;" (b) the typical screen-printed planar format.

200 μm), followed by drying at 80°C for 30 minutes. This process was repeated by employing different oxide films in fabricating heterolayer sensors. The heterolayers AO/BO indicate the upper and lower oxide layer, respectively. Before the sensors were tested for the sensing properties, they were fired at 800°C for 2 h in air and then preheated at 700°C for 48 h to stabilize their gas-sensing characteristics. A scanning electron microscope (SEM, Hitachi, S-2250N) shows typical cross-sectional views of the heterolayer sensors in Figure 1.7 [98]. In M-SnO$_2$/ TiO$_2$ (M: noble metals) and SiO$_2$/WO$_3$ sensors fabricated by the slide-off transfer printing method, M-SnO$_2$ and SiO$_2$ act as filter for O$_2$ and NO$_2$, respectively. Hence, sensitivity of H$_2$ and NO for the sensors was enhanced in mixed gases of H$_2$ and NO$_X$ (NO$_2$, NO) [98].

Another new structure type, carbon nanotubes (CNTs), was discovered in 1991 [99]. Carbon nanotubes have high mechanical and chemical stability due to their unique structure and properties and can be used as modules in nanotechnology. Two types of carbon nanotubes, single-walled nanotubes (SWNT) and multiwalled nanotubes (MWNTs), were developed. Single-walled nanotubes (SWNT) consist of

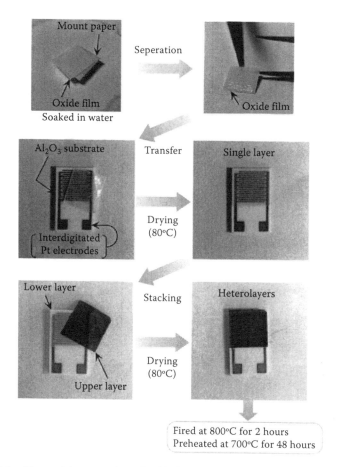

FIGURE 1.6 The stacking procedure of oxide films by slide-off transfer printing. (Reprinted from *Sens. Actuators B*, 77, Hyodo T. et al., Gas sensing properties of semi conductor heterolayer sensors fabricated by slide-off transfer printing, 2001, with permission from Elsevier.)

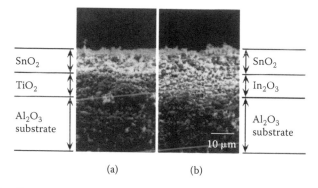

FIGURE 1.7 Cross-sectional SEM view of (a) SnO_2/TiO_2 and (b) SnO_2/In_2O_3 heterolayer sensors. (Reprinted from *Sens. Actuators B*, 77, Hyodo T. et al., Gas sensing properties of semi conductor heterolayer sensors fabricated by slide-off transfer printing, 2001, with permission from Elsevier.)

a honeycomb network of carbon atoms and can be imagined as a cylinder rolled from a graphitic sheet. Multiwalled nanotubes (MWNTs) are a coaxial assembly of graphitic cylinders generally separated by the plane space of graphite [100]. Carbon nanotubes can be used as gas sensors through controlling mechanical or chemical processes (e.g., nanotube bending or gas molecule adsorption) [101–105]. The electrical conductivities of an individual single-walled CNT or MWCNT ropes changed dramatically upon exposure to gaseous molecules, such as NO_2, NH_3, or water vapor [106,107]. Figure 1.8a showed a scheme of the CNTs NO_2 sensor layout for NO_2 detection with limits as low as 10 ppb [108]. It was prepared by a radio frequency plasma enhanced chemical vapour deposition (r.f. PECVD) on Si/Si_3N_4 substrates. The thin film (5 nm) of Ni catalyst was deposited onto Si_3N_4/Si substrates provided with platinum interdigital electrodes and a back-deposited thin-film platinum heater

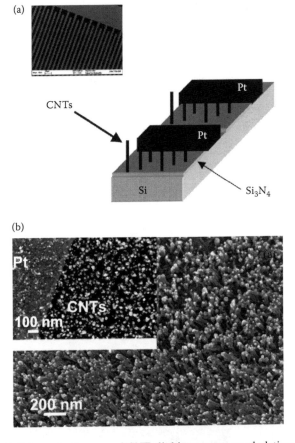

FIGURE 1.8 (a) Schematic diagram of CNTs linking prepatterned platinum contacts in a resistor geometry (b) SEM images of the as-grown structure of CNTs on Si/Si_3N_4; the inset shows the top view with the Pt electrode region highlighted and the as-grown structure of CNTs on a Si/Si_3N_4 substrate. (Reprinted from *Diamond. Relat. Mater.*, 13, Valentini L., Cantalini C., Armentano I., Kenny J. M., Lozzi L., and Santucci S., Highly sensitive and selective sensors based on carbon nanotubes thin films for molecular detection, 1301–1305, 2004, with permission from Elsevier.)

commonly used in gas-sensor applications. The scanning electron microscopy (LEO 1530) image is shown in Figure 1.8b.

1.2.1.3.3 Factors Affecting Sensor Performance

The important parameters of semiconducting metal oxide sensors include sensitivity, selectivity, stability, size, response time, reversibility, reliability, and dynamic range. Out of these, sensitivity and selectivity for the sensors are the most important. Sensitivity is the measure of a sensor ability to detect the presence of a gas in its environment and is often presented as the "detection limit." Various levels of sensitivity, from a few percent down to parts-per-billion (ppb) concentrations, may be required. Selectivity is the sensor ability to measure only one particular gas without responding to, or experiencing interference from, other chemical compounds present. An ideal sensor would have high "sensitivity" and be "selective." That is, it would show a large resistance change for a small change in the target gas concentration and have the ability to discriminate between different gases. It would also need to show a reproducible response over the required lifetime and be economically viable.

Besides the chemical factors, the grain size, the microporosity, and the film thickness and geometry can strongly affect the sensitivity of the sensors. The grain size of the materials affects surface area and conducting type of the materials. When the grain sizes of the materials decrease down to the nanosized range (or the actual grain size being smaller than two times the space-charge depth), the material will be dominantly of surface conductance and the sensitivity of the sensors is rapidly enhanced. The mechanism of surface conductance type is discussed in the next section. Physical characteristics such as the microporosity, the film thickness, and geometry also affect sensor response, especially if the sensing mechanism depends on diffusion or operating temperature. The response time is the minimum duration required for the gas to interact with the sensor and create a signal. It is defined that when an input signal goes through a well-defined change, the output response is defined by the time interval between 10% and 90% of the stationary value [109].

For instance, the described slide-off transfer printing method for fabricating multilayer sensors shows effective improvement in the sensitivity of the sensors from physical characteristics over those of thin or thick films. For thin- and thick-film sensors, diffusivity of a target gas is important for sensing performance and can be controlled by thickness and porosity of the sensor materials [110]. However, the thin-film sensors usually suffer from poor long-term stability during operation at elevated temperatures, whereas thick-film sensors by the screen-printing technology show excellent durability but are not adequate compared to the stacking technique for the multilayers on a substrate that can control both the reactivity and the diffusivity more precisely. Hence, the slide-off transfer printing method can overcome the defects of thin or thick film in the state-of-the-art in improving the sensitivity of the sensors. As mentioned earlier, $M-SnO_2/TiO_2$ (M: noble metals) and SiO_2/WO_3 sensors fabricated by the slide-off transfer printing method resulted in the enhancement of the sensitivity of H_2 and NO [98].

Apart from these, the detailed physical form of the sensor can also have an important influence on selectivity. The target gas has to diffuse through the microporous sensor material toward the sensing electrodes. The diffusion rate will depend on the mean free path of the gas molecules in relation to the diameter of the channels in the solid.

By controlling the graded microporosity specific gases can be selected and other gases that diffuse through the solid sensing material can be rejected. A graded microporosity, readily accomplished via screen printing, is a route to influencing selectivity. The electrode spacing is another important design parameter that can influence selectivity. The effective thickness of the sensing material overlying the electrodes will depend on the electrode spacing. By varying the electrode spacing the path of the gas molecules to the sensing electrodes can be regulated and hence specific gases can be selected.

Improving selectivity of the sensors usually can also be done through doping dopants or by adding filters such as Fe_2O_3-doped SnO_2 to sensitize for NO_2 or V_2O_5 and Pd-doped SnO_2 to sensitize for CO [111].

1.3 NANOSTRUCTURED GAS SENSORS

Usually, a material can be defined or considered as nanomaterial if one of its linear dimensions is less than 100 nm [112]. The new concept of nanostructure as applied to semiconducting metal oxide gas sensors has given rise to the accelerated development in surface conductance type sensors with low power, small size, and relatively low cost. In terms of the material grain sizes, semiconducting metal oxide gas sensors can be divided into bulk conductance (transducer function) and surface conductance (receptor function). When the grain sizes of the materials decrease down to the nanosized range (or the actual grain size is smaller than two times the space-charge depth, as in Figure 1.9), the material will be surface conductance type and their carriers diffuse on the grain surface with minimum energy [113]. Hence the nanostructured materials can function exceptionally well at very low temperature. At the same time, the surface-to-bulk ratio of nanostructured materials are much greater than that for coarse materials. This will result in more surface sites available at the solid-gas interface for the transduction reactions. Hence, it not only improves the sensitivity and response time but also effectively reduces the operating temperature

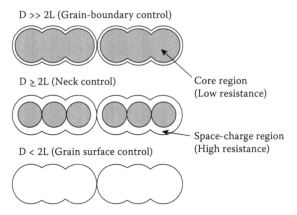

FIGURE 1.9 Schematic models for grain-size effects. D—the actual grain size, L—the space-charge depth.

and improves integrated circuit density for hybrid-array gas sensors. A typical example is shown in Figure 1.10 for In_2O_3 thin films prepared by sol-gel method. It can detect low levels (several hundreds ppb) of nitrogen dioxide in air due to the smaller nanosized material (down to 5 nm). Higher response and higher sensitivity for NO_2 in air are also the result of the decrease in the grain sizes in In_2O_3 [114].

Another example, by Y. Hu and O. K. Tan et al., shows encouraging results for nanostructured $SrTiO_3$ oxygen gas sensors with operating temperature down to near human-body temperature [115]. This is much lower than that of the conventional semiconducting metal oxide oxygen gas sensors (300°C–500°C) [5,116–118] and $SrTiO_3$ oxygen gas sensors (>700°C) [119]. The reduction of the grain sizes of the semiconducting metal oxide materials and synthesis of new materials are some of the possible techniques adopted to decrease the operating temperature.

As early as 1981, in fact, Ogawa et al. were using gas-phase evaporation of tin in an oxygen atmosphere to produce nanoscale SnO_2 crystals with an average size of 6 nm [120]. Up to now, nanosized powders have been generally prepared using various methods, including (a) chemical coprecipitation [121], (b) sol-gel process [122], (c) metalorganic deposition (MOD) [123], (d) plasma enhanced chemical vapor deposition (PECVD) [124], (e) atmospheric-pressure chemical vapor deposition (APCVD) [125], (f) physical vapor deposition (PVD) [126,127], (g) low-pressure flame deposition (LPFD) [128], (h) laser ablation [129], and (i) high-energy ball milling or mechanical alloying [130–141].

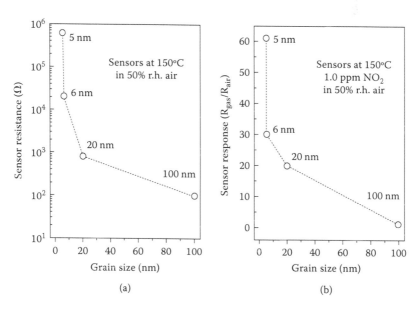

(a)

(b)

FIGURE 1.10 In_2O_3 thin-film sensor resistances (a) and response to 1.0 ppm NO_2 (b) in air with 50% relative humidity versus In_2O_3 grain size. (Reprinted from *Sens. Actuators B*, 44, Gurlo A., Ivanovskaya M., Bârsanb N., Schweizer-Berberich M., Weimar U., Göpel W., and Diéguez A., Grain size control in nanocrystalline In_2O_3 semiconductor gas sensors, 327–333, 1997, with permission from Elsevier.)

Methods (a) through (e) are basically chemical processing techniques aimed at obtaining homogeneous structure on an extremely fine scale of a few nanometers at the molecular level. The others are physical processing techniques. In these processing techniques, the high-energy ball milling has the advantages of high product yield, simple technology, low cost, mass production, and the ability to synthesize high melting metal, alloy, or semiconducting metal oxides. Table 1.2 shows a large variety of gases in the atmospheric environment, such as NO_x, CO_2, CO, SO_2, H_2S, C_2H_5OH, O_2, H_2, CH_4, and so on, that were detected by nanostructured semiconducting metal oxide gas sensors through various synthesis and production routes.

TABLE 1.2
Nanosized Semiconducting Oxides Used as Gas Sensors

Synthesis Route	Sensing Materials	Detected Gases
High-energy ball milling	ZrO_2–α-Fe_2O_3, SnO_2– α-Fe_2O_3, TiO_2–α-Fe_2O_3, $SrTiO_3$	C_2H_5OH, O_2 [115,130,131, 134–137,142–154]
Hydrothermal synthesis	SnO_2	CO, H_2 [155,156]
Chemical coprecipitation	WO_3-TiO_2	NO_2 [44]
	SnO_2-PdO	CO [157]
	In_2O_3, MoO_3-In_2O_3	NO_2, O_3 [158]
	ZrO_2–α-Fe_2O_3	CO, CH_4, H_2 [159]
	WO_3	NO_2 [160]
	α-Fe_2O_3	CO, SO_2, C_3H_8, CH_4, LPG [121]
Sol-gel	SnO_2	H_2, NO_2 [161,162]
	$La_{0.8}Sr_{0.2}FeO_3$	CO, NO_2 [156]
	SnO_2, MoOx–SnO_2	NO_2, CO [163,164]
	In_2O_3–NiO	CO [165]
	In_2O_3, In_2O_3-NiO, and In_2O_3-MoO_3	O_3, O_5, NO_2 [114,166]
	MoO_3–TiO_2	O_2, CO, NO_2 [167]
	TiO_2, Nb_2O_5–TiO_2	CO [155], O_2 [168,169]
	MoO_3, MoO_3–WO_3	O_2 [170]
	In_2O_3, In_2O_3–Pt, In_2O_3–Au	CH_4, CO, C_2H_5OH, NH_3 [171]
	In_2O_3	O_3 [172], NO_2 [173]
	In_2O_3–NiO_2	NO_2, CO [174,175]
	MoO_3–TiO_2	O_2 [176]
Sputtering deposition	ZnO	O_3 [177]
	Ti-W-O, Ti-O-Mo, Mo-W-O	CO, NO_2 [43,127,178–180]
	SnO_2, ZnO, In_2O_3, TiO_2, WO_3	NO_2, C_2H_5OH [181–186]
	TiO_2, Fe_2O_3	CO, NO_2 [187,188]
Aerosol pyrolysis	CeO_2	O_2 [189]
	SnO_2, SnO_2–PdO, SnO_2–CuO	CO [190]
	TiO_2	CO, NO_2 [178,191–193]
Laser ablation	SnO_2, In_2O_3, WO_3	O_3, NO_2 [194–196]
	SnO_2	CO, H_2, CH_4 [129,196]

In the next section, a variety of solid state environment gases sensors (NO_x, CO_2, CO, SO_2, O_2, etc.) are reviewed, and attention is also paid to semiconducting metal oxide type. Also discussed are the extension of the operating temperature to the near-human temperature regimes and better sensing properties derived from the nanostructured semiconducting metal oxide gas sensors.

1.4   SOLID-STATE ENVIRONMENTAL GAS SENSORS

Gases formed as a result of combustion processes have become a public concern. The main products in a combustion process of oil, coal, and peat are carbon oxides, nitrogen oxides, sulphur trioxide, hydrocarbons, and so on. They are dangerous to human life and health. The primary emissions are CO, NO, and unburnt hydrocarbons. Upon interaction with atmospheric oxygen (O_2) and sunlight, secondary pollutants, such as NO_2 and O_3, are generated. The lower exposure limits (LEL) of such gas pollutants in air are, for example, 35–50 ppm for CO, 3 ppm for NO_2, and 25 ppm for NO as stated in the international regulations for environmental pollution in air [197]. Carbon monoxide (CO), a result of fuel combustion in air (for example in engines), can poison blood by taking the place of oxygen molecules. This gas is very toxic and even lethal in high quantity. Sulfur dioxide (SO_2), created by the combustion of fossil combustibles containing sulfur and by industrial processes, causes lung irritation and, in the presence of humidity, forms acid rains, which are noxious for constructions and vegetation. Nitrogen oxides (NO_X: NO and NO_2) come from car exhausts and combustion processes, and in recent years the concentration of these gases in atmosphere has increased with increase in traffic and the number of buildings. Nitrogen oxides cause lung irritations, decrease the fixation of oxygen molecules on red blood corpuscles, contribute to acid rains, and generate the increase of ozone rate in the low atmosphere. Nitrogen monoxide is unstable and quickly forms NO_2, which is an oxidizing gas [198]. Although oxygen gas (O_2) is not a pollution gas in the environment, the oxygen content monitoring is of upmost importance for human beings in specific ambience, such as in submarine, space capsule, mine, and incubator. Specifically, low-temperature oxygen gas sensors are needed in medical, environmental, and domestic fields.

1.4.1   SENSOR ARRAY FOR A VARIETY OF GASES

Sensor array systems for environment can be designed for single- or multicomponent sensing. The techniques can employ silicon integrated circuits (ICs) or hybrid technique. Hybrid techniques are a combination of thick film, thin film, and integrated circuit technology [3]. The basic hybrid circuits are fabricated by the thick-film processing to produce complex patterns of insulating, resistive, conducting, and capacitive layers all on an electrically insulating or resisting substrate, usually alumina. Hybrid technology has the advantages in production scale, size of device, and encapsulation. These systems generally use component-specific sensors of various types, each responding to a single gas, to capture and store the row vector of responses. The row vector of responses can be expressed as the product of a matrix of regression coefficients and a column vector of concentrations. The equations can be solved using parametric methods or neural network techniques [199–201]. Figure 1.11 shows the

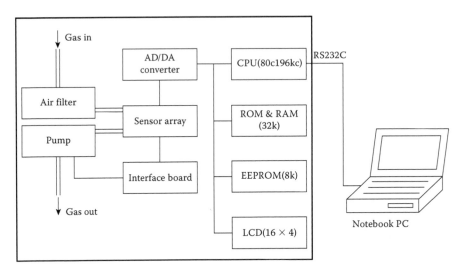

FIGURE 1.11 Schematic diagram of the portable electronic nose system. (Reprinted from *Sens. Actuators B*, 66, Hong H.-K., Kwon C. H., Kim S.-R., Yun D. H., Lee K., and Sung Y. K., Portable electronic nose system with gas sensor array and artificial neural network, 49–52, 2000, with permission from Elsevier.)

schematic diagram of a portable electronic nose system [202]. This system was fabricated and characterized using a semiconducting metal oxide gas sensor array and back-propagation artificial neural network. The sensor array consists of six thick-film gas sensors. The portable electronic nose system consists of an Intel 80c196kc, an EEPROM containing optimized connection weights of artificial neural network, and an LCD for displaying gas concentrations. As an application, the system can identify 26 carbon monoxide/hydrocarbon (CO/HC) car exhausting gases in the concentration range of CO 0%/HC 0 ppm to CO 7.6%/HC 400 ppm [202].

In the sensor array systems, the key factors are increasing the gas sensitivity (that is related to the magnitude of its response to a variation of the gas concentration) and decreasing the cross-sensitivity. Decreasing the cross-sensitivity is required for improving the gas selectivity, which is the ability to recognize one particular gas in the presence of other gases [203]. Unfortunately, most sensing materials used in semiconducting metal oxide gas sensors are sensitive to many gases [204,205].

In the single-component sensing, for example, one way to improve the selectivity and sensitivity of semiconducting metal oxide gas sensors is to use a multisensor array based on sintered and thick tin oxide films and to analyze the whole response using pattern recognition methods, such as artificial neural network models (ANN). It not only detects the individual components of the gas mixture (NO_2 and CO) but also measures the concentration of both gases with sufficient accuracy [206]. On the other hand, in multicomponent sensing, selectivity of semiconducting metal oxide gas sensors can be improved by several ways: modification of sensing material, for example, by doping with catalytic and electroactive admixtures [49,207,208]; optimization of working conditions to detect only a target gas; and also use of filtering membranes [209–217]. One reported sensor array consists of 16 discrete sensing elements formed by tin oxide thin layers deposited by sputtering [218]. The Pt doped is suitable for CO detection, Al doped is chosen for NO detection [219–221], Pd doped is selected for chlorinated- and amino-VOC, and SO_2 annealing is used to improve the response to both aromatic-VOC and SO_2 [7,219,222,223]. In another report, the nanocrystalline $LaFeO_3$ thick films with Au electrodes [198] were sensitive to CO at 250°C–270°C, CH_4 at 420°C–450°C, and NO_2 at 350°C at different concentrations in air. Different temperatures can be used to render the films sensitive to different gases. This gives a possibility to use an array of films for a selective detection in the case of a gas mixture. The use of filtering membranes to separate interfering gas molecules in terms of the different molecules size with different diffusion rate is also reported [211]. It makes selective interactions with specific gas molecules [212,213]. The membranes can increase stability of semiconducting metal oxide gas sensors by filtering corrosive and irreversibly adsorbing gases. In some cases they can increase sensitivity of sensor element to a target gas [165,215]. For example, in the filtering membranes with pure and Pt or Ru doped Al_2O_3 deposited on the surface of SnO_2 (Pd) films by aerosol pyrolysis method, all membranes reduce significantly the sensitivity to CO and increase the sensitivity to CH_4 at 200°C, while Ru-doped membranes significantly reduces the sensitivity to H_2 [224]. In other cases, the filtering organic membranes made by plasma polymers are also used [225]. The array sensors with thermally isolated structures based on silicon integrated circuits (ICs) are achieved as shown in Figure 1.12 [49]. Tin oxide sensitive layers have been deposited by reactive sputtering technique owing to its compatibility with Si-based IC technology. The active area has a size of 500 x 500 μm^2 and is supported by a membrane of silicon nitride. Polysilicon is used as heating material, and the power consumption is below 50 mW at the operating temperature of 350°C for every sensor prepared. Very low concentrations around 1 ppm of NO_2 have been detected with a 3-minute response time.

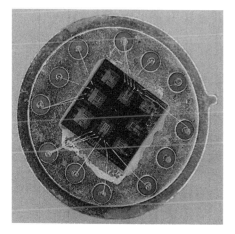

FIGURE 1.12 Chip assembled on standard TO-8 package. (Reprinted from *Sens. Actuators B*, 58, Horrillo M.C., Sayago I., Arés L., Rodrigo J., Gutiérrez J., Götz A., Gràcia I., Fonseca L., Cané C., and Lora-Tamayo E., Detection of low NO_2 concentrations with low power micromachined tin oxide gas sensors, 325–329, 1999, with permission from Elsevier.)

1.4.2 SO$_X$ Gas Sensors

In the global atmosphere, SO$_X$ (SO$_2$ and SO$_3$) gas is a major source of acid rain. The major industrial sources of SO$_2$ emissions are coal-fired power plants, oil and gas productions, and nonferrous smelting. Currently, chemical analyses (West-Gaeke coulometric technique and hydrogen peroxide method) and instrumental analyses (flame photometric detection and UV fluorescence technique) are used to determine SO$_2$ gas in stack gases. However, these methods are very complicated [1]. In order to seek more suitable methods for the continuous monitoring of SO$_2$ gases, various solid electrolytes have been developed, such as alkali metal sulfates [11,12], Ag-β″-alumina [13], Na-β-alumina [14], NASICON (Na$_2$Zr$_2$Si$_3$PO$_{12}$) [15], MgO-stabilized [16,17] and sulfate-based solid electrolytes (Type III sensors, combining NASICON or another Na$^+$ conductor (solid electrolyte) and Na$_2$SO$_4$ (auxiliary phase)) [15]. These sensors have shown a high sensitivity and linearity to SO$_2$ gas concentrations and have reached the levels for practical applications [226]. On the other hand, the semiconducting metal oxide SO$_2$ sensors are also used for their advantages stated earlier in this chapter [5].

Since the electronic interactions between SO$_2$ and SnO$_2$ were first studied by Lalauze et al. [7] in 1984, CeO$_2$ [8], SnO$_2$ [9,10], and doped-WO$_3$ [227] sensors for SO$_2$ have been developed. SO$_2$ sensing properties of several semiconducting metal oxides revealed complex temperature- and time-dependent surface states of SO$_2$-related adsorbates and the oxides. For example, the higher sensing property of 1.0 wt.% Ag-doped WO$_3$ sensor for SO$_2$ at 450°C is shown in Figure 1.13 [227]. The addition of Ag led to changes in both the surface states of SO$_2$-related adsorbates

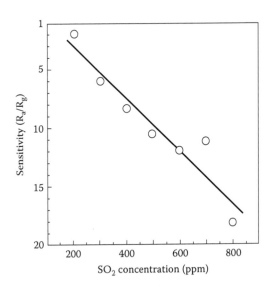

FIGURE 1.13 Correlation between sensitivity (R_a/R_g, R_a, R_g: the dc sensor resistance in a sample gas and air respectively) of (1.0 wt.%) Ag doped-WO$_3$ and SO$_2$ concentration at 450°C. (Reprinted from *Sens. Actuators B: Chem.*, 77, Shimizu Y., Matsunaga N., Hyodo T., and Egashira M., Improvement of SO$_2$ sensing properties of WO$_3$ by noble metal loading, 35–40, 2001, with permission from Elsevier.)

and the electronic interactions between the adsorbates and WO_3. The resistance of 1.0 wt.% Ag-doped WO_3 increases at 450°C with the formation of SO_4^{2-} as a result of reaction of a gaseous SO_2 molecule and two O_{ad}^{2-} ions on the Ag. This mechanism is different from pure WO_3 sensors for SO_2, where the resistance of WO_3 sensors increases in SO_2 ambient at 400°C owing to the formation of SO_2^- at sites different from those for oxygen adsorbates.

1.4.3 CO₂ GAS SENSORS

CO_2 is the main cause of the greenhouse effect, yet, by itself, it is harmless. CO_2 sensing is necessary for the autoventilation of air in living rooms and automobiles as well as for measuring or controlling bio-related activities [228–230].

Since the first solid-state CO_2 sensor based on electrochemical principles (Type II) was reported in 1977 by Gauthier and Chamberland [23], the researches on solid electrolyte sensor have been rather active, such as K_2CO_3, NASICON/Na_2CO_3, NASICON/Na_2CO_3-$BaCO_3$, NASICON/Li_2CO_3-$CaCO_3$, NASICON/$NdCoO_3$, and so on [23–28]. CO_2 is determined by the voltage difference between CO_2 gas and the alkaline carbonate coated on working electrode. However, the structure of the sensor is complicated, and the alkaline carbonate is deliquescent and strongly affected by water vapor, and the requirements of the electrode preparation are very strict. Hence, the other types of solid-state CO_2 gas sensors are needed.

While it was thought that the semiconducting metal oxide sensors are difficult to test CO_2 gas due to the stable chemical property of CO_2 in nature, the semiconducting metal oxide SnO_2 and In_2O_3 gas sensors for CO_2 have been fabricated [18–22]. The transition metal oxides, such as La_2O_3, Nd_2O_3, and SrO_2, were used as catalyst on it to improve the sensing properties. Among these sensors, La_2O_3-added Sn_2O_3 sensor showed the most superior sensitivity to CO_2 gas. The La_2O_3-added SnO_2 sensor was investigated using various methods such as powder mixing, soaking, impregnating, coating, and so on [18,22]. The coating method was proved to show better sensitivity to CO_2 gas than any other methods.

Another complex oxide compound of $BaTiO_3$ and PbO was developed for the capacitive-type sensor for CO_2 in 1990 [29]. From then on, more mixtures of $BaTiO_3$ and metal oxide for CO_2 gas have been studied, such as NiO-$BaTiO_3$, PbO-$BaTiO_3$, Y_2O_3-$BaTiO_3$, CuO-$BaTiO_3$, like CeO-$BaCO_3$/CuO, and so on [30–34]. These sensors exhibit a high sensitivity to CO_2 plus selectivity within a concentration range of 100–100,000 ppm. Figure 1.14 shows the sensitivity changes of the La_2O_3-doped SnO_2 thick-film sensor with increasing CO_2 concentration contained in an artificial air (80% N_2 + 20% O_2) [19]. The CO_2 gas sensor was fabricated using the La (0.01 M)-coated SnO_2 film heat treated at 1000°C for 5 minutes in the air. The sensitivity increased almost linearly with the increase of CO_2 concentration, and the detected range is between 0 and 2500 ppm.

1.4.4 CO GAS SENSORS

Carbon monoxide is one of the most common and dangerous pollutants present in the environment due to emissions from automated vehicles, aircraft, natural gas

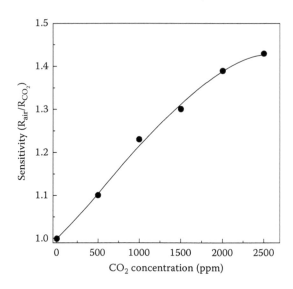

FIGURE 1.14 CO_2 sensitivity as a function of CO_2 concentration for La_2O_3-doped SnO_2 sensor. (Reprinted from *Sens. Actuators B*, 62, Kim D. H., Yoon J. Y., Park H. C., and Kim K. H., CO_2 sensing characteristics of SnO_2 thick film by coating lanthanum oxide, 61–66, 2000, with permission from Elsevier.)

emission, industrial wastage, sewage leaking, mines, and so on. Its poisonous effects on human life are well known. Among several kinds of chemical sensors, a compact CO gas sensor is required for monitoring carbon monoxide generated from automobile exhaust and the natural gas due to incomplete combustion [231]. Semiconducting metal oxide CO sensors are researched worldwide because of their advantages of simple structure, low cost, and time-saving properties over other techniques such as gas chromatography, chemical analysis, and infrared absorption, and so on.

For the semiconducting metal oxide CO sensors, improving their sensitivity and selectivity to CO among the various coexistent gases such as H_2, hydrocarbon, and water vapor are key factors [232]. In_2O_3, SnO_2, and ZnO were found to exhibit rather high sensitivities to CO at 300°C and above. Noble metals or Bi_2O_3 doped SnO_2 and Ti^{4+} doped α-Fe_2O_3 with Au effectively improve the CO sensitivity [35–38]. In_2O_3 appeared to be the most attractive in terms of the selectivity to CO over H_2. Specifically, alkali metal carbonates, rare-earth metal oxide, and some transitional metal doped-In_2O_3 are very effective in enhancing the sensitivity and the selectivity to CO [233]. For Rb-In_2O_3 and Co-In_2O_3 sensors, Co (0.5 wt.%)-doped In_2O_3 sensor by sputtering improved the sensitivity to CO at a lower temperature of 200°C compared to that of Rb-doped In_2O_3 to CO at 300°C; the detected limit was to 250 ppm [233].

The basic mechanism of the semiconducting metal oxide CO sensors relies on the conductivity changes experienced by the n-type semiconducting metal oxide material when surface chemisorbed oxygen reacts with reducing gases such as carbon monoxide (CO) or methane (CH_4) at elevated temperatures. The overall reactions that occur on the SnO_2 surface can be written simply, and in the case of CO as follows:

$$2e^- + O_2 \rightarrow 2O^- \tag{1.5}$$

$$O^- + CO \rightarrow CO_2 + e^- \tag{1.6}$$

where e^- represents a conduction band electron. In the absence of a reducing gas, electrons are removed from the semiconductor conduction band via the reduction of molecular oxygen, leading to a buildup of O^- species; consequently, the SnO_2 becomes very resistive. When CO is introduced, it undergoes oxidation to CO_2 by surface oxygen species and subsequently electrons are reintroduced into the conduction band, leading to a decrease in the resistance. For the FET type, the combination of a catalytic metal (Pd, Pt, or Ag) and adsorptive oxide (SnO_2 or ZnO) as gas-sensitive layers in a CAIS structure (catalytic gate-adsorptive oxide-insulator-semiconductor) gives a practical sensor for the detection of oxidizing (O_2) and reducing (CO, H_2) gases. The device can operate with a good performance in a low-temperature range (50°C–100°C) [73].

1.4.5 NO$_X$ GAS SENSORS

NO_X (NO and NO_2) gas is known to be very harmful to humans and is one of the main causes of acid rain. They are typical noxious gases released from combustion facilities and automobiles. The lower exposure limits (LEL) have been provided in the previous section. Solid-state NO_X sensors are widely demanded for monitoring NO_X in the environmental atmosphere as well as in combustion exhausts. Particularly, NO_X monitoring is indispensable for the feedback control of combustion systems or de-NO_X systems. The NO and NO_X have quite different properties from each other.

Various types of solid-state NO_2 sensors have been proposed based on semiconducting metal oxides (including heterocontact materials) [42–50,58,59,234–238], solid electrolytes [1,239,240], metal phthalocyanine [241], and SAW devices [242]. Among these NO_2 sensors, the semiconducting metal oxides and solid electrolytes appear to be the best. Specifically, semiconducting metal oxide gas sensors are most attractive because they are compact, sensitive, of low cost, and have low-power consumption. Their basic mechanism is that the NO_2 gas is adsorbed on the surface of the material; this decreases the free electron density into the space-charge layer and results in a resistance increase [243].

The dopants (Pd, Pt, Au) have a promotional effect on the speed of response to NO_2 for WO_3 thin films at low temperature [42]. The selectivity is also enhanced with respect to other reducing gases (e.g., CO, CH_4, H_2, etc.). Mixed oxides [43–50] have emerged as promising candidates for NO_2 gas detection due to the optimal combination of the various sensing properties of their pure components, such as TiO_2-WO_3 [43–45] by coprecipitation and precipitation of SiO_2-WO_3 [46] by spin-coating method, as well as Sn-W-O system [47,48], Al_2O_3-V_2O_5, and NiO-CuO [49,50]. Figure 1.15 shows that the sensitivity of nanocrystalline TiO_2-WO_3 sensors for NO_2

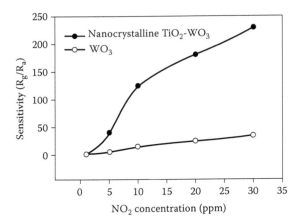

FIGURE 1.15 Sensitivity of TiO$_2$–WO$_3$ and WO$_3$ gas sensors as a function of NO$_2$ gas concentration. (Reprinted from *IEEE Sensors J.*, 1, Lee D.-D. and Lee D.-S., Environmental gas sensors, 214–224, 2001, with permission from Elsevier.)

is much better than that of normal WO$_3$ sensor, and that NO$_2$ can be detected above 1 ppm in air at an operating temperature range of 250°C–350°C [1]. The grain size and heterocontact structure rapidly improve the sensitive property of nanocrystalline TiO$_2$-WO$_3$ sensors for NO$_2$.

WO$_3$-based sensors mixed with different metal oxides (SnO$_2$, TiO$_2$, and In$_2$O$_3$) and doped with noble metals (Au, Pd, and Pt) show high sensitivity for NO$_2$ [244]. Figure 1.16 shows that WO$_3$-/SnO$_2$-Au as the sensing material from ball milling for NO$_2$ has good sensitivity and selectivity properties operating at 300°C (response and recovery times less than 2 minutes) [244]. Especially, the nanosized tellurium thin-film NO$_2$ sensor obtained by vacuum thermal evaporation exhibits high sensitivity to nitrogen dioxide at room temperature. The resistance of the tellurium films decreases reversibly in the presence of NO$_2$. The sensitivity of this device depends on the gas concentration. It improves for lower concentrations less than 3 ppm. The response time is considerably short and in the range of 2–3 minutes [245].

For NO$_x$ (NO and NO$_2$) gas, the test process becomes complicated. Some semi-conducting metal oxides used in solid-state gas sensor devices, such as SnO$_2$, WO$_3$, or In$_2$O$_3$, are far more sensitive to NO$_2$ than to NO, even when exposed to relatively low concentrations of NO$_2$. Acceptable sensor responses to NO are also obtained with these oxides. However, in NO/NO$_2$ mixtures, or in oxygen-rich atmospheres, where a NO to NO$_2$ conversion exists, the small percentages of NO$_2$ can result in similar or larger sensor signals than the remaining NO component, complicating the detection process [246,247]. In combustion exhaust control, the quantitative detection of monoxide can be considered more important than that of NO$_2$ due to NO being the primary created and having a larger concentration [83,234]. Therefore, a highly selective material for NO detection is needed. The Bi$_2$O$_3$ sensors were obtained by precipitation as a highly selective oxide for NO detection in presence of the NO$_2$ as an interfering species between 200°C and 500°C [248]. There are also Nb-doped α-Fe$_2$O$_3$ sensors for 0–100 ppm NO$_x$ (NO and NO$_2$) operating at

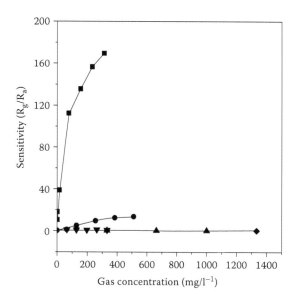

FIGURE 1.16 Effect of sensor cross-sensitivity on the sensing characteristics of sensor using WO_3-SnO_2-Au as sensing material. (■) NO_2, (●) C_2H_5OH, (▲) CH_4, (▼) CO, (♦) C_4H_{10}. (Reprinted from *Talanta*, 59, Su P.-G., Wu R.-J., and Nieh F.-P., Detection of nitrogen dioxide using mixed tungsten oxide-based thick film semiconductor sensor, 667–672, 2003, with permission from Elsevier.)

150°C–300°C [249] and $BaWO_4$-$BaCO_3$ (2:1 in molar ratio) for the range of 0–400 ppm NO and 0–200 ppm NO_2 at 600°C [250]. Solid-electrolyte NO_2 sensors, using Na^+ as conductor (solid electrolyte) and $NaNO_3$ as auxiliary phase, can detect limit down to less than 0.2 ppm [1]. Electrochemical planar sensors based on yttria-stabilized zirconia (YSZ) with semiconducting oxides (WO_3 and $LaFeO_3$) and mixed conductors ($La_{0.8}Sr_{0.2}FeO_3$) as sensing electrodes detect different concentrations of NO_2 and CO in air in the range 20–1000 ppm operating at 450°C–700°C [251].

1.4.6 O_2 Gas Sensors

Detecting the oxygen content in the environment is very important because of the severe effect that the oxygen level has on human life, as shown in Table 1.3. Hence, near-room-temperature and moderate-temperature oxygen sensors attract a lot of attention owing to their wide prospective application in medical, environmental, and domestic fields [252–254], especially in life support: area monitoring, hospital oxygen, breathing gases, physiological testing, and so forth. In the market, the oxygen sensors available in this temperature range are the electrochemical cells [255], but they have the shortcoming of high cost, periodic maintenance, and a liquid electrolyte. The semiconducting metal oxide materials have always received the most attention owing to the advantages of low cost, small size, simple structure, and ease of integration [5]. They do not require any reference electrode. However, they still pose the problems of relative high operating temperature and have no long-term stability.

TABLE 1.3
The Effect of Oxygen Level on Human Life

Oxygen Level (%)	Effect
21–19.5	Normal atmospheric oxygen level
19.5–16	Danger; the minimum safe level is 19.5%
16–12	Disturbed respiration; fatigue or exertion
11–10	Increased respiration; disturbed heart coordination; headaches
10–6	Nausea and vomiting; possible loss of consciousness, collapse
<6	Gasping respiration, cardiac arrest; death

These not only limit their applications but also reduce their reliability. One possible way to enhance the sensitivity and decrease the operating temperature is by decreasing the grain size of the materials [113]. When the grain size is small enough, that is, smaller than two times the space-charge depth, the semiconducting metal oxide materials behave like the surface conducting type (Figure 1.9 [113]) and the carriers move along the surface; hence they are able to operate at low temperature. On the other hand, for the semiconducting metal oxide material that is the bulk conducting type with grain-boundary control, the carriers need enough energy to overcome the Schottky barriers, and hence need to operate at high temperature (>700°C). For current surface conducting of semiconducting metal oxide oxygen gas sensors (such as SnO_2, ZnO, TiO_2, CeO_2, Nb_2O_5, WO_3, Ga_2O_3, and x-Fe_2O_3-(1-x)ZrO_2 etc. [5,149]), the operating temperature is between 300°C to 500°C.

Strontium Titanate ($SrTiO_3$) is a very important semiconducting metal oxide material for oxygen gas sensors owing to the advantages of low cost and the stability of the perovskite structure in thermal (up to 1200°C) and chemical atmospheres [119]. Most research work in $SrTiO_3$ focus on bulk conduction type [119,256–269], or grain-boundary control, with high operating temperatures (700°C–1000°C) [5] due to the conventional high temperature solid-state reaction synthesis method [263,270–273]. Recently, through high-energy ball milling and screen-printing techniques, nano-sized $SrTiO_3$ oxygen sensors operating independently of the relative humidity and operating at near human-body temperatures were obtained [115]. Their operating temperatures are much lower than that of the conventional surface conduction type of semiconducting metal oxide oxygen sensors above 300°C and conventional $SrTiO_3$ oxygen sensors above 700°C [5]. This is significant for application of semiconducting metal oxide oxygen in medical, environmental, and domestic fields.

Figures 1.17 and 1.18 show that the nanosized synthesized $SrTiO_3$ material (about 30 nm) was obtained through high-energy ball milling technique [115,150]. Figure 1.19 shows that the optimal relative resistance ($R_{nitrogen}/R_{20\% \, oxygen}$) value of 6.35 is obtained for the synthesized $SrTiO_3$ sample annealed at 400°C and operating at 40°C [134]. It exhibits good repeatability, as the relative change of resistance value is consistent in each cycle, as shown in Figure 1.20. The response time is 1.6 minutes and the recovery time is 5 minutes for the sensors [115].

1.5 SUMMARY AND FUTURE TRENDS

With the increase in needs to monitor a variety of gases in our environment, the solid-state gas sensors based on electrical parameters are widely studied, including silicon-based chemical sensors, semiconducting metal oxide sensors, catalysis,

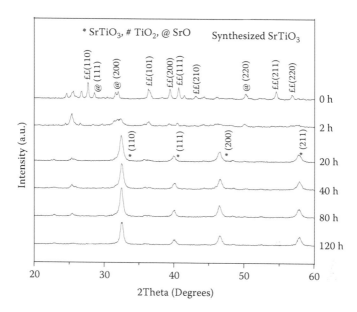

FIGURE 1.17 XRD patterns for synthesized $SrTiO_3$ after different milling times. (Reprinted *Chem. B*, 108, from Hu Y., Tan O. K., Pan J. S., and Yao X., A new form of nano-sized $SrTiO_3$ material for near human body temperature oxygen sensing applications, *J. Phys.* 11214–11218, 2004 with permission from Elsevier.)

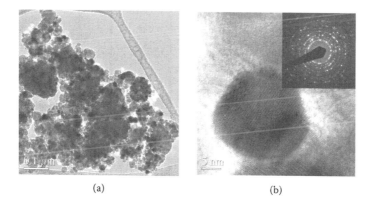

(a)   (b)

FIGURE 1.18 TEM bright-field images and SAD pattern for synthesized $SrTiO_3$ (120 h milling) (a) and (b). (Reprinted from *Sensors J.*, 5, Hu Y., Tan O. K., Cao W., and Zhu W., Fabrication & characterization of nano-sized $SrTiO_3$-based oxygen sensor for near-room temperature operation, *IEEE* 825–832, 2005 with permission from Elsevier.)

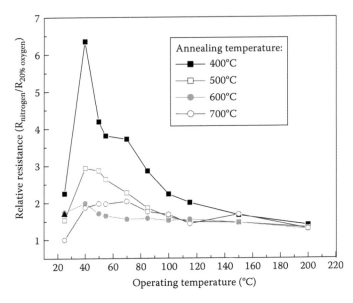

FIGURE 1.19 Relative resistance of synthesized nanosized SrTiO₃-based sample for different annealing temperatures. (Reprinted from *Sens. Actuators B*, 108, Hu Y., Tan O. K., Pan J. S., Huang H., and Cao W., The effects of annealing temperature on the sensing properties of low temperature nano-sized SrTiO₃ oxygen gas sensor, 244–249, 2005, with permission from Elsevier.)

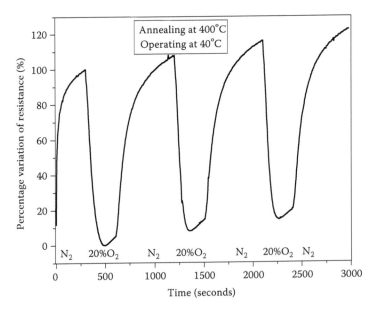

FIGURE 1.20 Response times of synthesized nanosized SrTiO₃-based sample with annealing at 400°C and operating at 40°C for several cycles. (Reprinted from *Sens. Actuators B*, 108, Hu Y., Tan O. K., Pan J. S., Huang H., and Cao W., The effects of annealing temperature on the sensing properties of low temperature nano-sized SrTiO₃ oxygen gas sensor, 244–249, 2005, with permission from Elsevier.)

solid electrolyte sensors, and membranes [4]. Even with so many types of solid-state gas sensors, only the solid electrolyte, semiconducting metal oxide, and capacitor-type (one of EFT) gas sensors are normally utilized. This chapter has systematically illustrated these three types of representative sensors from principles, structures, and state-of-the-art. Special attention is paid to the semiconducting metal oxide type as well as new state-of-the-art owing to its advantages of low cost, small size, simple structure, ease of integration, and no reference electrode [5]. The effect factors of the sensitivity and selectivity for the semiconducting metal oxide type have been analyzed from physical characteristics (such as the grain size, the microporous, the thickness, and geometry) and chemical composition (dopants).

The new concept of nanostructure as applied to semiconducting-metal-oxide gas sensors and the different synthesis techniques for the nanostructure semiconducting-metal-oxide materials (including single or sensor array) for varied gases sensors (NO_X, CO_2, CO, SO_2, O_2, etc.) in atmospheric environment are also reviewed in this chapter. Specifically, the nanostructured synthesized $SrTiO_3$ oxygen gas sensor shows better sensing properties and extends the operating temperature from above 300°C for the conventional metal oxide semiconducting oxygen gas sensors to the near-human temperature regimes. This is significant for the extension of the applications in medical, environmental, and domestic fields that are not achievable with the conventional coarse materials.

At present, there have been some developments in improving the sensitivity and selectivity for solid-state environment gas sensors. These include the use of organic filtering membrane coverings on solid-state gas sensors [274–277], using photo-activation to test the gas-sensing property [278,279], as well as the combination structure of conventional metal-oxide-semiconductor field-effect transistors with organic film [280,281]. An array of charge-flow transistors (CFT) incorporating polyaniline films [280] were reported that can detect NO and SO_2 down to 1 ppm. The electrical dc characteristics were similar to those of conventional metal-oxide-semiconductor field-effect transistors. The advantage of the CFT is that the currents measured in the microampere to millampere range greatly reduce the problems associated with noise. Additional advantages include the substantial reduction in the device size, the use of well-established semiconductor processing techniques, and the possibility of integrating the detection circuitry with the sensor device.

REFERENCES

1. Lee D.-D. and Lee D.-S., Environmental gas sensors, *IEEE Sensors J.*, 1, 214–224, 2001.
2. Seiyama T., Kato A., Fujiishi K., and Nagatani M., A new detector for gaseous components using semiconducting thin film, *Anal. Chem.*, 34, 1502–1503, 1962.
3. Belford R. E., Kelly R. G., and Owen A. E., Thick film devices, in *Chemical Sensors*, ed. T. E. Edmonds (Blacki and Son Ltd., USA by Chapman and Hall, New York, 1988), Chapter 11.
4. Morrison R. S., Selectivity in semiconductor gas sensors, *Sens. Actuators*, 12, 425–440, 1987.
5. Xu Y. L., Zhou X. H., and Sørensen O. T., Oxygen sensors based on semiconducting metal oxides: An overview, *Sens. Actuators B*, 65, 2–4, 2000.
6. Yamazoe N. and Miura N., New approaches in the design of gas sensors, in *Gas Sensors—Principles, Operation and Developments*, ed. G. Sberveglieri (Dordrecht, The Netherlands: Kluwer Academic, 1992), 1.
7. Lalauze R., Bui N., and Pijolat C., Interpretation of the electrical properties of a SnO_2 gas sensor after treatment with sulfur dioxide, *Sens. Actuators*, 6, 119–125, 1984.

8. Várhegyi E. B., Gerblinger J., Réti F., Perczel I. V., and Meixner H., Study of the behavior of CeO_2 in SO_2-containing environment, *Sens. Actuators B*, 24/25, 631–634, 1995.

9. Giraradin D., Berger F., Chambaudet A., and Planade R., Modelling of SnO_2 detection by tin dioxide gas sensors, *Sens. Actuators B*, 43, 147–153, 1997.

10. Berger F., Fromm M., Chambaudet A., and Planade R., Tin dioxide-based gas sensors for SO_2 detection: A chemical interpretation of the increase in sensitivity obtained after a primary detection, *Sens. Actuators B*, 45, 175–181, 1997.

11. Imanaka N., Yamaguchi Y., Adachi G., and Shiokawa J., Sulfur dioxide gas detection with Na_2SO_4-Li_2SO_4-$Y_2(SO_4)_3$-SiO_2 solid electrolyte by a solid reference electrode method, *J. Electrochem. Soc.*, 134, 725–728, 1987.

12. Adachi G. and Imanaka N., Development of an SO_x sensor based on metal sulfates, in *Chemical Sensor Technology*, ed. N. Yamazoe, Vol. 3 (Tokyo: Kodansha, 1991), 31.

13. Yang P. H., Yang J. H., Chen C. S., Peng D. K., and Meng G. Y., Performance evaluation of SO_x (x = 2, 3) gas sensors using Ag-β''-alumina solid solution, *Solid State Ionics*, 86/88, 1095–1099, 1996.

14. Akila R. and Jacob K. T., An SO_x (x = 2, 3) sensor using β-alumina/Na_2SO_4 couple, *Sens. Actuators*, 16, 311–323, 1989.

15. Choi S.-D., Chung W.-Y., and Lee D.-D., SO_2 sensing characteristics of Nasicon electrolytes, *Sens. Actuators B*, 35/36, 263–266, 1996.

16. Yan Y., Shimizu Y., Miura N., and Yamazoe N., Characteristics and sensing mechanism of SO_x sensor using stabilized zirconia and metal sulphate, *Sens. Actuators B*, 12, 77–81, 1993.

17. Yan Y., Shimizu Y., Miura N., and Yamazoe N., High-performance solid-electrolyte SO_x sensor using MgO-stabilized zirconia tube and Li_2O_4-$CaSO_4$-SiO_2 auxiliary phase, *Sens. Actuators B*, 20, 81–87, 1994.

18. Yoshioka T., Mizuno M., and Iwamato M., La_2O_3-loaded SnO_2 element as a CO_2 gas sensor, *Chem. Lett.*, 19, 1249–1252, 1991.

19. Kim D. H., Yoon J. Y., Park H. C., and Kim K. H., CO_2 sensing characteristics of SnO_2 thick film by coating lanthanum oxide, *Sens. Actuators B*, 62, 61–66, 2000.

20. Tamaki J., Akiyama M., and Xu C., Conductivity change of SnO_2 with CO_2 adsorption, *Chem. Lett.*, 18, 1243–1246, 1990.

21. Mizuno N., Kato K., Yoshioka T., and Iwamoto M., A remarkable sensitivity of CaO-loaded In_2O_3 element to CO_2 gas in the presence of water vapor, *Chem. Lett.*, 20, 1683–1684, 1992.

22. Mizuno N., Yoshioka T., Kato K., and Iwamoto M., CO_2-sensing characteristics of SnO_2 element modified by La_2O_3, *Sens. Actuators B*, 13/14, 473–475, 1993.

23. Gauthier M. and Chamberland A., Solid-state detectors for the potentiometric determination of gaseous oxides, *J. Electrochem. Soc.*, 124, 1579–1583, 1977.

24. Saito Y., Maruyamu T., and Sasaki S., *Gas Sensors Using NASICON as a Solid Electrolyte*, Report of the Research Laboratory of Engineering Materials, Tokyo Institute of Technology, 1984.

25. Yao S., Shimizu Y., Miura N., and Yamazoe N., Solid electrolyte CO_2 sensor using binary carbonate electrode, *Chem. Lett.*, 18, 2033–2036, 1990.

26. Yao S., Hosohara S., Shimizu Y., Miura N., Futata H., and Yamazoe N., Solid electrolyte CO_2 sensor using NASICON and Li-based binary carbonate electrode, *Chem. Lett.*, 19, 2069–2072, 1991.

27. Shimizu Y. and Yamashita N., Solid electrolyte CO_2 sensor using NASICON and perovskite-type oxide electrode, *Sens. Actuators B*, 64, 102–106, 2000.

28. Seo M.-G., Kang B.-W., Chai Y.-S., Song K.-D., and Lee D.-D., CO_2 gas sensor using lithium ionic conductor with inside heater, *Sens. Actuators B*, 65, 346–348, 2000.

29. Ishihara T., Kometani K., Hashida M., and Yakita Y., Mixed oxide capacitor of $BaTiO_3$-PbO as a new type CO_2 gas sensor, *Chem. Lett.*, 18, 1163–1166, 1990.

30. Matsubara S., Kaneko S., Morimoto S., Shimizu S., Ishihara T., and Takita Y., A practical capacitive type CO_2 sensor using $CeO_2/CaCO_3/CuO$ ceramics, *Sens. Actuators B*, 65, 128–132, 2000.

31. Ishihara T., Kometani K., Hashida M., and Yakita Y., Application of mixed oxide capacitor to the selective carbon dioxide sensor, *J. Electrochem. Soc.*, 138, 173–176, 1991.

32. Ishihara T., Kometani K., Hashida M., and Yakita Y., Mixed oxide capacitor of CuO-$BaTiO_3$ as a new type CO_2 gas sensor, *J. Am. Ceram. Soc.*, 75, 613–618, 1992.

33. Ishihara T. and Matsubara S., Capacitive type gas sensors, *J. Electroceram.*, 2, 215–228, 1998.

34. Liao B., Wei Q., Wang K., and Liu Y., Study on CuO-$BaTiO_3$ semiconductor CO_2 sensor, *Sens. Actuators B*, 80, 208–214, 2001.

35. Coles G. S. V., Williams G., and Smith B., Selectivity studies on tin oxide-based semiconductor gas sensors, *Sens. Actuators B*, 3, 7–14, 1991.

36. Kobayashi T., Haruta M., Sano K., and Nakane N., A selective CO sensor using Ti-doped Fe_2O_3 with coprecipitated ultrafine particles of gold, *Sens. Actuators B*, 13, 339–349, 1988.

37. Yamaura H., Jinkawa T., Tamaki J., Moriya K., Miura N., and Yamazoe N., Indium oxide-based gas sensor for selective detection of CO, *Sens. Actuators B*, 35/36, 325–332, 1996.

38. Yamazoe N., Kurokawa Y., and Seiyama T., Effects of additive on semiconductor gas sensor, *Sens. Actuators B*, 4, 283–289, 1983.

39. Chang S. C. and Hicks D. B., Tin oxide microsensors, *in Proc. 3rd Int. Conf. Solid State Sensors and Actuators (Transducers '85)*, Philadelphia, PA, USA, 381, 1985.

40. Sberveglieri G., Groppelli S., and Nelli P., Highly sensitive and selective NO_x and NO_2 sensor based on Cd-doped SnO_2 thin films, *Sens. Actuators B*, 4, 445–450, 1991.

41. Sugai T., Matsuzawa T., Murayama Y., Sato M., and Sakaguchi M., In_2O_3-based NO_x gas sensors, in *Digest 9th Chem. Sens. Symp.* Tokyo, Japan, 101, 1989.

42. Penza M., Martucci C., and Cassano G., NO_x gas sensing characteristics of WO_3 thin films activated by noble metals (Pd, Pt, Au) layers, *Sens. Actuators B*, 50, 52–59, 1998.

43. Nelli P., Depero L. E., Ferroni M., Groppelli S., Guidi V., Ronconi F., Sangaletti L., and Sberveglieri G., Sub-ppm NO_2 sensors based on nanosized thin films of titanium-tungsten oxides, *Sens. Actuators B*, 31, 89–92, 1996.

44. Lee D. S., Han S. D., Lee S. M., Huh J. S., and Lee D. D., The TiO_2-adding effects in WO_3-based NO_2 sensors prepared by coprecipitation and precipitation method, *Sens. Actuators B*, 65, 331–335, 2000.

45. Depro L. E., Ferroni M., Guidi V., Marca G., Martinelli G., Nelli P., Sangaletti L., and Sberveglieri G., Preparation and micro-structural characterization of nanosized thin film of TiO_2-WO_3 as a novel material with high sensitivity towards NO_2, *Sens. Actuators B*, 35/36, 381–383, 1996.

46. Wang X., Sakai G., Shimanoe K., Miura N., and Yamazoe N., Spin-coated thin films of SiO_2–WO_3 composites for detection of sub-ppm NO_2, *Sens. Actuators B*, 45, 141–146, 1997.

47. Solis J. L. and Lantto V., Gas-sensing properties of Sn_xWO_{3+x} mixed oxide thick films, *Sens. Actuators B*, 48, 322–327, 1998.

48. Solis J. L., Lantto V., Haggstrom L., Kalska B., Frantti J., and Saukko S., Synthesis of new compound semiconductors in the Sn–W–O system for gas-sensing studies, *Sens. Actuators B*, 68, 286–292, 2000.

49. Horrillo M.C., Sayago I., Arés L., Rodrigo J., Gutiérrez J., Götz A., Gràcia I., Fonseca L., Cané C., and Lora-Tamayo E., Detection of low NO_2 concentrations with low power micromachined tin oxide gas sensors, *Sens. Actuators B*, 58, 325–329, 1999.

50. Banno S., Imanaka N., and Adachi G., Selective nitrogen dioxide sensor based on nickel copper oxide mixed with rare earths, *Sens. Actuators B*, 24/25, 619–622, 1995.

51. Hotzel G. and Weppner W., Application of fast ionic conductors in solid state galvanic cells for gas sensors, *Solid State Ionics*, 18/19, 1223–1227, 1986.

52. Shimizu Y., Okamoto Y., Yao S., Miura N., and Yamazoe N., Solid electrolyte NO_2 sensors fitted with sodium nitrate and/or barium nitrate electrodes, *Denki Kagaku*, 59, 465, 1991.

53. Yao S., Shimizu Y., Miura N., and Yamazoe N., Development of high performance solid electrolyte NO_x sensor using sodium nitrite-based auxiliary phase, in *Digest 16th Chem. Sens. Symp.* Tokyo, Japan, 81, 1993.

54. Lu G., Miura N., and Yamazoe N., High-temperature sensors for NO and NO_2 based on stabilized zirconium and spinal-type oxide electrodes, *J. Mater. Chem.*, 7, 1445–1449, 1997.

55. Shimizu Y., Nishi H., Suzuki H., and Maeda K., Solid-state NO_x sensor combined with NASICON and Pb-Ru-based pyrochlore-type oxide electrode, *Sens. Actuators B*, 65, 141–143, 2000.

56. Tatsumi I., Shinobu S., and Yusaku T., Capacitive-type sensors for the selective detection of nitrogen oxides, *Sens. Actuators B*, 25, 392–395, 1995.

57. Sberveglieri G., Groppelli S., and Coccoli G., Radio frequency magnetron sputtering growth and characterization of indium-tin oxide (ITO) thin films for NO_2 gas sensors, *Sens. Actuators B*, 15, 235–242, 1988.

58. Akiyama M., Tamaki J., Harada T., Miura N., and Yamazoe N., Tungsten oxide-based semiconductor sensor highly sensitive to NO and NO_2, *Chem. Lett.*, 19, 1611–1614, 1991.

59. Akiyama M., Zhang Z., Tamaki J., Harada T., Miura N., and Yamazoe N., Development of high sensitivity NO_x sensor using metal oxides, in *Tech. Digest, 11th Sensor Symp.* Tokyo, Japan, 181, 1992.

60. Tomchenko A. A., Khatko V. V., and Emelianov I. L., WO_3 thick-film gas sensors, *Sens. Actuators B*, 46, 8–14, 1998.

61. Yao S., Shimizu Y., Miura N., and Yamazoe N., Use of sodium nitrite auxiliary electrode for solid electrolyte sensor to detect nitrogen oxides, *Chem. Lett.*, 20, 587–590, 1992.

62. Ishihara T., Fujita H., Nishiguchi H., and Takita Y., $SrSnO_3$-WO_3 as capacitive-type nitrogen oxide sensors for monitoring at high temperature, *Sens. Actuators B*, 65, 319–324, 2000.

63. Takada T. and Komatsu K., O_3 gas sensor of thin film semiconductor in In_2O_3, in *Proc. Transducers 87*, Tokyo, Japan, 693, 1987.

64. Takada T., Suzuki K., and Nakane M., Highly sensitive ozone sensors, in *Tech. Digest, 4th Int. Meeting Chem. Sens.*, Fukuoka, Japan, 470, 1992.

65. Cantalini C., Wlodarski W., Ki Y., Pasacantando M., Santucci S., Comini E., Faglia G., and Sberveglieri G., Investigation on the O_3 sensitivity properties of WO_3 thin films prepared by sol-gel, thermal evaporation and r.f. sputtering techniques, *Sens. Actuators B*, 64, 182–188, 2000.

66. Miyata T., Hikosaka T., and Minami T., Ozone gas sensors with high sensitivity using $Zn_2In_2O_5$-$MgIn_2O_4$ multicomponent oxide thin films, *Surf. Coat. Technol.*, 126, 219–224, 2000.

67. Logothetis E. M., *Chemical Sensor Technology*, ed. N. Yamazoe, Vol. 3 (New York: Elsevier Science, 1991), 89.

68. Weppner W., Solid-state electrochemical gas sensors, in *Proc. 2nd Int. Meet. Chemical Sensors*, Bordeauq, France, 59, 1986.

69. Yamazoe N. and Miura N., Environmental gas sensing, *Sens. Actuators B*, 20, 95–102, 1994.

70. Armgarth M. and Nylander C., Field effect gas sensors, in *Sensors—A Comprehensive Survey, Chemical and Biochemical Sensors*, ed. W. Gopel, T. A. Jones, M. Kleitz, I. Lundström, and T. Seiyama, Vol. 2 (Weinheim: VCH, 1991), 467.

71. Lundström I., Shivaraman S., Svensson C., and Lundiwist L., A hydrogen-sensitive MOS field effect transistor, *Appl. Phys. Lett.*, 26, 55–57, 1975.

72. Lundström I., Shiiaraman S., and Svensson C., A hydrogen-sensitive Pd-gate MOS transistor, *J. Appl. Phys.*, 46, 3876–3881, 1975.

73. Kang W. P. and Kim C. K., Performance and detection mechanism of a new class of catalyst (Pd, Pt, or Ag)-adsorptive oxide (SnO_2 or ZnO)-insulator-semiconductor gas sensors, *Sens. Actuators B*, 22, 47–55, 1994.

74. Dobos K. and Zimmer G., Performance of carbon monoxide-sensitive MOSFET's with metal-oxide semiconductor gates, *IEEE Trans. Electron Devices*, 32, 1165–1169, 1985.

75. Nicollean E. H. and Brews J. R., *MOS Physics and Technology* (New York: Wiley, 1982), 176.

76. Bogner M., Fuchs A., Scharnagl K., Winter R., Doll T., and Eisele I., Electrical field impact on the gas adsorptivity of thin metal oxide films, *Appl. Phys. Lett.*, 73, 2524–2526, 1998.

77. Lee D. D. and Chung W. Y., Gas-sensing characteristics of SnO_{2-x} thin film with added Pt fabricated by the dipping method, *Sens. Actuators B*, 20, 301–305, 1989.

78. Kang W. P., Xu J. F., Lalevic B., and Poteat T. L., Pd-SnO_2 MIS capacitor as a new type of oxygen sensor, *IEEE Electron Device Lett.*, 8, 211–213, 1987.

79. Kang W. P., Xu J. F., Lalevic B., and Poteat T. L., Sensing behavior of Pd-SnO_2 MIS structure used for oxygen detection, *Sens. Actuators B*, 12, 349–366, 1987.

80. Lundström I. and Svensson C., Gas-sensitive metal gate semiconductor devices, in *Solid-State Chem. Sens.*, ed. J. Janata and R. J. Huber (New York: Academic Press, 1985).

81. Ishihara T., Kometani K., Nishi Y., and Takita Y., Improved sensitivity of CuO-$BaTiO_3$ capacitive-type CO_2 sensor by additives, *Sens. Actuators B*, 28, 49–54, 1995.

82. Endres H. E., Hartinger R., Schwaiger M., Gmelch G., and Roth M., A capacitive CO_2 sensor system with suppression of the humidity interference, *Sens. Actuators B*, 57, 83–87, 1999.

83. Taguchi N., Gas detection device, *British Patent 1280809*, 1970.

84. Shaver P. J., Activated tungsten oxide gas detectors, *Appl. Phys. Lett.*, 11, 255–257, 1967.

85. Seiyama T. and Yamazoe N., *Fundamentals and Applications of Chemical Sensors*, Vol. 309, ed. D. Schuetzle and R. Hammerle (Washington, D.C.: American Chemical Society, 1986), 39.

86. Yamazoe N. and Miura N., *Chemical Sensor Technology*, Vol. 4, ed. S. Yamauchi (New York: Kodansha-Elsevier, 1992), 19.

87. Kohl D., Surface processes in the detection of reducing gases with SnO_2-based devices, *Sens. Actuators B*, 18, 71–113, 1989.

88. McAleer J. F., Moseley P. T., Norris J. O. W., and Williams D. E., Tin dioxide gas sensors. Part 1.-Aspects of the surface chemistry revealed by electrical conductance variations, *J. Chem. Soc., Faraday Trans.1*, 83, 1323–1346, 1987.

89. McAleer J. F., Moseley P. T., Norris J. O. W., Williams D. E., and Tofield B.C., Tin dioxide gas sensors. Part 2.-The role of surface additives, *J. Chem. Soc., Faraday Trans. 1*, 84, 441–457, 1988.

90. Park C. O. and Akbar S. A., Ceramics for chemical sensing, *J. Mats. Sci.* 38, 4611–4637, 2003.

91. Yamazoe N., Fuchigami J., Kishikawa M., and Seiyama T., Interactions of tin oxide surface with O_2, H_2O and H_2, *Surf. Sci.*, 86, 335–344, 1979.

92. Egashira M., Matsumoto T., Shimizu Y., and Iwanaga H., Gas-sensing characteristics of tin oxide whiskers with different morphologies, *Sens. Actuators B*, 14, 205–213, 1988.

93. Kawahara A., Yoshihara K., Katsuki H., Shimizu Y., and Egashira M., *Technical Digest of the 7th International Meeting on Chemical Sensors* (Beijing: International Academic, 1998), 364.

94. Takao Y., Miyazaki K., Shimizu Y., and Egashira M., High ammonia sensitive semiconductor gas sensors with double-layer structure and interface electrodes, *J. Electrochem. Soc.*, 141, 1028–1033, 1994.

95. Takao Y., Nakanishi M., Kawaguchi T., Shimizu Y., and Egashira M., Semiconductor dimethylamine gas sensors with high sensitivity and selectivity, *Sens. Actuators B*, 24/25, 375–379, 1995.

96. Cammann K., Institut für Chemo- und Biosensorik, Münster, *Chemical and Biochemical Sensors*, Ullmann's Encyclopedia of Industrial Chemistry, Wiley-VCH Verlag GmbH & Co. KGaA, 2002.

97. Moulson A. J. and Herbert J. M., *Electroceramics*, 2nd ed. (Greater New York City Area, USA: John Wiley & Sons, 2003), 212.

98. Hyodo T., Mori T., Kawahara A., Katsuki H., Shimizu Y., and Egashira M., Gas sensing properties of semiconductor heterolayer sensors fabricated by slide-off transfer printing, *Sens. Actuators B*, 77, 41–47, 2001.

99. Iijima S., Helical microtubules of graphitic carbon, *Nature*, 354, 56–58, 1991.

100. Dresselhaus M. S., Dressehaus G., and Eklund P. C., *Science of Fullenrenes and Carbon Nanotubes* (New York: Academic, 1996), Chapter 19.

101. Mickelson E. T., Huffman C. B., Rinzler A. G., Smalley R.E., Hauge R.H., and Margrave J.L., Fluorination of single-wall carbon nanotubes, *Chem. Phys. Lett.*, 296, 188–194, 1998.

102. Dillon A. C., Jones K. M., Bekkedahl T. A., Kiang C. H., Bethune D. S., and Heben M. J., Storage of hydrogen in single-walled carbon nanotubes, *Nature*, 386, 377–379, 1997.

103. Stan G., Bojan M. J., Curtarolo S., Gatica S. M., and Cole M. W., Uptake of gases in bundles of carbon nanotubes, *Phys. Rev. B*, 62, 2173–2180, 2000.

104. Gatica S. M., Bojan M. J., Stan G., and Cole M. W., Quasi-one- and two-dimensional transitions of gases adsorbed on nanotube bundles, *J. Chem. Phys.*, 114, 3765–3769, 2001.

105. Kong J., Franklin N. R., Zhou C., Chapline M. G., Peng S., Cho K., and Dai H., Nanotube molecular wires as chemical sensors, *Science*, 287, 622–625, 2000.

106. Zhao J., Buldum A., Han J., and Lu J. P., Gas molecule adsorption in carbon nanotubes and nanotube bundles, *Nanotechnology*, 13, 195–200, 2002.

107. Ong K. G., Zeng K., and Grimes C. A., A wireless, passive carbon nanotube-based gas sensor, *IEEE Sensors J.*, 2, 82–88, 2002.

108. Valentini L., Cantalini C., Armentano I., Kenny J. M., Lozzi L., and Santucci S., Highly sensitive and selective sensors based on carbon nanotubes thin films for molecular detection, *Diamond. Relat. Mater.*, 13, 1301–1305, 2004.

109. D'amico A., Natale C. D., and Taroni A., Sensors parameters, *Sensors for Domestic Applications: Proceedings of the First European School on Sensors (ESS'94)* (Lee, Italy: Castro Marina, 1994), 3.

110. Shimizu Y. and Egashira M., Basic aspects and challenges of semiconductor gas sensors, *MRS Bull.*, 24, 18–24, 1999.

111. Moulson A. J. and Herbert J. M., *Electroceramics: Materials, Properties, and Applications*. 2nd ed. (Greater New York City Area, USA: John Wiley & Sons, 2003).

112. Jayadevan K. P. and Tseng T. Y., *Encyclopedia of Nanotechnology*, ed. H. S. Nalwa, Vol. 8 (USA: American Scientific Publishers, 2004), 333.

113. Xu C., Tamaki J., Miura N., and Yamazoe N., Grain size effects on gas sensitivity of porous SnO_2-based elements, *Sens. Actuators B*, 3, 147–155, 1991.

114. Gurlo A., Ivanovskaya M., Bârsanb N., Schweizer-Berberich M., Weimar U., Göpel W., and Diéguez A., Grain size control in nanocrystalline In_2O_3 semiconductor gas sensors, *Sens. Actuators B*, 44, 327–333, 1997.

115. Hu Y., Tan O. K., Pan J. S., and Yao X., A new form of nano-sized $SrTiO_3$ material for near human body temperature oxygen sensing applications, *J. Phys. Chem. B*, 108, 11214–11218, 2004.

116. Sberveglieri G., Recent developments in semiconducting thin-film gas sensors, *Sens. Actuators B*, 23, 103–109, 1995.

117. Moseley P. T., Materials selection for semiconductor gas sensors, *Sens. Actuators B*, 6, 149–156, 1992.

118. Meixner H. and Lampe U., Metal oxide sensors, *Sens. Actuators B*, 33, 198–202, 1996.

119. Meneskou W., Schreiner H.-J., Härdtl K. H., and Ivers-Tiffée E., High temperature oxygen sensors based on doped $SrTiO_3$, *Sens. Actuators B*, 59, 184–189, 1999.

120. Ogawa H., Abe A., Nishikawa M., and Hayakawa S., Preparation of tin oxide films from ultrafine particles, *J. Electrochem. Soc.*, 128, 685–689, 1981.

121. Nakatani Y. and Matsuoka M., Effects of sulfate ion on gas sensitive properties of α-Fe_2O_3 ceramics, *Jpn. J. Appl. Phys.*, 21, 1758–1762, 1982.

122. Liu X. Q., Tao S. W., and Shen Y. S., Preparation and characterization of nanocrystalline of α-Fe_2O_3 by a sol-gel process, *Sens. Actuators B*, 40, 161–165, 1997.

123. Xue S., Ousi-Benomar W., and Lessard R. A., α-Fe_2O_3 thin films prepared by metal organic deposition (MOD) from Fe (III) 2-ethylhexanoate, *Thin Solid Films*, 250, 194–201, 1994.

124. Liu Y., Zhu W., Tan O. K., and Shen Y., Structural and gas sensing properties of ultrafine Fe_2O_3 prepared by plasma enhanced chemical vapor deposition, *Mater. Sci. Eng. B*, 47, 171–176, 1997.

125. Chai C. C., Peng J., and Yan B. P., Preparation and gas sensing properties of alpha-Fe_2O_3 thin films, *J. Electron. Mater.*, 24, 799–804, 1995.

126. Edelman F., Rothshild A., Komem Y., Mikhelashvili V., Chack A., and Cosandey F., E-Gun sputtered and reactive ion sputtered TiO_2 thin films for gas sensors, *Inst. Electron Technol.*, 33, 89–107, 2000.

127. Ferroni M., Boscarino D., and Comini E., Nanosized thin films of tungsten-titanium mixed oxides as gas sensors, *Sens. Actuators B*, 58, 289–294, 1999.

128. Cosandey F., Skandan G., and Singhal A., Materials and processing issues in nanostructured semiconductor gas sensors, *JOM-e*, 52(10), 2000.

129. Williams G. and Cole G. S. V., The gas-sensing potential of nanocrystalline tin dioxide produced by a laser ablation technique, *MRS Bull.*, 24, 25–29, 1999.

130. Tan O. K., Cao W., and Zhu W., Alcohol sensor based on a non-equilibrium nanostructured $xZrO_2$-$(1-x)\alpha$-Fe_2O_3 solid solution system, *Sens. Actuators B*, 63, 129–134, 2000.

131. Cao W., Tan O. K., Zhu W., and Jiang B., Mechanical alloying and thermal decomposition of $(ZrO_2)_{0.8}$-$(\alpha$-$Fe_2O_3)_{0.2}$ powder for gas sensing applications, *J. Solid. State. Chem.*, 155, 320–325, 2000.

132. Jiang J. Z., Lu S. W., Zhou Y. X., Mørup S., Nielsen K., Poulsen E. W., Berry F. J., and McMannus J., Correlation of gas sensitive properties with Fe_2O_3-SnO_2 ceramic microstructure prepared by high energy ball milling, *Mater. Sci. Forum*, 235/238, 941–946, 1997.

133. Jiang J. Z., Lin R., Lin W., Nielsen K., Mørup S., Dam-Johansen K., and Clasen R., Gas sensitive properties and structure of nanostructured $(\alpha Fe_2O_3)_x$-$(SnO_2)_{1-x}$ materials prepared by mechanical alloying, *J. Phys. D: Appl. Phys.*, 30, 1459–1467, 1997.

134. Hu Y., Tan O. K., Pan J. S., Huang H., and Cao W., The effects of annealing temperature on the sensing properties of low temperature nano-sized $SrTiO_3$ oxygen gas sensor, *Sens. Actuators B*, 108, 244–249, 2005.

135. Tan O. K., Hu Y., and Pan J. S., Electrical conduction properties of nano-sized $SrTiO_3$ semiconductor for low temperature resistive oxygen gas sensor, *the 10th International Meeting on Chemical Sensors: Chemical Sensors*, Vol. 20, Supplement B, Tsukuba, Japan, 16, 2004.

136. Cao W., Tan O. K., Zhu W., Jiang B. and Gopal Reddy C. V., An amorphous-like x-αFe_2O_3-$(1-x)ZrO_2$ solid solution system for low temperature resistive-type oxygen sensing, *Sens. Actuators B*, 77, 421–426, 2001.

137. Tan O. K., Cao W., Hu Y., and Zhu W., Nanostructured oxides by high-energy ball milling technique: Application as gas sensing materials, *Solid State Ionics*, 172, 309–316, 2004.

138. Chen Y., Fitz Gerald J., Chadderton L. T., and Chaffron L., Investigation of nanoporous carbon powders produced by high energy ball milling and formation of carbon nanotubes during subsequent annealing, *J. Metastable Nanocrystal. Mater.*, 2–6, 375–380, 1999.

139. Colin S. B., Caër G. Le, Villieras F., Devaux X., Simonnit M. O., Girot T., and Weisbecker P., From high-energy ball milling to surface properties of TiO₂ powders, *J. Metastable Nanocrystal. Mater.*, 12, 27–36, 2002.

140. Hahn J. D., Wu F., and Bellon P., Cr-Mo Solid solutions forced by high-energy ball milling, *Metall. Mater. Trans. A.*, 35A, 1105–1111, 2004.

141. Zieliński P. A., Schulz R., Kaliaguine S., and Van Neste A., Structural transformations of alumina by high-energy ball milling, *J. Mater. Res.*, 8, 2985–2992, 1993.

142. Sakai G., Baik N. S., Miura N., and Yamazoe N., Gas sensing properties of tin oxide thin films fabricated from hydrothermally treated nanoparticles: Dependence of CO and H₂ response on film thickness, *Sens. Actuators B*, 77, 116–121, 2001.

143. Baik N. S., Sakai G., Miura N., and Yamazoe N., Hydrothermally treated sol solution of tin oxide for thin-film gas sensor, *Sens. Actuators B*, 63, 74–79, 2000.

144. Mor G. K., Carvalho M. A., Varghese O. K., Pishko M. V., and Grimes C. A., A room-temperature TiO₂-nanotube hydrogen sensor able to self-clean photoactively from environmental contamination, *J. Mater. Res.*, 19, 628–634, 2004.

145. Shimizu Y., Kuwano N., Hyodo T., and Egashira M., High H₂ sensing performance of anodically oxidized TiO₂ film contacted with Pd, *Sens. Actuators B*, 83, 195–201, 2002.

146. Varghese O. K., Gong D. W., Paulose M., Ong K. G., Grimes C. A., and Dickey E. C., Highly ordered nanoporous alumina films: Effect of pore size and uniformity on sensing performance, *J. Mater. Res.*, 17, 1162–1171, 2002.

147. Dickey E. C., Varghese O. K., Ong K. G., Gong D. W., Paulose M., and Grimes C. A., Room temperature ammonia and humidity sensing using highly ordered nanoporous alumina films, *Sensors*, 2, 91–110, 2002.

148. Tan O. K., Cao W., and Zhu W., Alcohol sensor based on a non-equilibrium nanostructured xZrO₂–(1−x)α-Fe₂O₃ solid solution system, *Sens. Actuators B*, 63, 129–134, 2000.

149. Cao W., Tan O. K., Zhu W., Jiang B. and Pan J. S., Study of xα-Fe₂O₃-(1-x)ZrO₂ solid solution for low temperature resistive oxygen gas sensors, *IEEE Sensors J.*, 3, 421–434, 2003.

150. Hu Y., Tan O. K., Cao W., and Zhu W., Fabrication & characterization of nano-sized SrTiO₃-based oxygen sensor for near-room temperature operation, *IEEE Sensors J.*, 5, 825–832, 2005.

151. Tan O. K., Cao W., Hu Y., Zhu W. and Yao X., Nano-structured oxide semiconductor materials for gas sensing applications, *Ceram. Int.*, 30, 1127–1133, 2004.

152. Hu Y., Tan O. K., Cao W., and Zhu W., A low temperature nano-structured SrTiO₃ thick film oxygen gas sensor, *Ceram. Int.*, 30, 1819–1822, 2004.

153. Hu Y., Tan O. K., Zhu W., Characterization of nano-sized SrTi$_{1±x}$O$_{3-\delta}$ metal oxide semiconducting oxygen gas sensors for near human-body temperature application, *IEEE Sensors 2005*, Irvine, CA, 136, 2005.

154. Hu Y., Tan O. K., and Ong C. Y., The effects of doping dopants on the characterization of low temperature nano-sized SrTiO$_{3-\delta}$ oxygen gas sensors, *The 6th East Asia Conference on Chemical Sensors*, Guilin, China, 74, 2005.

155. Traversa E., DiVona M. L., Licoccia S., Sacerdoti M., Carotta M. C., Crema L., and Martinelli G., Sol-gel processed TiO₂-based nano-sized powders for use in thick-film gas sensors for atmospheric pollutant monitoring, *J. Sol-Gel. Sci. Technol.*, 22, 167–179, 2001.

156. Bartolomeo E. Di, Kaabbuathong N., D'Epifanio A., Grilli M. L., Traversa E., Aono H., and Sadaoka Y., Nano-structured perovskite oxide electrodes for planar electrochemical sensors using tape casted YSZ layers, *J. Eur. Ceram. Soc.*, 24, 1187, 2004.

157. Li H. T., Cui D. Y., Chen X. Y., Dong H. Q., and Cui Z. W., *Proceedings of the SPIE-The International Society for Optical Engineering*, 4077, 284, 2000.

158. Gurlo A., Barsan N., Ivanovskaya M., Weimar U., and Gopel W., In₂O₃ and MoO₃–In₂O₃ thin film semiconductor sensors: Interaction with NO₂ and O₃, *Sens. Actuators B*, 47, 92–99, 1998.

159. Reddy C. V. G., Akbar S. A., Cao W., Tan O. K., and Zhu W., Preparation and characterization of iron oxide-zirconia nano powder for its use as an ethanol sensor material, in *Chem. Sens. Hostile Environ., Ceram. Trans.*, 130, 67, 2002.

160. Lee D. S., Han S. D., Huh J. S., and Lee D. D., Nitrogen oxides-sensing characteristics of WO_3-based nanocrystalline thick film gas sensor, *Sens. Actuators B*, 60, 57–63, 1999.

161. Shukla S., Patil S., Kuiry S. C., Rahman Z., Du T., Ludwig L., Parish C., and Seal S., Synthesis and characterization of sol–gel derived nanocrystalline tin oxide thin film as hydrogen sensor, *Sens. Actuators B*, 96, 343–353, 2003.

162. Dieguez A., Romano-Rodriguez A., Morante J. R., Kappler J., Barsan N., and Gopel W., Nanoparticle engineering for gas sensor optimisation: Improved sol–gel fabricated nanocrystalline SnO_2 thick film gas sensor for NO_2 detection by calcination, catalytic metal introduction and grinding treatments, *Sens. Actuators B*, 60, 125–137, 1999.

163. Chiorino A., Ghiotti G., Prinetto F., Carotta M. C., Gallana M., and Martinelli G., Characterization of materials for gas sensors. Surface chemistry of SnO_2 and MoO_x–SnO_2 nano-sized powders and electrical responses of the related thick films, *Sens. Actuators B*, 59, 203–209, 1999.

164. Davis S. R., Chadwick A. V., and Wright J. D., The effects of crystallite growth and dopant migration on the carbon monoxide sensing characteristics of nanocrystalline tin oxide based sensor materials, *J. Mater. Chem.*, 8, 2065–2071, 1998.

165. Ivanovskaya M. and Bogdanov P., Effect of NiII ions on the properties of In_2O_3-based ceramic sensors, *Sens. Actuators B*, 53, 44–53, 1998.

166. Ivanovskaya M., Gudo A., and Bogdanov P., Mechanism of O_3 and NO_2 detection and selectivity of In_2O_3 sensors, *Sens. Actuators B*, 77, 264–267, 2001.

167. Galatsis K., Li Y. X., Wlodarski W., Comini E., Faglia G., and Sberveglieri G., Semiconductor MoO_3–TiO_2 thin film gas sensors, *Sens. Actuators B*, 77, 472–477, 2001.

168. Atashbar M. Z., Sun H. T., Gong B., Wlodarski W., and Lamb R., XPS study of Nb-doped oxygen sensing TiO_2 thin films prepared by sol-gel method, *Thin Solid Films*, 326, 238–244, 1998.

169. Li M. and Chen Y., An investigation of response time of TiO_2 thin-film oxygen sensors, *Sens. Actuators B*, 32, 83–85, 1996.

170. Galatsis K., Li Y. X., Wlodarski W., and Kalantar-Zadeh K., Sol–gel prepared MoO_3–WO_3 thin-films for O_2 gas sensing, *Sens. Actuators B*, 77, 478–483, 2001.

171. Romanovskaya V., Ivanovskaya M., and Bogdanov P., A study of sensing properties of Pt- and Au-loaded In_2O_3 ceramics, *Sens. Actuators B*, 56, 31–36, 1999.

172. Epifani M., Capone S., Rella R., Siciliano P., and Vasanelli L., In_2O_3 thin films obtained through a chemical complexation based sol-gel process and their application as gas sensor devices, *J. Sol-Gel. Sci. Technol.*, 26, 741–744, 2003.

173. Gurlo A., Ivanovskaya M., Pfau A., Weimar U., and Gopel W., Sol-gel prepared In_2O_3 thin films, *Thin Solid Films*, 307, 288–293, 1997.

174. Ivanovskaya M., Bogdanov P., Faglia G., and Sberveglieri G., The features of thin film and ceramic sensors at the detection of CO and NO_2, *Sens. Actuators B*, 68, 344–350, 2000.

175. Bogdanov P., Ivanovskaya M., Comini E., Faglia G., and Sberveglieri G., Effect of nickel ions on sensitivity of In_2O_3 thin film sensors to NO_2, *Sens. Actuators B*, 57, 153–158, 1999.

176. Li Y. X., Galatsis K., Wlodarski W., Ghantasala M., Russo S., Gorman J., Santucci S., and Passacantando M., Microstructure characterization of sol-gel prepared MoO_3–TiO_2 thin films for oxygen gas sensors, *J. Vacuum Sci. & Technol., A-Vacuum Surfaces Films*, 19, 904–909, 2001.

177. Bender M., Fortunato E., Nunes P., Ferreira I., Marques A., Martins R., Katsarakis N., Cimalla V., and Kiriakidis G., Highly sensitive ZnO ozone detectors at room temperature, *Jpn. J. Appl. Phys.*, 42, L435–L437, 2003.

178. Guidi V., Carotta M. C., Ferroni M., Martinelli G., Paglialonga L., Comini E., and Sberveglieri G., Preparation of nanosized titania thick and thin films as gas-sensors, *Sens. Actuators B*, 57, 197–200, 1999.

179. Rickerby D. G., Horrillo M. C., Santos J. P., and Serrini P., Microstructural characterization of nanograin tin oxide gas sensors, *Nano Struct. Mater.*, 9, 43–52, 1997.

180. Comini E., Ferroni M., Guidi V., Faglia G., Martinelli G., and Sberveglieri G., Nanostructured mixed oxides compounds for gas sensing applications, *Sens. Actuators B*, 84, 26–32, 2002.

181. Kim T. W., Lee D. U., and Yoon Y. S., Microstructural, electrical, and optical properties of SnO_2 nanocrystalline thin films grown on InP (100) substrates for applications as gas sensor devices. *J. Appl. Phys.*, 88, 3759–3761, 2000.

182. Ryzhikov A. S., Vasiliev R. B., Rumyantseva M. N., Ryabova L. I., Dosovitsky G. A., Gilmutdinov A. M., Kozlovsky V. F., and Gaskov A. M., Microstructure and electrophysical properties of SnO_2, ZnO and In_2O_3 nanocrystalline films prepared by reactive magnetron sputtering, *Mat Sci. Eng. B-Solid State Mat. Adv. Technol.*, 96, 268–274, 2002.

183. Karthigeyan A., Gupta R. P., Scharnagl K., Burgmair M., Zimmer M., Sharma S. K., and Eisele I., Low temperature NO_2 sensitivity of nano-particulate SnO_2 film for work function sensors, *Sens. Actuators B*, 78, 69–72, 2001.

184. Comini E., Sberveglieri G., and Guidi V., Ti–W–O sputtered thin film as n- or p-type gas sensors, *Sens. Actuators B*, 70, 108–114, 2000.

185. Rembeza S. I., Rembeza E. S., Svistova T. V., and Borsiakova O. I., Electrical resistivity and gas response mechanisms of nanocrystalline SnO_2 films in a wide temperature range, *Phys. Status. Solidi. A.*, 179, 147–152, 2000.

186. Santos J., Serrini P., OBeirn B., and Manes L., A thin film SnO_2 gas sensor selective to ultra-low NO_2 concentrations in air, *Sens. Actuators B*, 43, 154–160, 1997.

187. Comini E., Sberveglieri G., Ferroni M., Guidi V., Frigeri C., and Boscarino D., Production and characterization of titanium and iron oxide nano-sized thin films, *J. Mater. Res.*, 16, 1559–1564, 2001.

188. Ferroni M., Guidi V., Martinelli G., Faglia G., Nelli P., and Sberveglieri G., Characterization of a nanosized TiO_2 gas sensor, *Nanostruct. Mater.*, 7, 709–718, 1996.

189. Izu N., Shin W., and Murayama N., Fast response of resistive-type oxygen gas sensors based on nano-sized ceria powder, *Sens. Actuators B*, 93, 449–453, 2003.

190. Safonova O. V., Rumyantseva M. N., Ryabova L. I., Labeau M., Delabouglise G., and Gaskov A. M., Effect of combined Pd and Cu doping on microstructure, electrical and gas sensor properties of nanocrystalline tin dioxide, *Mat. Sci. Eng. B-Solid State Mat. Adv. Technol.*, 85, 43–49, 2001.

191. Ferroni M., Carotta M. C., Guidi V., Martinelli G., Ronconi F., Sacerdoti M., and Traversa E., Preparation and characterization of nanosized titania sensing film, *Sens. Actuators B*, 77, 163–166, 2001.

192. Carotta M. C., Ferroni M., Gnani D., Guidi V., Merli M., Martinelli G., Casale M. C., and Notaro M., Nanostructured pure and Nb-doped TiO_2 as thick film gas sensors for environmental monitoring, *Sens. Actuators B*, 58, 310–317, 1999.

193. Ferroni M., Carotta M. C., Guidi V., Martinelli G., Ronconi F., Richard O., VanDyck D., and VanLanduyt J., Structural characterization of $Nb–TiO_2$ nanosized thick-films for gas sensing application, *Sens. Actuators B*, 68, 140–145, 2000.

194. Starke T. K. H. and Coles G. S. V., High sensitivity ozone sensors for environmental monitoring produced using laser ablated nanocrystalline metal oxides, *IEEE Sensors J.*, 2, 14–19, 2002.

195. Starke T. K. H., Coles G. S. V., and Ferkel H., High sensitivity NO_2 sensors for environmental monitoring produced using laser ablated nanocrystalline metal oxides, *Sens. Actuators B*, 85, 239–245, 2002.

196. Starke T. K. H. and Coles G. S. V., Laser-ablated nanocrystalline SnO_2 material for low-level CO detection, *Sens. Actuators B*, 88, 227–233, 2003.

197. Brunet J., Talazac L., Battut V., Pauly A., Blanc J. P., Germain J. P., Pellier S., and Soulier C., Evaluation of atmospheric pollution by two semiconductor gas sensors, *Thin Solid Films*, 391, 308–313, 2001.

198. Toan N. N., Saukko S., and Lantto V., Gas sensing with semiconducting perovskite oxide $LaFeO_3$, *Physical B*, 327, 279–282, 2003.

199. Gardner J. W., Microsensor array devices, in *Microsensors: Principles and Applications* (Chichester: Wiley, 1994), 279.

200. Gardner J. W. and Barlett P. N., Pattern recognition in gas sensing, in *Techniques and Mechanisms in Gas Sensing*, ed. P. T. Moseley, J. Norris and D. E. Williams (Bristol: Adam Hilger, 1991), 347.

201. Vaihinger S. and Göpel W., Multi-component analysis in chemical sensing, in *Sensors. A Comprehensive Survey*, ed. W. Gopel, J. Hesse and J. N. Zemel, Vol. 2, Part I (Weinheim: VCH, 1991), 192.

202. Hong H.-K., Kwon C. H., Kim S.-R., Yun D. H., Lee K., and Sung Y. K., Portable electronic nose system with gas sensor array and artificial neural network, *Sens. Actuators B*, 66, 49–52, 2000.

203. Moseley P. T., Norris J. O. W., Williams D. E. (eds.), *Techniques and Mechanisms of Gas Sensing* (Bristol: Adam Hilger, 1991).

204. Göpel W., Hesse J., and Zemel J. N. (eds.), *Sensors. A Comprehensive Survey*, Vol. 1, 2nd ed. (Weinheim: VCH, 1991).

205. Grattan K. T. V. and Augousti A. T. (eds.), *Sensors and Technology: Systems and Applications* (Bristol: Adam Hilger, 1991), 121.

206. Martín M. A., Santos J. P., and Agapito J. A., Application of artificial neural networks to calculate the partial gas concentrations in a mixture, *Sens. Actuators B*, 77, 468–471, 2001.

207. Rumyantseva M. N., Safonova O. V., Boulova M. N., Ryabova L. I., Gaskov A. M., Chenevier B., Labeau M., Hazemann J. L., and Lucazeau G., Doping effects in tin dioxide in relation with gas sensing phenomena, *Recent Res. Dev. Mater. Sci. Eng.*, 1, 85–115, 2002.

208. McAleer J. F., Moseley P. T., Norris J. O. W., Williams D. E., and Tofield B. C., Tin dioxide gas sensors-Part 2, *J. Chem. Soc., Faraday Trans.*, 84, 441–457, 1988.

209. Abbas M. N., Moustafa G. A., and Gopel W., Multicomponent analysis of some environmentally important gases using semiconductor tin oxide sensors, *Anal. Chim. Acta.*, 431, 181–194, 2001.

210. Kocemba I. and Paryjszak T., Metal films on a SnO_2 surface as selective gas sensors, *Thin Solid Films*, 272, 15–17, 1996.

211. Althainz P., Dahlke A., Frietsch-Klarhof M., Goschnick J., and Ache H. J., Reception tuning of gas-sensor microsystems by selective coatings, *Sens. Actuators B*, 24/25, 366–369, 1995.

212. Montmeat P., Pijolat C., Tournier G., and Viricelle J.-P., The influence of platinum membrane on the sensing properties of a tin dioxide thin film, *Sens. Actuators B*, 84, 148–159, 2004.

213. Kwon C. H., Yun D. H., Hong H.-K., Kim S.-R., Lee K., Lim H. Y., and Yoon K. H., Multi-layered thick film gas sensor array for selective sensors by catalytic filtering technology, *Sens. Actuators B*, 65, 327–330, 2000.

214. Strakova M., Matisova E., Simon P., Annus J., and Lisy J. M., Silicon membrane measuring system with SnO_2 gas sensor for on-line monitoring of volatile organic compounds in water, *Sens. Actuators B*, 52, 274–282, 1998.

215. Park S. O., Akbar S. A., and Hwang J., Selective gas detection with catalytic filters, *Mater. Chem. Phys.*, 75, 56–60, 2002.

216. Fleischer M., Kornely S., Weh T., Frank J., and Meixner H., Selective gas detection with high-temperature operated metal oxides using catalytic filters, *Sens. Actuators B*, 69, 205–210, 2000.

217. Cabot A., Arbiol J., Cornet A., Morante J. R., Chen F., and Liu M., Mesoporous catalytic filters for semiconductor gas sensors, *Thin Solid Films*, 436, 64–69, 2003.
218. Getino J., Gutiérrez J., Arés L., Robla J. I., Horrillo M. C., Sayago I., and Agapito J. A., Integrated sensor array for gas analysis in combustion atmospheres, *Sens. Actuators B*, 33, 128–133, 1996.
219. Schierbaum K. D., Weimar U., and Göpel W., Comparison of ceramic thick-film and thin-film chemical sensors based upon SnO_2, *Sens. Actuators B*, 7, 709–716, 1992.
220. Horrillo M. C., Gutiérrez J., Arés L., Robla J. I., Sayago I., Getino J., and Agapito J., The influence of the tin oxide deposition technique on the sensitivity to CO, *Sens. Actuators B*, 25, 507–511, 1995.
221. Sayago I., Gutiérrez J., Arés L., Robla J. I., Horrillo M. C., Getino J., and Agapito J., The interaction of different oxidizing agents on doped tin oxide, *Sens. Actuators B*, 25, 512–515, 1995.
222. Lalauze R. and Pijolat C., A new approach to selective detection of gas by an SnO_2 solid-state sensor, *Sens. Actuators B*, 5, 55–63, 1984.
223. Coles G. S., Gallagher K. S., and Watson J., Fabrication and preliminary results on tin(IV)-oxide-based gas sensors, *Sens. Actuators B*, 7, 89–96, 1985.
224. Ryzhikov A., Labeau M., and Gaskov A., Al_2O_3 (M = Pt, Ru) catalytic membranes for selective semiconductor gas sensors, *Sens. Actuators B*, 109, 91–96, 2005.
225. Nehlsen S., Hunte T., Muller J., Gas permeation properties of plasma polymerized thin film siloxane-type membranes for temperatures up to 350°C, *J. Med. Screen.*, 106, 1–7, 1995.
226. Yamazoe N. and Miura N., Potentiometric gas sensors for oxidic gases, *J. Electroceram.*, 2/4, 243–255, 1998.
227. Shimizu Y., Matsunaga N., Hyodo T., and Egashira M., Improvement of SO_2 sensing properties of WO_3 by noble metal loading, *Sens. Actuators B: Chem.*, 77, 35–40, 2001.
228. Yahiro H. and Katayama S.-I., Application of metal ion-exchanged zeolites as materials for carbon dioxide sensor, *Denki Kagaku*, 61, 451–452, 1993.
229. Mutschall D. and Obermeier E., A capacitive CO_2 sensor with on-chip heating, *Sens. Actuators B*, 24/25, 412–414, 1995.
230. Ogura K. and Shiigi H., A CO_2 sensing composite film consisting of base-type poly-aniline and poly (vinyl alcohol), *Electrochem. Soli-State Lett.*, 2, 478–480, 1999.
231. Misra S. C. K., Mathur P., and Srivastava B. K., Vacuum-deposited nanocrystalline polyaniline thin film sensors for detection of carbon monoxide, *Sens. Actuators A*, 114, 30–35, 2004.
232. Murakami N., Takahata K., and Seiyama T., Selective detection of CO by SnO_2 gas sensor using periodic temperature change, in *Proceedings of the Digest of Technical Papers for 4th International Conference on Solid-State Sensors and Actuators (Trandducers'87)*, Tokyo, 1987, 618.
233. Lee H.-J., Song J.-H., Yoon Y.-S., Kim T.-S., Kim K.-J., and Choi W.-K., Enhancement of CO sensitivity of indium oxide-based semiconductor gas sensor through ultra-thin cobalt adsorption, *Sens. Actuators B*, 79, 200–205, 2001.
234. Satake K., Katayama A., Ohkoshi H., Nakahara T. T., and Takeuchi T., Titania NO_x sensors for exhaust monitoring, *Sens. Actuators B*, 20, 111–117, 1994.
235. Gurlo A., Barsan N., Ivanovskaya M., Weimar U., and Gopel W., In_2O_3 and MoO_3–In_2O_3 thin film semiconductor sensors: Interaction with NO_2 and O_3, *Sens. Actuators B*, 47, 92–99, 1998.
236. Ferroni M., Guidi V., Martinelli G., Sacerdoti M., Nelli P., and Sberveglieri G., MoO_3-based sputtered thin films for fast NO_2 detection, *Sens. Actuators B*, 48, 285–288, 1998.
237. Sberveglieri G., Depero L., Groppelli S., and Nelli P., WO_3 sputtered thin films for NOx monitoring, *Sens. Actuators B*, 26/27, 89–92, 1995.
238. Cantalini C., Pelino M., Sun H. T., Faccio M., Santucci S., Lozzi L., and Passacantando M., Cross sensitivity and stability of NO_2 sensors from WO_3 thin film, *Sens. Actuators B*, 35/36, 112–118, 1996.

239. Miura N., Ono M., Shimanoe K., and Yamazoe N., A compact solid-state amperometric sensor for detection of NO_2 in ppb range, *Sens. Actuators B*, 49, 101–109, 1998.

240. Tierney M. J., Kim H.yun.-O.k L., Madou M., and Otagawa T., Microelectrochemical sensor for nitrogen oxides, *Sens. Actuators B*, 13/14, 408–411, 1993.

241. Sadaoka Y., Jones T. A., and Gopel W., Effect of heat treatment on the electrical conductance of lead phthalocyanine films for NO_2 gas detection, *J. Mater. Sci. Lett.*, 8, 1095–1097, 1989.

242. Nieuwenhuizen M. S. and Barendu A. W., Processes involved at the chemical interface of a SAW chemosensor, *Sens. Actuators B*, 11, 45–62, 1987.

243. Chirino A., Boccuzzi F., and Ghiitti G., Surface chemistry and electronic effects of O_2, NO and NO/O_2 on SnO_2, *Sens. Actuators B*, 5, 189–192, 1991.

244. Su P.-G., Wu R.-J., and Nieh F.-P., Detection of nitrogen dioxide using mixed tungsten oxide-based thick film semiconductor sensor, *Talanta*, 59, 667–672, 2003.

245. Tsiulyanu D., Marian S., Miron V., and Liess H.-D., High sensitive tellurium based NO_2 gas sensor, *Sens. Actuators B*, 73, 35–39, 2001.

246. Pijolat C., Pupier C., Sauvan M., Tournier G., and Lalauze R., Gas detection for automotive pollution control, *Sens. Actuators B*, 59, 195–202, 1999.

247. Becker T., Mühleberger S., Braunmühl Chr. Bosch-v., Müller G., Ziemann T., and Hechtenberg K. V., Air pollution monitoring using tin-oxide-based microreactor systems, *Sens. Actuators B*, 69, 108–119, 2000.

248. Cabot A., Marsal A., Arbiol J., and Morante J. R., Bi_2O_3 as a selective sensing material for NO detection, *Sens. Actuators B*, 99, 74–89, 2004.

249. Cantalini C., Sun H. T., Faccio M., Ferri G., and Pelino M., Niobium-doped α-Fe_2O_3 semiconductor ceramic sensors for the measurement of nitric oxide gases, *Sens. Actuators B*, 24–25, 673–677, 1995.

250. Tamaki J., Fujii T., Fujimori K., Miura N., and Yamazoe N., Application of metal tungstate-carbonate composite to nitrogen oxides sensor operative at elevated temperature, *Sens. Actuators B*, 24–25, 396–399, 1995.

251. Bartolomeo E. D., Kaabbuathong N., Grilli M. L., and Traversa E., Planar electrochemical sensors based on tape-cast YSZ layers and oxide electrodes, *Solid State Ionics*, 171, 173–181, 2004.

252. Watson J., The tin oxide gas sensor and its application, *Sens. Actuators B*, 5, 29–42, 1984.

253. Schipper E. F., Kooyman R. P. H., Heidman R. G., and Greve J., Feasibility of optical waveguide immunosensors for pesticide detection: Physical aspects, *Sens. Actuators B*, 24–25, 90–93, 1995.

254. Brailsford A. D. and Logothetis E. M., Selected aspects of gas sensing, *Sens. Actuators B*, 52, 195–203, 1998.

255. Takeuchi T., Oxygen sensors, *Sens. Actuators B*, 14, 109–124, 1988.

256. Gerblinger J., Hardtl K. H., Meixner H., and Aigner R., High-temperature microsensors, in *Sensors, A Comprehensive Survey*, ed. W. Gopel, Vol. 8 (Weinheim: VCH, 1995), 181.

257. Lampe U., Gerblinger J., and Meixner H., Comparison of transient response of exhaust gas sensors based on thin films of selected metal oxides, *Sens. Actuators B*, 7, 787–791, 1992.

258. Cho S.-S. and Kim H.-G., Dynamic characteristics of $SrTiO_3$ thick film as an oxygen sensor, *4th International Conference on Electronic Ceramics and Applications, Proceedings 2: Electroceramics IV*, Aachen, 1994, 749.

259. Wernicke R., The kinetics of equilibrium restoration in barium titanate ceramics, *Philips Res. Rep.*, 31, 526–543, 1976.

260. Moseley P. T., Solid state gas sensors, *Meas. Sci. Technol.*, 8, 223–237, 1997.

261. Yu C., Shimizu Y., and Arai H., Investigations on a lean-burn sensor using perovskite type oxides, *Chem. Lett.*, 14, 563–566, 1986.

262. Moseley P. T. and Williams D. E., Gas sensors based on oxides of early transition metals, *Polyhedron*, 8, 1615–1618, 1989.

263. Moos R., Menesklou W., Schreiner H.-J., and Hardtl K. H., Materials for temperature independent resistive oxygen sensors for combustion exhaust gas control, *Sens. Actuators B*, 67, 178–183, 2000.

264. Steinsvink S., Bugge R., Gjønnes J., Taftø J., and Norby T., Defect structure of $SrTi_{1-x}Fe_xO_{3-y}$ (x = 0–0.8) investigated by electrical conductivity measurements and electron energy loss spectroscopy (EELS), *J. Phys. Chem. Solids*, 58, 969–976, 1997.

265. Jurado J. R., Figueiredo F. M, Gharbage B., and Frade J. R., Electrochemical permeability of $Sr_{0.7}(Ti,Fe)O_{3-\delta}$ materials, *Solid State Ionics*, 118, 89–97, 1999.

266. Abrantes J. C. C., Labrincha J. A., and Frade J. R., Evaluation of $SrTi_{1-y}Nb_yO_{3+\delta}$ materials for gas sensors, *Sens. Actuators B*, 56, 198–205, 1999.

267. Denk I., Claus J., and Maier J., Electrochemical investigations of $SrTiO_3$ boundaries, *J. Electrochem. Soc.*, 144, 3526–3535, 1997.

268. Vollmann M., Hagenbeck R., and Waser R., Grain-boundary defect chemistry of acceptor-doped titanates: Inversion layer and low-field conduction, *J. Am. Ceram. Soc.*, 80, 2301–2314, 1997.

269. Costa M. E. V., Jurado J. R., Colomer M. T., and Frade J. R., Effects of humidity on the electrical behavior of $Sr_{0.97}Ti_{0.97}Fe_{0.03}O_{3-\delta}$, *J. Eur. Ceram. Soc.*, 19, 769–772, 1999.

270. Zhou X. H., Sørensen O. T., and Xu Y. L., Defect structure and oxygen sensing properties of Mg-doped $SrTiO_3$ thick fi lm sensors, *Sens. Actuators B*, 41, 177–182, 1997.

271. Zhou X. H., Sørensen O. T., Cao Q. X., and Xu Y. L., Electrical conduction and oxygen sensing mechanism of Mg-doped $SrTiO_3$ thick film sensors, *Sens. Actuators B*, 65, 52–54, 2000.

272. Feighery A. J., Abrantes J. C. C., Labrincha J. A., Ferreira J. M. F., and Frade J. R., Microstructural effects on the electrical behavior of $SrTi_{0.95}Nb_{0.05}O_{3+\delta}$ materials on changing from reducing to oxidizing conditions, *Sens. Actuators B*, 75, 88–94, 2001.

273. Ding T. and Jia W., Electrophoretic deposition of $SrTi_{1-x}Mg_xO_{3-\delta}$ films in oxygen sensor, *Sens. Actuators B*, 82, 284–286, 2002.

274. Nehlsen S., Hunte T., and Müller J., Gas permeation properties of plasma polymerized thin film siloxane-type membranes for temperatures up to 350°C, *J. Med. Screen.*, 106, 1–7, 1995.

275. Gu C., Sun L., Zhang T., Li T., and Zhang X., High-sensitivity phthalocyanine LB film gas sensor based on field effect transistors, *Thin Solid Films*, 327–329, 383–386, 1998.

276. Hu W., Liu Y., Xu Y., Liu S., Zhou S., Zhu D., Xu B., Bai C., and Wang C., The gas sensitivity of a metal-insulator-semiconductor field-effect-transistor based on Langmuir-Blodgett films of a new asymmetrically substituted phthalocyanine, *Thin Solid Films*, 360, 256–260, 2000.

277. Xie D., Jiang Y., Pan W., Jiang J., Wu Z., and Li Y., Study on bis [phthalocyaninato] praseodymium complex/silicon hybrid chemical field-effect transistor gas sensor, *Thin Solid Films*, 406, 262–267, 2002.

278. Comini E., Cristalli A., Faglia G., and Sberveglieri G., Light enhanced gas sensing properties of indium oxide and tin dioxide sensors, *Sens. Actuators B*, 65, 260–263, 2000.

279. Comini E., Faglia G., and Sberveglieri G., UV light activation of tin oxide thin films for NO_2 sensing at low temperatures, *Sens. Actuators B*, 78, 73–77, 2001.

280. Barker P. S., Bartolomeo C. Di, Monkman A. P., Petty M. C., and Pride R., Gas sensing using a charge-flow transistor, *Sens. Actuators B*, 24–25, 451–453, 1995.

281. Xie D., Jiang Y., Pan W., and Li Y., A novel microsensor fabricated with charge-flow transistor and a Langmuir-Blodgett organic semiconductor film, *Thin Solid Films*, 424, 247–252, 2003.

2 Conducting Polymer Nanocomposite Membrane as Chemical Sensors

Manoj K. Ram
Eclipse Energy Systems Inc.
St. Petersburg, Florida

Ozlem Yavuz, Matt Aldissi
Fractal Systems Inc.
Safety Harbor, Florida

CONTENTS

2.1   INTRODUCTION

The Environmental Protection Agency (EPA) Office of Air Quality Planning and Standards (OAQPS) has set National Ambient Air Quality Standards for pollutants like carbon monoxide (CO), nitric oxide (NO_x), lead, sulphur dioxide (SO_2), ozone (O_3), aromatic hydrocarbons, particulate matters, and so forth. Some of these pollutants are listed in Table 2.1 [1]. EPA has defined the primary standard (with an adequate margin of safety, to protect the public health) and secondary standard (air quality defines levels of air quality necessary to protect the public welfare from any known or anticipated adverse effects of a pollutant) for keeping the quality of air. Such standards are subject to revision and are promulgated with the need to protect the public health. So, it is necessary to measure the pollutants in air below the standard set by EPA (U.S.) and also the European Union (EU) for such hazardous gases (Table 2.1).

Gas sensors based on metal oxide sensing layers—tin oxide (SnO_2), zirconium oxide (ZrO_2), tungsten oxide (WO_3), titanium oxide (TiO_2), indium oxide (In_2O_3), tantalum oxide (Ta_2O_5), and so on and in particular SnO_2—are widely known for such combustion products [2–5]. Although SnO_2 sensors do not primarily exhibit selectivity toward any of these species in general, a certain degree of selectivity is obtained by forming arrays of sensors that are distinguished by their cross-sensitivity and doping with metals such as platinum (Pt), palladium (Pd), ruthenium (Ru), rhodium (Rh), silver (Ag), copper (Cu), and nickel (Ni) [6]. A second concern is that metal oxide sensors tend to suffer from baseline drifts upon interaction with poisoning species such as SO_2 and nitrogen dioxide (NO_2). The third concern is due to the dual response of oxides used in the automotive field, particularly SnO_2, to oxidizing (NO_2) or reducing (CO) gases [7,8], but it, however, reveals some disadvantages such as lack of selectivity and sensitivity at ambient humidity and the higher (300°C–500°C) operating temperature [9,10]. To meet the need for analyzing gas mixtures, and to overcome stability, selectivity, and cost problems, other classes of thin-film sensors are being developed [11]. The electronic conductors (WO_3, TiO_2) [12,13]; mixed conductors $La_2Ni_{0.9}Co_{0.1}O_{4+\delta}$ [14,15]; ionic conductors such as ZrO_2, lanthanum fluoride (LaF_3) [16,17], phthalocyanines (PbPc, LuPc) [18–20]; and conducting polymers (polypyrroles, polythiophenes, and polyanilines) have been used for the development of NO_x, hydrocarbon, organic vapor, and CO gases, respectively [21–32]. Besides, Fourier Transform Infrared (FTIR) spectroscopy has also been used for the detection of NO_x type of gases, with cost being an issue [33–35]. WO_3 thin films, used in micro-NO gas sensors for portable detectors, show a good sensitivity but lack selectivity and operate at 300°C [36]. However, the presence of humidity affects the performance and sensitivity of metal oxide type of sensors for practical use [37].

So, there has also been considerable interest in the use of conducting polymers, particularly polypyrrole (PPy), polythiophene, and polyaniline (PANI), in the form of thin films or blends with conventional polymers as sensors for airborne volatiles such as alcohols, ethers, halogens, ammonia (NH_3), NO_2, and warfare simulants

TABLE 2.1
Standard Set for Maintaining the Quality of Air [1,62]

Pollutant	Standard Value	Standard Set by NAAQS for U.S.	EU Standard Attention Level
Carbon monoxide (CO)			12.5 ppm
8-hour average	9 ppm (10 mg/m³)	Primary	
1-hour average	35 ppm (40 mg/m³)	Primary	
Nitrogen dioxide (NO₂)	0.053 ppm (100 µg/m³)	Primary and secondary	0.1 ppm
Annual arithmetic mean			
Ozone (O₃)			0.09 ppm
1-hour average	0.12 ppm (235 µg/m³)	Primary and secondary	
8-hour average	0.08 ppm (157 µg/m³)	Primary and secondary	
Lead (Pb)	1.5 µg/ m³	Primary and secondary	
Quarterly average			
Sulfur dioxide (SO₂)	0.03 ppm (80 µg/m³)	Primary	
Annual arithmetic mean	0.14 ppm (365 µg/m³)	Primary	
24-hour average			
3-hour average	0.50 ppm (1300 µg/m³)	Secondary	
Particulate matter (PM 10)	150 µg/m³	Primary	
24 hours			
Particulate matter (PM 2.5)			
Annual arithmetic mean	15.0 µg/m³	Primary	
8 hours	35 µg/m³	Primary	

[38–43]. Polymer-based sensors are relatively low-cost materials, their fabrication techniques are quite simple, and they can be deposited on different types of substances [44,45]. The gas-sensing properties of PPy by exposing PPy-impregnated filter paper to ammonia vapor have been measured [46]. It has been observed that nucleophilic gases such as NH_3, methanol (CH_3OH), and ethanol (C_2H_5OH) vapors cause a decrease in conductivity with electrophilic gases [NO_x, phosphorous trichloride (PCl_3), SO_2] having the opposite effect [47–49]. Most of the widely studied conducting polymers in gas-sensing applications are polythiophenes [50,51], PPys [52,53], PANIs, and their composites [50,54–59] films. The gas sensors have also been studied using electrically conducting composite such as polyacrylonitrile (PAn)/PPy, polythiophene (PTh)/polystyrene (PS), polythiophene/polycarbonate (PC), PPy/polystyrene (PS), PPy/polycarbonate (PC), and PPy/poly(methyl methacrylate) (PMMA) films [60–62]. The interaction of this polymer with gas molecules decreases the polaron density in the band-gap of the polymer. It has been observed that PANI–PMMA composite coatings are sensitive to very low concentrations of NH_3 gas (10 ppm). The acrylic acid-doped PANI has been shown to measure the ammonia vapor over a broad range of concentrations of 1–600 ppm [63]. Our recent experience has shown that functionalization of PPy, that is, poly(nitrotoluene pyrrole), reveals selectivity to NO_x [64]. Electrical and electrochemical characterization of LB films of PANIs and polythiophenes have shown good reproducibility in sensing characteristics [64–67]. The Langmuir-Blodgett (LB) films of poly(ortho-anisidine) detect protonic acids hydrochloric acid (HCl) and sulphuric acid (H_2SO_4) in water at a sensitivity levels less than 0.1 ppm [32].

The main advantages of such films are highlighted with examples of high sensitivity and fast response time. Emphasis is also placed on the synergistic combination of distinct materials in the films, in cases where control of molecular architecture is essential for the sensing ability. Following a short description of the two methods for producing the films, sections are devoted to gas sensors, taste sensors, and biosensors, all based on either LB or LBL films [17,18,32,66]. Recently, the nanocomposite membranes have been used in the detection of volatile gases (NO_2, CO, and ammonia etc.) sensitively and selectively [67–77]. In this chapter, we describe the fabrication and characterization of gas sensors based on highly organized ultrathin films of conducting polymers and their nanocomposite films.

2.2 TECHNIQUES FOR FILM FABRICATION

We have concentrated solely on the sensors produced from organic compounds. There are two main methods to produce nanostructured organic films, namely the LB [78–80] and the self-assembly or layer-by-layer (LBL) technique [81–83].

2.2.1 LANGMUIR-BLODGETT OR LANGMUIR-SCHAEFER TECHNIQUES

The LB technique is widely used for the deposition of amphiphilic molecules and macromolecules with amphiphilic segments. LB technique denotes the transfer of monolayers/multilayers from air-water interface onto a solid substrate. The molecular film at the water-air interface is known as Langmuir film [84–90]. This technique has been used for preparing organic thin films in which the functional parts are arranged in ordered states. Salts of fatty acids are classic objects of LB technique. The PANI, poly(o-anisidine) (POAS), poly(o-toluidine) (POT), and poly(o-ethoxyaniline) (PEOA). Langmuir monolayers were studied and fabricated by Ram et al. [88–91] (Figure 2.1). The stability of a Langmuir monolayer is usually associated with a high collapse pressure, a steep increase in the pressure curve in the condensed phase, and a small hysteresis in the compression–expansion cycle. The solid support is vertically dipped and raised. If the even number of monolayers is deposited on the substrate with continuous insertion and withdrawal, Y-type LB films are formed. If the monolayers are deposited by inserting the substrate, X-type LB films are produced. If deposition occurs while substrate is taken out of water, Z-type LB films are formed.

2.2.2 MOLECULAR-LEVEL PROCESSING OF CONJUGATED POLYMERS USING LAYER-BY-LAYER MANIPULATION

Layer-by-layer (LBL) nanostructured films based on electrostatic interactions between charged species have been extensively employed as vapor sensors. The self-assembly is the spontaneous assembly of sets of comparatively simple subunits

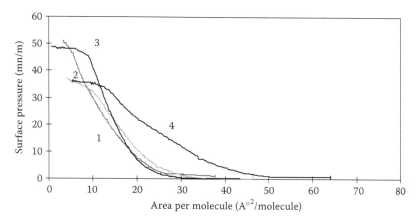

FIGURE 2.1   Pressure-area isotherm of Langmuir monolayer in aqueous subphase at pH 1. 1—PANI, 2—POT, 3—POAS, and 4—PEOA [91].

(molecular or otherwise) into highly complex supramolecular or molecular species of defined structure. Recently, the "molecular self-assembly" phenomenon has been applied to prepare materials with novel optical and electrical properties. It is believed that self-assembled monolayers (SAM), which have desired control on the order at the molecular level, should be considered as a potential technique for the construction of future advanced materials [80–83]. The most commonly employed LBL assembly interactions are electrostatic, where a film is created by combining a polycation and a polyanion layer accurately. The most substantial advantages of the LBL self-assembly are the quite accurately controlled average thickness of the polyelectrolytes layers, where the macroscopic properties of molecular film can be controlled by microscopic structure. In this approach, thin films are built up on a LBL basis by alternatively exposing a substrate to positive and negative polyelectrolytes—polyanion and polycation. On each deposition cycle the charge on the exposed surface is compensated and reversed by adsorbed polymers. The amount of material deposited on each cycle approaches a constant and reproducible value, permitting any number of layers to be incorporated. The instrument required is minimized, and the technique has been proven to be a powerful method to produce thin structures with defined composition. The schematic of LBL multilayer film formation of polyanions and the polycations is schematically depicted in Figure 2.2. This technique was, then, successfully applied to at least more than 20 water-soluble, linear polyanions including conducting polymers, ceramics, dyes, porphyrin, DNA, and proteins [83 and cross-references]. Molecular-level processing of conjugated polymers [i.e., PPy, PANI, poly(phenylene vinylene), poly(o-anisidine)] by LBL technique was also fabricated by LBL self-assembled technique. The self-assembly LBL technique was used for the sequential adsorption of polycation, poly(diallyldimethylammonium chloride) (PDDA), and polyanion, sulfonated PANI (SPANI) on various substrate [92,93].

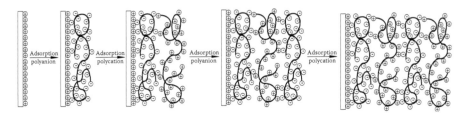

FIGURE 2.2 Schematic of the multilayer assemblies by consecutive adsorption of polyanion and polycation.

2.2.3 IN-SITU SELF-ASSEMBLED TECHNIQUE

In-situ self-assembled LBL technique has been introduced by Rubner et al. for deposition of PPy, poly(thiophene acetic acid), and PANI conducting polymer films [84]. It is an extension technique of LBL self-assembly. The in-situ molecularly oriented film of conducting polymer on activated surfaces can be deposited while the monomers are chemically polymerized in a solution containing electrolyte and oxidizing agents [93–95]. The monolayers of electrically conducting polymers are spontaneously adsorbed onto a substrate from the dilute solutions and subsequently built up into multilayer thin films by alternating deposition with a soluble polyanion or nonionic polymer. Multilayered structures are fabricated by alternatively dipping a substrate into in-situ polymerized conducting polymeric solution and a polyanion solution. We have fabricated the in-situ self-assembled films of PPy. A schematic drawing of the procedure is sketched in Figure 2.3a. A single layer of PPy was obtained in 5 minutes by the in-situ polymerization. The alternate bilayers of polystyrene sulfonate/polypyrrole (PSS/PPy) were fabricated by dipping the protonated substrates in PSS solution for 10 minutes, and for 5 minutes in PPy active solution followed by washing and drying in each step of deposition. The schematic of LBL deposition of PSS and PPy films has been shown in Figure 2.3b. Various in-situ self-assembled films of PANI and substituted thiophenes have been fabricated [93].

2.2.3.1 Fabrication of Conducting Polymer-SnO₂ Composite Films

The SnO_2 particles were prepared using a standard procedure [64–69]. $SnCl_4$ was made soluble in 1 M HCl and then added to 400 ml of deionized water. Aqueous ammonia was added dropwise to this solution to obtain a dispersion of fine SnO_2 particles. The solution was centrifuged to remove excess ammonia and unreacted $SnCl_4$. The pH of the dispersion medium was adjusted by addition of water, followed by the appropriate amounts of aniline monomer and $(NH_4)_2S_2O_8$ (ammonium persulfate) oxidant to initiate the polymerization of aniline in the aqueous medium. The polystyrene sulphonates (PSS) (Mw = 70,000) treated substrate was then introduced in the resulting solution for deposition of self-assembled film.

2.2.3.2 Sulfonated Self-Assembled Films

Sulfonated polyaniline (SPAn) is of interest because of its unusual physical properties and improved processability, and it is easy to process into supramolecular films

(a) Schematic of in-situ self-assembly of PPy films as a function of time on PSS surface

(b) Schematic of in-situ self-assembly of PPy/PSS films by layer by layer technique

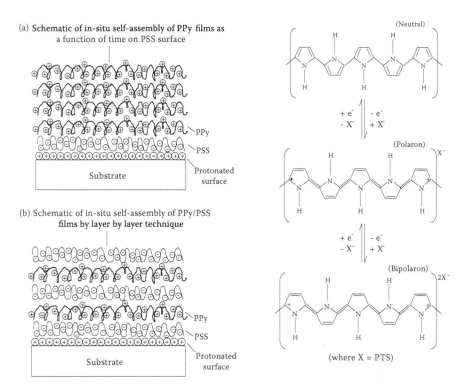

FIGURE 2.3 (a) Schematic of in-situ self-assembly of PPy on PSS surface as function of time. (b) Schematic of in-situ self-assembled LBL films of PPy with PSS.

[64–69]. SPAn is the first self-doped water-soluble conducting PANI derivative and a prime model for dopant and secondary dopant induced processability in parent PANI doped with HCl. The fabrication of SPAn LBL films using the polycation, poly(diallyldimethylammonium chloride) (PDDA), and SPAn as the polyanion were deposited on various surfaces activated with PSS solution for 15 minutes (prepared by using 2 mg ml-1 of PSS in water), which provided the charges necessary to adsorb the first layer of PDDA. The commercial SPAn (Aldrich) was used for deposition of SPAn self-assembled layers. SPAn solution (2 ml) was dissolved in 40 ml water and pH was adjusted to 3, and the resulting solution was filtered to remove any trace of undissolved SPAn particles. Later, this solution was adjusted to pH 1.2 by the dropwise addition of 1 M HCl solution. The multilayer structure was fabricated by alternate dipping of treated substrates in the PDDA and SPAn solution for 10 minutes each by rigorously washing in a solution of pH 2 and drying in a nitrogen gas flow. The alternating layers of PDDA and SPAn were deposited onto various PSS-treated substrates.

2.2.3.3 The UV-Vis Study on PANI/PSS LBL Films

Figure 2.4 shows the optical spectra of polystyrene sulfonate/polyaniline (PSS/PANI) as a function of a number of bilayers from 1 to 25. It reveals the two sharp absorption

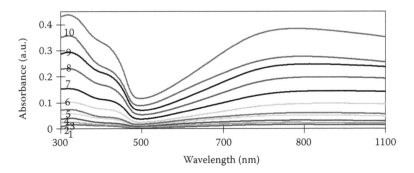

FIGURE 2.4 The optical spectra of PSS/PANI as a function of the number of bilayers made in 2.8 MeSA vs (1) 1 bilayer, (2) 2 bilayers, (3) 4 bilayers, (4) 5 bilayers, (5) 8 bilayers, (6) 11 bilayers, (7) 15 bilayers, (8) 18 bilayers, (9) 20 bilayers, and (10) 25 bilayers [63,89].

bands at 340 nm and 800–850 nm for the film made at pH 2.8 using MeSA. The observed peak at 340 nm can be attributed to a π-π* transition centered on the benzoid ring (interband transition), and the band seen at 850 nm is due to the dopant incorporated, when the films were formed at pH 2.8 (partial doped state of PANI). A small band at 450 nm can also be seen in Figure 2.4, which could be due to the polarons when the films are formed at pH 2.8 using MeSA [94]. The UV-visible absorbance increases gradually with the increase in the number of bilayers of PANI/PSS LBL films from 1 to 25 bilayers, which reveals the uniformity in deposition. The films were undoped using NaOH solution for 10 minutes each, and the respective UV-visible spectra of the films were recorded. There is an induced absorption peak at 330 nm and a broad band is located near 620 nm, which characterizes an emeraldine base (giving the familiar blue color) form of PANI. The uniformity of the film could be well maintained while undoping the film for the emeraldine base form of PANI. It demonstrates a linear increase in the measured absorption magnitude till 25 bilayers [94].

2.3 GAS SENSORS BASED ON CONDUCTING NANOCOMPOSITE FILMS

The chemical–physical properties of nanocomposite and membrane finds unique place in sensor application due to combinational properties. The basic use of nanocomposite is to the products, which show many folds of improvement on the physical and mechanical properties or on the processing properties upon addition of very minute quantity of nanomaterials [99]. Nanoscale particles not only enhance the mechanical properties but also have wide potential in the field of electronic, magnetic, optical, and chemical field. The polymer nanocomposites provide improvement over other known composites in thermal, mechanical, electrical, and even air barrier properties [64–70]. Formulation of nanocomposite membranes with suitable polymer, suitable nanoparticles, and the processing technology of the nanocomposite are critical to success factor to dominate the gas sensor product in the market.

2.3.1   ENVIRONMENTAL SENSORS (HEAVY METALS AND ACIDS)

Environmentally polluting gases like HF, HCl, Cl_2, and so on, have been detected with sensitivity down to 0.1 ppm with response times varying from milliseconds to seconds, which is at the forefront of the existing technology. The presence of heavy metals, even at low concentrations, may also cause severe hazards to the normal functioning of the aquatic ecosystem [32]. PPy-coated carbon nanofibers (CNFs) were fabricated using one-step vapor deposition polymerization (VDP). The PPy-coated CNFs showed an excellent performance as a toxic gas sensor detecting ammonium (NH^{4+}) and hydrochloric acid (HCl) [95]. The LBL multiwalled carbon nanotubes (MWNTs) and PANI multilayer films on gassy carbon (GC) electrodes have been shown to be excellent amperometric sensors for H_2O_2 from +0.2 V over a wide range of concentrations [96]. The biosensors for choline are also developed using LBL assembled functionalized MWNTs and PANI multilayers [96]. The response signal of PPy-coated CNFs was dependent on the thickness of PPy layer. Recent studies on using polythiophene for the detection of metals can be found in literature, and we have also seen the detection of heavy metals ions with high accuracy using conducting polymers. POAS LB films of 40 monolayers on interdigitated electrode were used to detect protonic acid inside water [93]. The films were undoped in aqueous ammonia for 5 minutes and later washed in water and dried by blowing nitrogen gas. Each of the LS films on interdigitated electrode was dipped at different concentrations of HCl, H_2SO_4, and CH_3COOH solutions for 5 minutes and dried by blowing nitrogen gas [32]. Consequently, current-voltage measurements of the films on interdigitated electrode were performed, and the magnitude of current was measured at 0.5 V for each acid-doped POAS LS film [32]. Significant changes in the current magnitude for various concentrations of HCl, H_2SO_4, and CH_3COOH ions showed a continuous increase in log (current magnitude) versus log (concentration of acid). The magnitude of current was found to be different for various concentrations of acid: CH_3COOH-treated interdigitated electrode/POAS LS films showed a decrease in magnitude of current of 2 orders compared to HCl^- and $H_2SO_4^-$ treated films. The presence of a small amount of acid in the solution causes diffusion into the films, which can be sensed by LS film. The DPV study reveals that heavy metal ions in the solution can be recognized due to the well-separated oxidation potential of various metal ions [97]. Several heavy metal cations (Pb^{2+}, Hg^{2+}, Cd^{2+}, and Cu^{2+}) in water have been measured using conducting polymer and composite membrane [97].

2.3.2   CARBON MONOXIDE SENSOR

Much of the welfare of modern societies relies on the combustion of fossil fuels. To a greater or lesser extent, all processes for energy are associated with the production of toxic by-products such as CO, NO_x, and aromatic hydrocarbons [98]. Because their chemical and physical properties may be tailored over a wide range of characteristics, the use of polymers and composite are finding a permanent place in sophisticated electronic measuring devices such as sensors [100,101]. A significant part of CO and NO_x emission originates from exhaust of motor vehicles due to their increasing number each year. The interaction of CO and NO_x with sunlight tends to produce O_3, which, due to its strongly oxidizing behavior, is believed to be harmful to plants and to the respiratory system of human beings [100]. The CO gas sensor has been fabricated using organized films of PANI-SnO$_2$ and PANI-TiO$_2$ nanocomposites systems.

The gas-sensing system consists of a Parr pressure chamber equipped with the necessary valves to achieve vacuum, introduce gases, and perform in situ four-probe conductivity measurements [68,69]. Resistance values obtained for three types of different nanocomposite PANI-metal oxide films are shown in Figures 2.5 and 2.6. CO acts as an oxidant for PANI-metal oxide nanocomposite films and reveals a

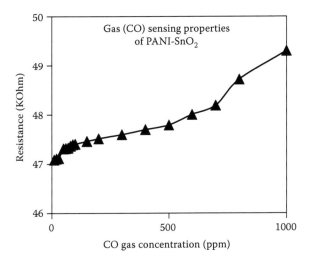

FIGURE 2.5 Resistance change of self-assembled PANI-SnO$_2$ films versus CO gas concentration.

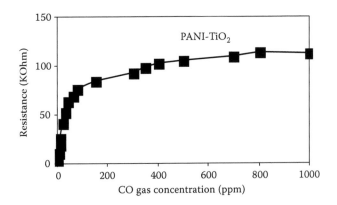

FIGURE 2.6 Resistance change of self-assembled PANI-TiO$_2$ films versus CO gas concentration.

decrease in the resistance value (in kΩ range) for increases in concentration of CO gas. Besides, CO gas-treated films show better conductivity and electrochemical activity compared to untreated PANI-metal oxide nanocomposite films [68]. This is probably due to the fact that interaction between the polymer and SnO_2 nanoparticles resulted in a reduced form by becoming an easily oxidizable PANI. It is important to note that the reversibility of gas adsorption could easily take place under vacuum at room temperature in few minutes. Similar results are obtained with TiO_2 film as shown in Figure 2.6.

2.3.3 NITROGEN DIOXIDE SENSOR

The LBL, in-situ self-assembled LBL films of PPy, PANI, polyhexylthiophene, and poly(ethylene dioxythiophene) (PEDT) and hybrid cross-linked nanocomposite thin film for NO_x gas sensing were fabricated [68,70,101–106] for NO_2 sensing application. Layered films of conducting polymer nanocomposite films were fabricated similar to the one explained in Sections 2.2.1 and 2.2.2.

We attempted to investigate the detection capability of such thin films in the ppb range. The plot shown in Figure 2.7 for a 6-layer polyhexylthiophene (PHTh) film clearly reveals the high sensitivity. The self-assembled films of PANI show the increase in resistance from few KΩ to 110 MΩ upon exposure to 100–200 ppm of NO_2, with a concomitant structural and color change to the emeraldine base [68]. The response of PHTh-TiO_2 (Figure 2.8) films to a mixture of NO_2 and CO in equal amounts indicates high specificity for NO_2, as a sharp decrease in resistance is observed when NO_2 is introduced in the chamber, whereas no significant change takes place when CO is introduced. The gases are introduced into the chamber with 10-ppm increments each time. The sharp decrease in resistance indicates an excellent sensitivity of such films. In addition, total reversibility was observed upon removal

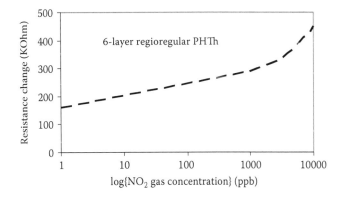

FIGURE 2.7   Resistance change of NO_2 gas at ppb levels using 6-layer PHTh films.

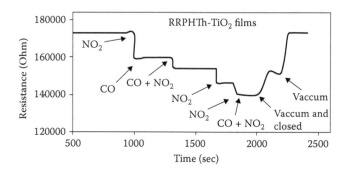

FIGURE 2.8 Resistance change of PHTh-TiO$_2$ films versus time in the presence of CO and NO$_2$.

TABLE 2. 2
Comparisons between Different Materials Tested for the Various Gases [68,70]

Structure	NO$_2$ (ppm)	Comments
SnTiO$_2$ thin films (at 400°C)	1.7	Thin films are sensitive than thick films
SnO$_2$ (crystalline, RF sputtered) (300°C)	5000	Highly selective but effected due to influence of water
SnO$_2$-reactive evaporated, measured at 450°C	0.92	Poor selectivity and stability
Pd-doped SnO$_2$	2.5	Stability
WO$_3$ thin films at 300°C	2.6	1–2 minute response time
Pt–Pd/Si/Al	6.0	No response for detection
Pt/SnO$_2$/n-Si/p-Si/A	1000	The barrier height change or conductivity change of SnO$_2$
CoO/SiO$_2$ nanocomposite	–	Selective to NO
SWCNT/SnO$_2$ nanocomposite material	5 to 60 ppm	High sensitivity, and fast recovery, while working at a relatively low temperature
RRPHTh and SnO$_2$ composite	<100 ppb	<1 s response time, material-based selectivity

of the gas mixture by vacuum. The resistance change of nanocomposite and conducting polymer membranes due to NO$_2$ gas are tabulated in Table 2.2.

2.3.4 ETHANOL SENSOR

Ethanol is a common chemical material in the industry for alimental manufacturing, cosmetic production, and organic synthesis [107]. The ethanol sensors find application in various areas, such as drunken driving, monitoring of fermentation, and other processes in chemical industries [108,109]. Ethanol is also one of the major causes of traffic accidents in the advanced countries, so the laws and regulations are made to improve the fitness of their people by changing the allowed alcohol level from time to time. So the new, simple, and cost-effective techniques are always required for understanding the blood alcohol level (BAC) in the body [110]. The concentration of ethanol is estimated using gas chromatography, liquid chromatography, solid-state sensor, electrochemical method, infrared spectroscopy, and fuel cell techniques [111–114]. Conventional ethanol sensors are mostly based on SnO_2, zinc oxide (ZnO), SnO_2-Nb_2O_5, lanthanum iron oxide ($LaFeO_3$), cadmium oxide (CdO)—iron oxide (Fe_2O_3), Cd-doped SnO_2, WO_3 and γ-Fe_2O_3, and barium stannate but suffer from high sensitivity and specificity to ethanol vapor [115–121]. The ethanol is also determined using alcohol oxidase sensors based on the detection of oxygen and hydrogen peroxide or using microbial sensor measuring oxygen. As regards alcohol dehydrogenase, it is possible to directly detect the cofactor nicotinamide adenine dinucleotide, but this involves the use of a high oxidizing potential, thus increasing the likelihood of interferences due to oxidizable substances present in real samples [121]. The conductance of single-walled semiconducting carbon nanotubes in field-effect transistor (FET) geometry is also recently attempted for alcoholic vapors [122]. The conducting polymer resistors are of increasing interest in the field of gas and odor sensing due to their ease of fabrication and ability to operate at room temperature [123–131]. PANI and its substituted derivatives are found to be sensitive to different alcohol such as methanol, ethanol, propanol, and heptanol vapors [132]. Unfortunately, two disadvantages to their use have been reported by some conducting polymers, that is, they not only have a significant batch-to-batch variation in resistance of 35% but also show a significant (null gas) temperature coefficient of resistance of 0°C. The Pd-PANI nanocomposite, $ZnFe_2O_4$/ZnO polymer nanocomposite, vanadium oxide-PANI nanocomposite, and thionine-carbon nanofiber nanocomposite membranes have been used as alcohol sensors [132–135]. So, we have used the conducting nanocomposite polymer films for detecting alcohol accurately and sensitively. We made an attempt to fabricate ethanol gas sensor based on highly organized ultrathin films of nanocomposite films based on our earlier approach [68–70]. Our approach has focused on the synthesis, characterization, and measurements of PANI-SnO_2 and PANI-TiO_2 nanocomposite film for ethanol gas sensing application.

After the assembly of sensing system, the resistance of the nanocomposite films deposited onto interdigitated electrodes was read by using an electrometer, a desired concentration of testing alcohol was added into the test chamber, and the resistance change was measured. The response current is equal to the measured current minus the background current. The response time of the sensor was estimated based on the volume of the test chamber, which induced a latent time between the gas introduction and the beginning of the sensor response. We have calculated ethanol concentration by converting the weight typically provided in micrograms (μg) to an in-stock concentration by using the following equation:

$$PPM = \frac{(\mu g / MW)}{(V_{std} / GC)} \qquad (2.1)$$

where μg is the total mass of analyte collected, MW is the molecular weight of analyte, V_{std} is the sample volume (i.e., volume of gas sampled) standardized to STP (68°F, 29.92 in Hg) in liters, and GC is the gas constant (4.056)—molar gas volume at STP (68°F, 29.92 in Hg).

We have used interdigitated electrodes in these experiments. Two interdigitated electrodes were spaced to 50 μm between any two pairs. Each track in the interdigitated electrode was 50 μm in width and 40 nm in height. This conductivity change of the sensor films is observed at room temperature. The film was connected to a digital multimeter, with the change in resistance being recorded as pressure varied due to insertion of alcohol. Resistance change is obtained for the introduction of 100 ppm of ethanol to PANI-SnO_2 and PANI-TiO_2 sensor films on the interdigitated gold electrodes. The decrease in the film resistance is observed due to the ethanol doping in the composite films. There is a sudden change in the conductivity value, but the saturation is observed after 1 minute of ethanol exposure. The removal of ethanol was observed as soon as the gas was removed by creating a vacuum in the measuring chamber. It took around 1 minute to see the complete reversibility in the systems as shown in Figure 2.9. As mentioned earlier, as soon as the film came in contact with ethanol vapor, the resistance started to decrease and took nearly 30–40 seconds to saturate, and whenever the vacuum was created, the resistance of PANI-SnO_2 film

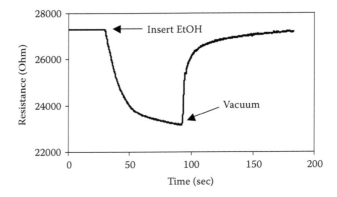

FIGURE 2.9 Resistance change of self-assembled PANI-SnO_2 films upon exposure to ethanol.

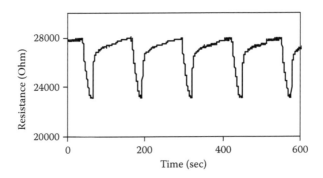

FIGURE 2.10 Resistance change of self-assembled PANI-TiO_2 films upon exposure to ethanol.

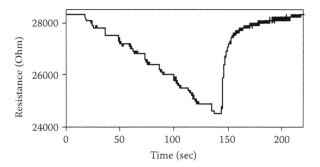

FIGURE 2.11   Reversibility in resistance change of self-assembled PANI-TiO$_2$ films upon exposure to ethanol (100 ppm).

started to increase until the initial point. We have also treated PANI-TiO$_2$ nanocomposite films with ethanol for their reversibility (Figure 2.10). This is a reversible process; resistance decreases with exposure of ethanol and, vice versa, increases without the exposure. The reversibility followed for the PANI-SnO$_2$ films is shown in Figure 2.11. The pattern of the ethanol sensing has been observed similarly by insertion of 100 ppm in equal time. This interesting result reveals that the absorption and desorption of ethanol show reversible behavior.

2.3.5   AMMONIA SENSOR

The detection of minor gas leaks in a hazardous work environment has been a challenging research problem for many decades as it concerns a lot of health, safety, and environmental risks. Tens of thousands of sites containing toxic chemical spills, leaking underground storage tanks, and chemical waste dumps require accurate characterization and long-term monitoring to reduce health risks and ensure public safety. Although a number of chemical sensors are commercially available for field measurements of chemical species (e.g., portable gas chromatographs, surface-wave acoustic sensors, optical instruments, etc.), few have been adapted for use in long-term monitoring applications [138–144]. Spectrometric analysis has, certainly, the richest information, but it is also the most complicated, and the realized devices are still cumbersome and expensive. Thin films of tin oxide-intercalated PANI nanocomposite have been deposited at room temperature through solution route technique. It was found that the PANI film resistance increases, while that of the nanocomposite (PANI-SnO$_2$) film decreases from the respective unexposed value. These changes on removal of ammonia gas are reversible in nature, and the composite films showed good sensitivity with relatively faster response/recovery time [145]. The behaviors, including sensitivity, reproducibility, and reversibility, to various ammonia gas concentrations ranging from 8 ppm to 1000 ppm are investigated using poly(pyrrole) thin films as sensitive layer [146]. A novel system for the detection of ammonia was developed by monitoring the conductance of inkjet printed or screen-printed PANI films with a radio frequency detector, and the sensor does not require an internal power source or associated circuitry and therefore may be a low-cost device suitable for smart packaging applications [147]. The change in resistance of PANI on exposure to aqueous ammonia has been utilized for the study of a prototype chemical sensor and discusses various aspects of PANI and its suitability as a chemical sensor, particularly with reference to aqueous ammonia solution. On the other hand, more compact, more simple, and inexpensive devices were born, but most of them often suffer from the same fault: lack of selectivity. Indeed, these sensors react more or less similarly to a collection of substances, and this could lead to false alerts, or worse, the molecules to be detected could be masked by some interfering compounds. Most of gas sensors available in the market are semiconducting sensors, measuring a conductivity alteration under gas adsorption. Table 2.3 shows the materials, method, and range of concentration of ammonia sensor available in literatures [138–144].

The ammonia sensor containing PANI as the sensitive element is proposed to show high sensitivity, wide range of measured concentrations (1–2000 ppm), and high stability of electrical parameters [149]. This reducing property of PANI is exploited to establish an ammonia gas sensor module for a concentrations range from 5 to 250 ppm with a response and recovery time of typically 50–1000 seconds [150]. The metal oxide sensor has a great effect especially with regard to water vapor, which increases its conductivity. In this context, it is important to identify operating conditions, which either avoids or minimizes sensor-ammonia gas poisoning problems. So, nanocomposite films are used to measure the ammonia gas sensitively and selectively.

2.3.5.1 Methods to Assess the Influence of Water Vapor on the Ammonia Sensor Signal

The ammonia concentration varies with various meteorological elements such as water vapor content changes with time in the environment. Also, because water

TABLE 2.3
Ammonia Sensor Materials and Characteristics [145–150]

Sensing Material	Process	Sensitivity (ppm)	Comments
Nb_2O_5	Conductivity	1	Depends on humidity
$Cr_2Ti_{2-x}O_{7-2x}$	Conductivity	–	Cross-sensitivity due to O_2 and/or H_2O
PPy doped with Cu and Pd	Conductivity	1	Partially reduced
PANI-polyacrylnitrile	Conductivity	1.5	Reversibility (20 times)
ZnO and aluminum oxide	Conductivity	2–5	–
Nafion-crystal violet	Optical fiber	0–20	Calibration on neural network
WO_3 doped with Au and Ru	Conductivity	–	Sensitivity depends on thickness and annealing temp.
Pd-ZNO		30	Tested with or without dopant, response 4 seconds
MoO_3/layer of Ti (50 nm)	Conductivity	–	Operating temp 200°C

vapor content affects the sensor section of the ammonia sensor, that influence must be eliminated. Ammonia is highly water soluble, so dehumidification also removes ammonia. Therefore, one has no choice but to measure the concentration of ammonia, including whatever water vapor is in the air. For that reason, in the laboratory we have synthesized air with varied ammonia concentrations and water vapor pressures, and have examined the ammonia sensor's output to investigate the effects of water vapor pressure on the sensor. Water vapor pressure is controlled by passing room air through a bubbling device (a humidifier–dehumidifier) that is capable of controlling the temperature of the intake air. Both humidity and ammonia concentration are thus controlled by mixing standard ammonia gas with the air after bubbling, as shown in the schematic in Figure 2.12. The flow rate of the synthesized air is kept at several liters per minute. Standard gas containing ammonia and humidified air are mixed in ratios of 0:10, 2:8, 4:6, 6:4, 8:2, and 10:0. Measurements were performed over several days using our sensors.

We changed the ammonia concentration and water vapor pressure and measured the ammonia sensor's output under various combinations of the two. The sensor reading response took several seconds because responses took time even if water vapor pressure was varied while keeping ammonia concentration at zero. We attributed this response time to the responsiveness of the system as a whole. The sensor's output was tested in ambient temperature, which varies with time. The response time curve of sensor has been observed at different concentration. Our ammonia sensor is quite capable of detecting even low ammonia concentrations of about 0.01 ppm.

Conductivity of the proposed materials is typically in the range of 0.1 to several hundreds or several thousands S/cm, depending on the type of polymer and the synthesis technique, and thus exceeding the conductivity of metal oxides. The highly conducting electrode is beneficial for the determination of electrode sensitivity because we can have a large change in conductivity with low concentrations of the gases being monitored, as shown in Figure 2.13. The change in conductivity is caused by electron transfer between adsorbed molecules and conducting

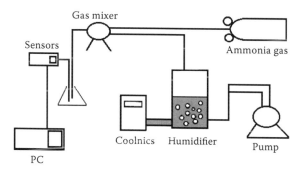

FIGURE 2.12 Schematic of apparatus for measurement of ammonia sensor's responses to NH_3 concentration and water vapor pressure.

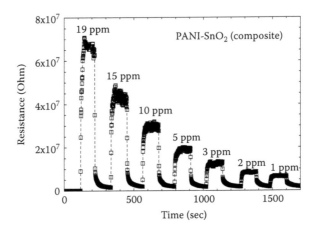

FIGURE 2.13 Ammonia sensing by PANI-SnO$_2$ nanocomposite membranes.

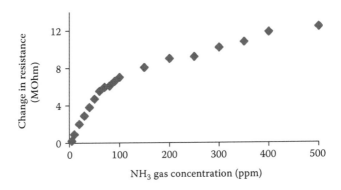

FIGURE 2.14 Ammonia sensing by PANI-SnO$_2$ nanocomposite membranes.

polymer chains. Fine-tuning of sensor properties occurs by the choice of dopants, contacts (intergranular), structure of nanosized particles, overall thickness, and temperature. The proposed inorganic-organic hybrid nanocomposite containing a conducting polymer and a metal oxide exhibits high stability in air and excellent sensing properties according to our recent experiments (Figure 2.14). Intrinsically conducting polymers (ICPs) are highly sensitive to gases and show saturation effects at higher gas concentrations. In terms of sensitivity, the measured sensitivity is less than 0.01 ppm resolution of ammonia [68–70]. The supramolecular approach is utilized to fabricate films of conducting materials via the LBL self-assembly technique for this particular application for the first time to emphasize sensitivity and selectivity issues that are otherwise unattainable with other fabrication methods or materials.

2.3.6   Chemical Warfare Sensor

Early detection of chemical and biological toxins is an important aspect of the defense against the use of such agents in military or terrorist activities. There are also multiple applications in food and water safety, wherein naturally occurring toxins can have devastating effects. Such sensors must enable real-time detection and accurate identification of different classes of pesticides (e.g., carbamates and organophosphates) as well as chemical warfare nerve agents. Present field analytic sensors are bulky with limited specificity, require specially trained personnel, and, in some cases, depend upon lengthy analysis time and specialized facilities. Especially discriminate between widely used organophosphate (OP) pesticides and G- and V-type another study chemical warfare agents are comparatively detected by different analytical techniques such as gas chromatography-infrared detection-mass spectral detection (GC-IR-MS); liquid chromatography-mass spectrometry (LC-MS); nuclear magnetic resonance (NMR) using the nuclei H, C and P; and gas chromatography-atomic emission detection (GC-AED) [151–153]. Nerve gas agents such as sarin, soman, tabun, and VX, along with blister agents such as mustard gas and lewisite were compared for their detection using techniques such as gas detection tube, flame photometric detector, ion mobility spectrometer, surface acoustic wave detector, photo ionization detector, FTIR, and gas chromatography-mass spectrometry for chemical warfare agents. However, these techniques, which are time-consuming and expensive and require highly trained personnel, are available only in sophisticated laboratories and are not amenable to online and rapid monitoring. Biological methods such as immunoassays, biosensors, and inhibition of cholinesterase activity for OP determination have also been reported [154–157]. Carbon black composite sensing arrays have been used to detect explosives and chemical warfare agents such as sarin and soman [158]. Quartz crystal microbalance with immobilized nanoscale ZSM-5 zeolite film has been developed as a sensor for nerve agent. A minimum concentration of 1 part per million (ppm) was detected using the stimulant dimethylmethylphosphonate [159]. Microsensor Systems Inc., a leading producer of surface acoustic wave (SAW) sensors have demonstrated the nerve gas agents and blister gas agents, with a sensitivity of 0.04 ppm in 20 seconds and 0.01 ppm in 120 seconds, respectively [160]. Chemical vapor detection and biosensor array based on flexural plate wave sensor have been demonstrated. A siloxane polymer coating 50-nm thick is applied on the surface for the detection of specific chemical agents. A sensitivity of 10 ppm was demonstrated in the experimental study [161]. EMS-based sensor arrays have been used to detect nerve agents such as tabun and sarin, with a sensitivity of 4 and 26 parts per billion (ppb) respectively. Blister agents such as sulphur mustard were detected with a sensitivity of 16 ppb [162]. Detection of dimethyl methylphosphonate (DMMP), a stimulant for the nerve agents, was demonstrated with ppt detection capability within 10 seconds of exposure using cantilever [163]. The detection of thiol-containing products of enzyme-catalyzed nerve agents is done at the carbon nanotube-modified screen-printed electrode. The sensitivity demonstrated for such sensors is 258 ppb [163]. The capacitive resistor based on dielectric properties has demonstrated the sensitivity of detection of 100 ppm for toxic industrial solvents and 1 ppm for chemical warfare agent as well as explosives

[164]. A polymer-carbon nanotube composite film is provided for use as a sensor for detecting chemical vapors. Nanocomposites of SnO_2 and In_2O_3 have been demonstrated as excellent material for semiconductor gas sensors for toxic industrial gases. Addition of additives is shown to have enhanced sensitivity and selectivity performance [165]. Tin, niobium, and vanadium oxide thin films have been developed for the detection of nerve gas agents sarin. For agglomerates of the size of 40 nm, a sensitivity of 70 ppb was demonstrated [166]. Little work has been devoted to develop the chemical warfare agents using the nanocomposite polymeric materials. There is no one device that meets the need of onsite detection of both chemical and warfare agents [167]. Shortcomings of chemical agent's sensors are that no polymer coating for sensors can display complete selectivity to all possible interferences [168]. The costs of using mass spectrometry, gas chromatography, and ion mobility spectrometry are high, and they are not as easy to use as small electronic nose-type devices [169–173]. We are fabricating chemical warfare sensor using the sensor array of conducting nanocomposite membrane for our future work.

2.4 CONCLUSIONS

We have performed numerous experiments using a variety of materials that have been well characterized using spectroscopic, microscopy, and electrical techniques. The use of ultrathin films for conducting polymers and their nanocomposites with certain inorganic oxides, processed via the LBL self-assembly technique, has proved to be an effective way for simple, fast, and cost-effective sensing of gases. The self-assembled films are typically deposited on precleaned surfaces that are then modified by a polyanion or polycation as a necessary step before the deposition of the active layer to ensure that the proper sequence of deposition is followed and for adhesion purposes. We have fabricated the films with different number of layers (thickness) of the same material and found that 2–10 monolayers are sufficient to form a cohesive film with excellent gas detection capability. The synergy between the conducting polymer and the inorganic oxide leads to the formation of novel materials with gas detection capability beyond that of the conducting polymer and the inorganic oxide used separately. With this technology, sensitivity and selectivity have been dramatically improved compared to commercial systems. For example, sensitivity levels exceed those of oxide-based materials by >2 orders of magnitude with good reversibility in most cases without requiring difficult and costly processes used in commercial systems. Selectivity has been addressed by looking into using different materials that respond differently to the gases. Furthermore, conducting polymer films are known to perform >106 of doping/undoping cycles in similar applications, such as capacitors and electrochromics. With gases that do not form a strong charge transfer complex with the polymers, cycling, and therefore, multiple uses of our sensors will be a natural outcome without using heat activation or deactivation. We have evaluated the stability of the films in situ and ex situ versus temperature and gases. The stability versus time and temperature of these films has been tested and found to be appropriate at temperatures >200°C and in some cases up to 400°C. We have optimized gas detection properties of several materials for several gases (NO_2, CO, ammonia, and ethanol). Recently, we used the nanocomposite for sensitively and selectively measuring the simulants of chemical warfare agents. The sensors are also tested in the presence of mixed gases for specificity and sensitivity studies with excellent results. For example, SO_2 gas did not change the conductivity of the nanocomposite films, indicating that our sensing layers that are specific to NO_2 or CO will not be affected by the presence of SO_2 gas. In order to address viability of this technology, we have fabricated the sensing layers onto interdigitated electrode arrays. The electrodes were characterized for their gas detection capability, which proved to be more effective than their planar electrode counterparts used at the beginning of this effort.

ACKNOWLEDGMENTS

The EPA and NASA Ames Research Center supported this project.

REFERENCES

1. National Ambient Air Quality Standards (NAAQS), http://www.epa.gov/air/criteria.html.
2. Becker T. H., Muhlberger S. T., Braunmuhl, Bosch-v Chr, Muller G., Ziemann T. H., and Hechtenberg K. V., Air pollution monitoring using tin-oxide-based microreactor systems, *Sens. Actuators B*, 69, 108–119, 2000.
3. Santos J., Serrini P., O'Beirn B., and Manes L., A thin film SnO_2 gas sensor selective to ultra-low NO_2 concentrations in air, *Sens. Actuators B,* 43, 154–160, 1997.
4. Pijolat C., Pupier C., Sauvan M., Tournier G., and Lalauze R., Gas detection for automotive pollution control, *Sens. Actuators B*, 59, 195–202, 1999.
5. Meixner H., Gerblinger J., Lampe U., and Fleischer M., Thin-film gas sensors based on semiconducting metal oxides, *Sens. Actuators B*, 23, 119–125, 1995.
6. Lim J.-W., Kang D.-W., Lee D.-S., Huh J.-S., and Lee D.-D., Heating power-controlled micro-gas sensor array, *Sens. Actuators B*, 77, 139–144, 2001.
7. Zhang W., de Vasconcelos E. A., Uchida H., Katsube T., Nakatsubo T., and Nishioka Y. A., A study of silicon Schottky diode structures for NO_x gas detection, *Sens. Actuators B*, 65, 154–156, 2000.
8. Rickerby D. G., Horrillo M. C., Santos J. P., and Serrini P., Microstructural characterization of nanograin tin oxide gas sensors, *Nanostructured Materials*, 9, 43–52, 1997.
9. Melendez J., de Castro A. J., Lopez F., and Meneses J., Spectrally selective gas cell for electrooptical infrared compact multigas sensor, *Sens. Actuators A*, 46(47), 417–421, 1995.
10. Koshizaki N., Yasumoto K., and Sasaki T., Mechanism of optical transmittance change by NO_x in CoO/SiO_2 nanocomposites films, *Sens. Actuators B*, 66, 122–124, 2000.
11. Kocache R. M. A., in *Solid State Gas Sensors*, ed. P. T. Moseley and B. C. Tofield, IOP Publ. Ltd., Bristol: Adam Hilger, 1987, p. 1.
12. Korotcenkov G., Metal oxides for solid-state gas sensors: What determines our choice? *Mater. Sci. Eng. B,* 139(1), 1–23, 2007.
13. Shimizu K., Kashiwagi K., Nishiyama H., Kakimoto S., Sugaya S., Hitoshi Yokoi H., and Satsuma A., Impedance metric gas sensor based on Pt and WO_3 co-loaded TiO_2 and ZrO_2 as total NO_x sensing materials, *Sens. Actuators B*, 130(2), 707–712, 2008.
14. Kharton V. V., Yaremchenko A. A., Tsipis E. V., Valente A. A., Patrakeev M. V., Shaula A. L., Frade J. R., and Rocha J., Characterization of mixed-conducting $La_2Ni_{0.9}Co_{0.1}O_{4+\delta}$ membranes for dry methane oxidation, *Appl. Catal. A*, 261(1), 25–35, 2004.

15. Yamada Y., Ueda A., Shioyama H., Maekawa T., Kanda K., Suzuki K., and Kobayashi T., A semiconductor gas sensor system for high throughput screening of heterogeneous catalysts for the production of benzene derivatives, *Meas. Sci. Technol.*, 16(1), 229–234, 2005.

16. Fergus J. W., A review of electrolyte and electrode materials for high temperature electrochemical CO_2 and SO_2 gas sensors, *Sens. Actuators B*, 134(2), 1034–1041, 2008.

17. Ding H., Erokhin V., Ram M. K., Paddeu S., and Nicolini C., Detection of hydrogen sulfide: The role of fatty acid salt Langmuir–Blodgett films, *Mater. Sci. Eng. C*, 11(2), 121–128, 2000.

18. Ding H., Erokhin V., Ram M. K., Paddeu S., Valkova L., and Nicolini C., A physical insight into the gas-sensing properties of copper (II) tetra-(*tert*-butyl)-5,10,15,20–tetraazaporphyrin Langmuir–Blodgett films, *Thin Solid Films*, 379, 279–286, 2000.

19. Umar A. A., Salleh M. M., and Yahaya M., Self-assembled monolayer of copper(II) meso-tetra(4-sulfanatophenyl) porphyrin as an optical gas sensor, *Sens. Actuators B*, 101(1–2), 231–235, 2004.

20. Agbor N. E., Petty M. C., and Monkman A. P., Polyaniline thin films for gas sensing, *Sens. Actuators B*, 28, 173–179, 1995.

21. Timmer B., Olthuis W., and van den Berg A., Ammonia sensors and their applications—a review, *Sens. Actuators B*, 107, 666–677, 2005.

22. Nicolas-Debarnot D. and Poncin-Epaillard F., Polyaniline as a new sensitive layer for gas sensors, *Anal. Chim. Acta*, 475, 1–15, 2003.

23. Maksymiuk K., Chemical reactivity of polypyrrole and its relevance to polypyrrole based electrochemical sensors, *Electroanalysis*, 18, 1537–1551, 2006.

24. Ameer Q., Adeloju S. B., Polypyrrole-based electronic noses for environment and industrial analysis, *Sens. Actuators B*, 106, 541–552, 2005.

25. MacDiarmid A. G., "Synthetic metals": A novel role for organic polymers, *Angew. Chem.*, 113, 2649–2659, 2001; *Angew. Chem. Int. Ed.*, 40, 2581–2590, 2001.

26. Bartlett P. N. and Ling-Chung K., Conducting polymer gas sensors Part III, *Sens. Actuators*, 20, 287–292, 1989; Gardner J. W. and Bartlett P. N., *Electronic Noses, Principles and Applications* (Oxford: Oxford Science Publications, 1999).

27. McQuade D. T., Pullen A. E., and Swager T. M., Conjugated polymer-based chemical sensors, *Chem. Rev.*, 100, 2537–2574, 2000.

28. Monkman A. P., Petty M. C., Agbor N. E., and Scully M. T., Polyaniline gas sensor, U.S. Patent., 5536473. July 16, 1996.

29. Torsi L., Pezzuto M., Siciliano P., Rella R., Sabbatini L., Valli L., and Zambonin P. G., Conducting polymer doped with metallic inclusions: New materials for gas sensors. *Sens. Actuators B*, 48, 362–367, 1998.

30. Chiang J. C. and MacDiarmid A. G., Polyaniline. Protonic acid doping of the emeraldine form to the metallic regime, *Synth. Met.*, 13, 193–197, 1986.

31. Dhawan S. K., Kumar D., Ram M. K., Chandra S., Trivedi D. C., Application of conducting polyaniline as sensor material for ammonia, *Sens. Actuators B*, 40(2–3), 99–103, 1997.

32. Paddeu S., Ram M. K., Carrara S., and Nicolini C., Langmuir-Schaefer films of poly (*o*-anisidine) conducting polymer for sensors and displays, *Nanotechnology*, 9, 228–236, 1998.

33. Siebert R. and Müller J., Infrared integrated optical evanescent field sensor for gas analysis: Part II. Fabrication, *Sens. Actuators A*, 119(2), 584–592, 2005.

34. Melendez J., De Castro A. J., Lopez F., and Meneses J., Spectrally selective gas cell for electrooptical infrared compact multigas sensor. *Sens. Actuators A*, 47(1), 417–421, 1995.

35. Werle P., Slemr F., Maurer K., Kormann R., Mucke R., and Janker B., Near- and mid-infrared *laser-optical* sensors for gas analysis, *Optics and Lasers in Engineering*, 37(2), 101–114, 2002.
36. Labidi A., Jacolin C., Bendahan M., Abdelghani A., Guérin J., Aguir K., and Maaref M., Impedance spectroscopy on WO_3 gas sensor. *Sens. Actuators B*, 106(2), 713–718, 2005.
37. Baraton M.-I., Metal oxide semiconductor nanoparticles for chemical gas sensors, *IEEJ Trans. Sensors Micromach.*, 126, 553–559, 2006.
38. Bartlett P. N., Archer P. B. M., and Ling-Chung S. K., Conducting polymer gas sensors. Part I: Fabrication and characterization, *Sens. Actuators*, 19, 125–140, 1989.
39. Bartlett P. N. and Ling-Chung S. K., Conducting polymer gas sensors. Part II: Response of polypyrrole to methanol vapour, *Sens. Actuators B*, 19, 141–150, 1989.
40. Bartlett P. N. and Ling-Chung S. K., Conducting polymer gas sensors. Part III: Results for four different polymers and five different vapours, *Sens. Actuators*, 20, 287–292, 1989.
41. Chabukswar V. V., Pethkar S., and Athawale A. A., Acrylic acid doped polyaniline as an ammonia sensor, *Sens. Actuators B*, 77, 657–663, 2001.
42. Jain S., Chakane S., Samui A. B., Krishnamurthy V. N., and Bhoraskar S. V., Humidity sensing with weak acid-doped polyaniline and its composites, *Sens. Actuators B*, 96, 124–129, 2003.
43. Hu H., Trejo M., Nicho M. E., Saniger J. M., and Garcia-Valenzuela A., Adsorption kinetics of optochemical NH_3 gas sensing with semiconductor polyaniline films, *Sens. Actuators B*, 82, 4–23, 2002.
44. Thust M., Schöning M. J., Frohnhoff S., Arens-Fischer R., Kordos P., and Lüth H. Porous silicon as a substrate material for potentiometric biochemical sensors, *Meas. Sci. Technol.*, 7, 26–29, 1996.
45. Li K., Diaz D. C., He Y., Campbell J. C., and Tsai C., Electroluminescence from porous silicon with conducting polymer film contacts, *Appl. Phys. Lett.*, 64, 2394–2396, 1994.
46. Nylander C., Armgarth M., and Lundstrom I., An ammonia detector based on a conducting polymer. In: (2nd edn. ed.), *Anal. Chem. Symp. Ser.*, 17, 203–207, 1983.
47. Adhikari B. and Majumdar S., Polymers in sensor applications. *Prog. Polym. Sci.*, 29(7), 699–766, 2004.
48. Amrani E. H., Ibrahim S., and Persaud K. C., Synthesis, chemical characterisation and multifrequency measurements of poly *N*-(2-pyridyl) pyrrole for sensing volatile chemicals, *Mater. Sci. Eng.*, C1, 17–22, 1993.
49. Severin E. J., Doleman B. J., and Lewis N. S., An investigation of the concentration dependence and response to analyte mixtures of carbon black/insulating organic polymer composite vapor detectors, *Anal. Chem.*, 72(4), 658–668, 2000.
50. Bartlett P. N. and Ling-Chung S. K., Conducting polymer gas sensors. Part III: Results for four different polymers and five different vapours, *Sens. Actuators*, 20, 287–292, 1989.
51. Marsella M. J., Carroll P. J., and Swager T. M., Design of chemoresistive sensory materials: Polythiophene-based pseudopolyrotaxanes, *J. Am. Chem. Soc.*, 117, 9832–9841, 1995.
52. Bruschi P., Cacialli F., Nannini A., and Neri B., Gas and vapour effects on the resistance fluctuation spectra of conducting polymer thin-film resistors, *Sens. Actuators B*, 18–19, 421–425, 1994.
53. Torsi L., Pezzuto M., Siciliano P., and Rella R., Sabbatini conducting polymers doped with metallic inclusions: New materials for gas sensors L., Valli L., Zambonin P. G., *Sens. Actuators B*, 48, 362–367, 1998.

54. Hirata M. and Sun L., Characteristics of an organic semiconductor polyaniline film as a sensor for NH_3 gas, *Sens. Actuators A*, 40, 159–163, 1994.

55. Unde S., Ganu J., and Radhakrishnan S., Conducting polymer-based chemical sensor: Characteristics and evaluation of polyaniline composite films, *Adv. Mater. Opt. Electron.*, 6, 151–157, 1996.

56. Ogura K. and Shiigi H. A., CO_2 sensing composite film consisting of base-type polyaniline and poly(vinyl alcohol), *Electrochem. Solid State Lett.*, 2, 478–480, 1999.

57. Ogura K., Shiigi H., Oho T., and Tonosaki T., A CO_2 sensor with polymer composites operating at ordinary temperature, *J. Electrochem. Soc.*, 147, 4351–4355, 2000.

58. Hong K. H., Oh K. W., Kang T. J., Polyaniline-nylon 6 composite fabric for ammonia gas sensor, *J. Appl. Polym. Sci.*, 92, 37–42, 2004.

59. Ogura K., Shiigi H., Nakayama M., and Ogawa A., Thermal properties of poly(anthranilic acid) (PANA) and humidity-sensitive composites derived from heat-treated PANA and poly(vinyl alcohol), *J. Polym. Sci. A*, 37, 4458–4465, 1999.

60. Ogura K., Saino T., Nakayama M., and Shiigi H., The humidity dependence of the electrical conductivity of a soluble polyaniline-poly(vinyl alcohol) composite film, *J. Mater. Chem.*, 7, 2363–2366, 1997.

61. Bhat N. V., Gadre A. P., and Bambole V. A., Investigation of electropolymerized polypyrrole composite film: Characterization and application to gas sensors, *J. Appl. Polym. Sci.*, 88(1), 22–29, 2003.

62. Wang H. L., Toppare L., and Fernandez J. E., Conducting polymer blends: polythiophene and polypyrrole blends with polystyrene and poly (bisphenol A carbonate), *Macromolecules*, 23, 1053–1059, 1990.

63. Li D., Jiang Y., Wu Z., Chen X., and Li Y., Self-assembly of polyaniline ultrathin films based on doping-induced deposition effect and applications for chemical sensors, *Sens. Actuators B*, 66, 125–127, 2000.

64. Ram M. K., Yavuz O., and Aldissi M., Gas sensors based on ultrathin films of conducting polymers, in *Colloidal Nanoparticles in Biotechnology*, ed. Abdelhamid Elaissari (New Jersey: John Wiley & Sons, 2008), 223–245.

65. Ram M. K. (ed.), *The Supramolecular Engineering of Conducting Materials* (Trivandrum, Kerela, India: Transworld Research Network, 2005).

66. Ram M. K., Adami M., Sartore M., Salerno M., Paddeu S., and Nicolini C., Comparative studies on Langmuir-Schaefer films of polyanilines. *Synth. Metals*, 1100, 249–259, 1999.

67. Yavuz O., Ram M.K., and Aldissi M., Electromagnetic Applications of Conducting and Nanocomposite Materials, *The New Frontiers of Organic and Composite Nanotechnology*, in ed. V. Erokhin, M. Ram, and O. Yavuz (New York: Elsevier, 2007).

68. Ram M. K., Yavuz O., Lahsangah V., and Aldissi M., CO gas sensing from ultrathin nano-composite conducting polymer film, *Sens. Actuators B*, 106(2), 750–757, 2005.

69. Ram M. K., Yavuz O., and Aldissi M., NO_2 gas sensing based on ordered ultrathin films of conducting polymer and its nanocomposite, *Synth. Met.*, 151, 77–84, 2005.

70. Bavastrello V., Stura E., Carrara S., Erokhin V., and Nicolini C., Poly(2,5-dimethylaniline)-MWNTs nanocomposite: A new material for conductimetric acid vapours sensor, *Sens. Actuators B*, 98, 247–253, 2004.

71. Bavastrello V., Ram M. K., and Nicolini C. Synthesis of multiwalled carbon nanotubes and poly(o-anisidine) nanocomposite material: Fabrication and characterization of its Langmuir-Schaefer films, *Langmuir*, 18, 1535–1541, 2002.

72. Lu J., Ma A., Yang S, and Ng K. M., Surfactant assisted solid-state synthesis and gas sensor application of a SWCNT/SnO_2 nanocomposite material. *J. Nanosci. Nanotechnol.*, Apr-May, 7(4–5), 1589–1595, 2007.

73. Tai H., Jiang Y., Xie G., Yu J., and Chen X., Fabrication and gas sensitivity of polyaniline–titanium dioxide nanocomposite thin film, *Sens. Actuators B*, 125(2), 644–650, 2007.

74. Sadek A. Z., Wlodarski W., Shin K., Kaner R. B., and Kalantar-zadeh K., A layered surface acoustic wave gas sensor based on a polyaniline/In$_2$O$_3$ nanofibre composite, *Nanotechnology*, 17, 4488–4492, 2006.

75. Santhanam K. S. V., Sangoi R., and Fuller L., A chemical sensor for chloromethanes using a nanocomposite of multiwalled carbon nanotubes with poly(3-methylthiophene) *Sens. Actuators B*, 106(2), 766–771, 2005.

76. Sharma S., Nirkhe C., Pethkar S., and Athawale A. A., Chloroform vapour sensor based on copper/polyaniline nanocomposite, *Sens. Actuators B*, 85(1), 131–136, 2002.

77. Vieira S. M. C., Beecher P., Haneef I., Udrea F., Milne W. I., Namboothiry M. A. G., Carroll D. L., Park J., and Maeng S., Use of nanocomposites to increase electrical "gain" in chemical sensors, *Appl. Phys. Lett.*, 91, 203111–203113, 2007.

78. Roberts G. (ed.), *Langmuir-Blodgett Films* (Plenum: New York, 1990).

79. Petty M. C., *Langmuir-Blodgett Films: An Introduction* (Cambridge University Press, 1996), 1–256.

80. Ulman A., *An Introduction to Ultrathin Organic Films from Langmuir-Blodgett to Self-Assembly* (San Diego: Academic Press, 1991).

81. Decher G., Templating, self-assembly and self-organization, *Comprehensive Supramolecular Chemistry*, vol. 9, ed. J.-P. Sauvage and M. W. Hosseini (Oxford: Pergamon, 1996), 507.

82. Decher G., Fuzzy nanoassemblies: Towards layered polymeric multicomposites, *Science*, 277, 1232–1237, 1997.

83. Erokhin V., Ram M. K., and Yavuz O., *New Frontiers of Organic and Composite Nanotechnology* (Elsevier Science, 2007), 1–477.

84. Rubner M. F. and Skotheim T. A., *Conjugated Polymers*, ed. J. L. Bredas and R. Silbey (Amsterdam: Kluwer, 1991), 363–403.

85. Punkka E. and Rubner M.F., Molecular Heterostructure Devices Composed of Langmuir-Blodgett Films of Conducting Polymers, *J. Electron. Mater.* 21, 1057–1063, 1992.

86. Mattoso L. H. C., Paterno L. G., Campana S. P., and Oliveira O. N. Jr., Kinetics of self-assembled films from doped poly (o-ethoxyaniline) *Synth. Met.*, 84, 123–124, 1997.

87. Ram M. K., Joshi M., Mehrotra M., Dhawan S. K., and Chandra S., *Thin Solid Films*, 304, 65, 1997; Choia H. J., Kima Ji W., Tob K., *Polymer*, 40, 2163, 1999.

88. Ram M. K., Paddeu S., Carrara S., Maccioni E., and Nicolini C., Poly (ortho-anisidine) Langmuir-Blodgett films: Fabrication and characterization, *Langmuir*, 13(10), 2760–2765, 1997.

89. Paddeu S., Ram M. K., and Nicolini C., Investigation of ultra-thin films poly (ortho)-anisidine conducting polymer obtained by the Langmuir-Blodgett technique *J. Phys. Chem. B.*, 101, 4759–4766, 1997.

90. Ram M. K., Maccioni E., and Nicolini C., Electrochromic Response of polyaniline and it's copolymeric systems, *Thin Solid Films*, 303, 27–33, 1997.

91. Ram M. K., Manuela Adami A., Marco Sartore A., Marco Salerno, Sergio Paddeu, and Nicolini C., Comparative studies on Langmuir–Schaefer films of polyanilines, *Synth. Met.*, 100, 249–259, 1999.

92. Ram M. K. and Nicolini C., Supramolecular engineering and applications of polyanilines, in *Supramolecular Engineering of Conducting Materials*, ed. M. K. Ram (Trivandrum, Kerela, India: Transworld Research Network, Research Signpost, 2005).

93. Nicolini C., Erokhin V., and Ram M. K., Supramolecular layer engineering for industrial nanotechnology, in *Nano-Surface Chemistry*, ed. M. Rosoff (New York: Marcel Dekker, 2001), 141–212.

94. Ram M. K., Salerno M., Adami M., Faraci P., and Nicolini C., Physical properties of polyaniline films: Assembled by the layer-by-layer technique, *Langmuir*, 15, 1252–1259, 1999.

95. Bertoncello P., Notargiacomo A., Riley D. J., Ram M. K., and Nicolini C., Preparation, characterization and electrochemical properties of Nafion® doped poly(ortho-anisidine) Langmuir–Schaefer films, *Electrochem. Commun.*, 5(9), 787–792, 2003.

96. Qu F., Yang M., Jiang J., Shen G., and Yu R. Amperometric biosensor for choline based on layer-by-layer assembled functionalized carbon nanotube and polyaniline multilayer film, *Anal. Biochem.*, 344, 108–114, 2005.

97. Ram M. K., Bertoncello P., and Nicolini C., Langmuir-Schaefer films of processable poly(o-ethoxyaniline) conducting polymer: Fabrication and characterization as sensor for heavy metals, *Electroanalysis*, 13, 574–581, 2001.

98. Becker T. H., Muhlberger S. T., Braunmuhl Bosch-v Chr, Muller G., Ziemann T. H., and Hechtenberg K. V., Air pollution monitoring using tin-oxide-based microreactor systems, *Sens. Actuators B*, 69, 108–119, 2000.

99. Adhikari B. and Majumdar S., Polymers in sensor applications, *Prog. Polym. Sci.*, 29, 699–766, 2004.

100. Chen Y.-S., Li Y., Wang H.-C., and Yang M.-J., Gas sensitivity of a composite of multiwalled carbon nanotubes and polypyrrole prepared by vapor phase polymerization, *Carbon*, 45, 357–363, 2007.

101. Santos J., Serrini P., O'Beirn B., and Manes L., A thin film SnO_2 gas sensor selective to ultra-low NO_2 concentrations in air, *Sens. Actuators B*, 43, 154–160, 1997.

102. Kaushik A., Khan R., Gupta V., Malhotra B. D., Ahmad S., and Singh S. P., Hybrid cross-linked polyaniline-WO_3 nanocomposite thin film for NO_x gas sensing, *J. Nanosci. Nanotechnol.*, 9(3), 1792–1796, 2009.

103. Tiwari A. and Gong S. Q., Electrochemical synthesis of chitosan-co-polyaniline/ $WO_3 \cdot nH_2O$ composite electrode for amperometric detection of NO_2 gas, *Electroanalysis*, 20(6), 1775–1781, 2008.

104. Chen L. Y., Bai S. L., Zhou G. J., Dianging L., Aifan C., and Liu C. C., Synthesis of ZnO-SnO_2 nanocomposites by microemulsion and sensing properties for NO_2, *Sens. Actuators B.*, 134(2), 360–366, 2008.

105. Sadek A. Z., Wlodarski W., Shin K., Kaner R. B., and Kalantar-zadeh K., A layered surface acoustic wave gas sensor based on a polyaniline/In_2O_3 nanofibre composite, *Nanotechnoly*, 17(17), 4488–4492, 2006.

106. Li D., Jiang Y. D., Wu Z. M., Chen X. D., and Liy. R., Self-assembly of polyaniline ultrathin films based on doping-induced deposition effect and applications for chemical sensors, *Sens. Actuators B*, 66(1–3), 125–127, 2000.

107. Pang C.-C., Chen M.-H., Lin T.-Y., and Chou T.-C., An amperometric ethanol sensor by using nickel modified carbon-rod electrode, *Sens. Actuators B*, 73(2–3), 221–227, 2001.

108. Oomman K., Varghese L., Malhotra K., and Sharma G. L., High ethanol sensitivity in sol–gel derived SnO_2 thin films, *Sens. Actuators B*, 55(2–3), 161–165, 1999.

109. Ge J.-P., Wang J., Zhang H.-X., Wang X., Peng Q., and Li Y.-D., High ethanol sensitive SnO_2 microspheres, *Sens. Actuators B*, 113(2), 937–943, 2006.

110. Swift R., Transdermal alcohol measurement for estimation of blood alcohol concentration, *Clinical and Experimental Research*, 24(4), 422–423, 2000.

111. Schuhmann W., Zimmermann H., Habermüller K., and Laurinavicius V., Electron-transfer pathways between redox enzymes and electrode surfaces: Reagentless

biosensors based on thiol-monolayer-bound and polypyrrole-entrapped enzymes, *Faraday Discuss*, 116, 245–255, 2000.

112. Yamamoto M., Iwai Y., Nakajima T., and Arai Y., Fourier transform infrared study on hydrogen bonding species of carboxylic acids in supercritical carbon dioxide with ethanol, *J. Phys. Chem. A*, 103, 3525–3529, 1999.

113. Pérez-Ponce A., Rambla F. J., Garrigues J. M., Garrigues S., and de la Guardia M., Partial least-squares-Fourier transforms infrared spectrometric determination of methanol and ethanol by vapour-phase generation. *The Analyst*, 123, 1253–1258, 1998.

114. Hwang B. J., Yang J. Y., and Lin C. W., A Microscopic Gas-sensing model for Ethanol sensors based on conductive polymer composite from polypyrrole and poly(ethylene oxide), *J. Electrochem.* Soc. 146, 1231–1236, 1999.

115. Hellegouarch F., AreKhonsari F., and Planade R., Amouroux PECVD prepared SnO_2 thin films for ethanol sensors, *Sens. Actuators B*, 73, 27–34, 2001.

116. Weber I. T., Andrade R., Leite E. R., and Longo E., A study of the $SnO_2 \cdot Nb_2O_5$ system for an ethanol vapour sensor: A correlation between microstructure and sensor performance, *Sens. Actuators B*, 72, 180–183, 2001.

117. Liu X., Xu Z., Liu Y., and Shen Y., A novel high performance ethanol gas sensor based on $CdO—Fe_2O_3$ semiconducting materials, *Sens. Actuators B*, 52, 270–273, 1998.

118. Zhang T., Shen Y., and Zhang R., Ilmenite structure-type β-$CdSnO_3$ used as an ethanol sensing material, *Mater. Lett.*, 23, 69–71, 1995.

119. Tan O. K., Zhu W., Yan Q., and Kong L. B., Size effect and gas sensing characteristics of nanocrystalline $xSnO_2$- $(1–x)$ α-Fe_2O_3 ethanol sensors. *Sens. Actuators B*, 65, 361–365, 2000.

120. Tao S., Gao F., Liu X., and Srensen T. O., Ethanol-sensing characteristics of barium stannate prepared by chemical precipitation, *Sens. Actuators B*, 71, 223–227, 2000.

121. Zhao S., Sin J. K. O., Xu B., Zhao M., Peng Z., and Cai H., A high performance ethanol sensor based on field-effect transistor using a $LaFeO_3$ nano-crystalline thin-film as a gate electrode, *Sens. Actuators B*, 64, 83–87, 2000.

122. Leca B. and Marty J.-L., Reagentless ethanol sensor based on a NAD-dependent dehydrogenase, *Biosens. Bioelectron.*, 12(11), 1083–1088, 1997.

123. Chi Q. and Dong S., Electrocatalytic oxidation of reduced nicotinamide coenzymes at methylene green-modified electrodes and fabrication of amperometric alcohol biosensors, *Anal. Chim. Acta*, 285, 125–133, 1994.

124. Torsi L., Tanese M. C., Cioffi N., Gallazzi M. C., Sabbatini L., and Zambonin P. G., Alkoxy-substituted polyterthiophene thin-film-transistors as alcohol sensors, *Sens. Actuators B*, 98, 204–207, 2004.

125. Gardner J. W., Vidic M., Ingleby P., Pike A. C., Brignell J. E., Scivier P., Bartlett P. N., Duke A. J., and Elliott J. M., Response of a poly(pyrrole) resistive micro-bridge to ethanol vapour, *Sens. Actuators B*, 48, 289–295, 1998.

126. Deng Z., Stone D. C., and Thompson M., Characterization of polymer films of pyrrole derivatives for chemical sensing by cyclic voltammetry, X-ray photoelectron spectroscopy and vapour sorption studies, *Analyst*, 122, 1229–11238, 1997.

127. Athawale A. A. and Kulkarni M. V., Polyaniline and its substituted derivatives as sensor for aliphatic alcohols, *Sens. Actuators B*, 67, 173–177, 2000.

128. Hwang B. J., Yang J.-Y., and Lin C.-W., A microscopic gas-sensing model for ethanol sensors based on conductive polymer composites from polypyrrole and poly(ethylene oxide), *J. Electrochem. Soc.*, 146(3), 1231–1236, 1999.

129. Ball I. J., Huang S.-C., Miller K. J., Wolf R. A., Shimano J. Y., and Kaner R. B., The pervaporation of ethanol/water feeds with polyaniline membranes and blends, *Synth. Met.*, 102, 1311–1312, 1999.

130. Gardner W., Vidic M., Ingleby P., Pike A. C., Brignell J. E., Scivier P., Bartlett P. N., Duke A. J., and Elliott J. M., Response of a poly(pyrrole) resistive micro-bridge to ethanol vapour, *Sens. Actuators B*, 48, 289–295, 1998.

131. Dupoet P. D., Miyamoto S., Murakami T., Kimura J., and Karube I., Direct electron transfer with glucose oxidase immobilized in an electropolymerized poly(Nmethylpyrrole) film on a gold electrode. *Anal. Chem.*, 235, 255–263, 1990.

132. Amaya T., Saio D., Hirao T., Versatile Synthesis of Polyaniline/Pd Nanoparticles and Catalytic Application, *Macromol. Symp.*, 270(1), 88–94.

133. Nicolas B., Céline M. L., Julien D., Hélène S., Florent C., Lahire B., and Rénal B., Integrative chemistry portfolio toward designing and tuning vanadium oxide macroscopic fibers sensing and mechanical properties, *Comptes Rendus Chimie*, In press, 2009.

134. Lina W. U.,Mclntosh M., Xueji Z., Huangxian J. U., Amperometric sensor for ethanol based on one-step electropolymerization of thionine-carbon nanofiber nanocomposite containing alcohol oxidase, Talanta, 74, 387–392, 2007.

135. Arshak K., Moore E., Cunniffe C., Nicholson M., and Arshak A., Preparation and characterisation of $ZnFe_2O_4$/ZnO polymer nanocomposite sensors for the detection of alcohol vapours, *Superlattices. Microstruct.*, 42(1–6), 479–488, 2007.

136. Dexmer J., Leroy C. M., Binet L., et al., Vanadium oxide-PANI nanocomposite-based macroscopic fibers: 1D alcohol sensors bearing enhanced toughness, *Chem. Mater.*, 20(17), 5541–5549, 2008.

137. Wu L. N., McIntosh M., Zhang X. J., et al., Amperometric sensor for ethanol based on one-step electropolymerization of thionine-carbon nanofiber nanocomposite containing alcohol oxidase, *Talanta*, 74(3), 387–392, 2007.

138. Clifford P. K. and Tuma D. T., Characteristics of semiconductor gas sensors I. Steady state gas response, *Sens. Actuators B*, 3, 233–254, 1983.

139. Srivastava R. K., Lal P., Dwivedi R., and Srivastatva S. K., Sensing mechanism in tin oxide-based thick-film gas sensors, *Sens. Actuators B*, 21, 213–218, 1994.

140. Lundström I., Sevensson C., Spetz A., Sundgren H., and Winquist F., From hydrogen sensors to olfactory images—twenty years with catalytic field-effect devices, *Sens. Actuators B*, 13–14, 16–23, 1993.

141. Spetz A., Armgath M., and Lundström I., Hydrogen and ammonia response of metal-silicon dioxide-silicon structures with thin platinum gates, *J. Appl. Phys.*, 64, 1274–1283, 1988.

142. Ghauch A., Rima J., Charef A., Suptil J., Fachinger C., and Martin-Bouyer M., Quantitative measurements of ammonium hydrogenophosphate and CU (II) by diffuse reflectance spectrometry, *Talanta*, 48, 385–392, 1999.

143. Jin Z., Su Y., and Duan Y., Development of a polyaniline-based optical ammonia sensor, *Sens. Actuators B*, 72, 75–79, 2001.

144. Simon K., Dasgupta P. K., and Vecera Z., Wet effluent denuder coupled liquid/ion chromatography systems, *Anal. Chem.*, 63, 1237–1242, 1991.

145. Deshpande N. G., Gudage Y. G., Sharma R., Vyas J. C., Kim J. B., and Lee Y. P., Studies on tin oxide-intercalated polyaniline nanocomposite for ammonia gas sensing applications, *Sens. Actuators B*, 138(1), 76–84, 2009.

146. Carquigny S., Sanchez J. B., Berger F., Lakard B., Lallemand F., Ammonia gas sensor based on electrosynthesized polypyrrole films, *Talanta*, 78(1), 199–206, 2009.

147. Clark N. B. and Maher L. J., Non-contact, radio frequency detection of ammonia with a printed polyaniline sensor, *React. Funct. Polym.*, 69, 594–600, 2009.

148. Nicolas-Debarnot D., Fabienne poncin-epaillard polyaniline as a new sensitive layer for gas sensors, *Anal. Chim. Acta.*, 475, 1–15, 2003.

149. Prasad G. K., Radhakrishnan T. P., Kumar D. S., and Krishna M. G., Ammonia sensing characteristics of thin film based on polyelectrolyte templated polyaniline, *Sens. Actuators B*, 106(2), 626–631, 2005.

150. Kukla A. L., Shirshov Y. M., and Piletsky S. A., Ammonia sensors based on sensitive polyaniline films, *Sens. Actuators B*, 37(3), 135–140, 1996.

151. Tu A. T. and Gaffield W. Editors, Natural and selected synthetic toxins, biological implications Washington, *American Chemical Society*, 426, 2000. (ACS symposium series, 745).

152. Black R. M., Clarke R. J., Read R. W., and Reid M. T. J., Application of gas chromatography-mass spectrometry and gas chromatography-tandem mass spectrometry to the analysis of chemical warfare samples, found to contain residues of the nerve agent sarin, sulphur mustard and their degradation products, *J. Chromatogr. A*, 662, 301–321, 1994.

153. Robinson J. P. P., *SIPRI Yearbook, World Armaments and Disarmament*, Chapter 6 (London: Taylor & Francis, 1985).

154. Karousos N. G., Aouabdi S., Way A. S., and Reddy S. M., Quartz crystal microbalance determination of organophosphorus and carbamate pesticides. *Anal. Chim. Acta*, 469, 189–196, 2002.

155. Wang J., Krause R., Block K., Musameh M., Mulchandani A., Mulchandani P., Chen W., and Schoning M. J., Dual amperometric-potentiometric biosensor detection system for monitoring organophosphorus neurotoxins, *Anal. Chim. Acta*, 469, 197–203, 2002.

156. Sadik O. A., Land W., and Wang J., Targeting chemical and biological warfare agents at the molecular level, *Electroanalysis*, 15(4), 1149–1159, 2003.

157. Seto Y., Kanamori-Kataoka M., Tsuge K., Ohsawa I., Maruko H., Sekiguchi H., Sano Y., Yamashiro S., Matsushita K., Sekiguchi H., Itoi T., and Iura K., Development of an on-Site Detection Method for Chemical and Biological Warfare Agents. *Toxins Review*, 29, 299–312, 2007.

158. Lewis N. S., Electronic nose chip microsensors for chemical agent and explosives detection, California Institute of Technology: Noyes Laboratory, Pasadena, and California 2001.

159. Xie H. F., Yang Q. D., Sun X. X., Yu T., Zhou J., and Huang Y. P., Gas sensors based on nanosized-zeolite films to identify dimethylmethylphosphonate. *Sens. Mater.*, 17, 21–28, 2005.

160. Cajigas J. C., Longworth T. L., Davis N., and Ong K. Y., Testing of HAZMATCAD Detectors Against Chemical Warfare Agents: Summary Report of Evaluation Performed at Soldier Biological and Chemical Command, Microsensor Systems, Inc, Bowling Green, Kentucky. pp. 1–12, 2003.

161. Cunningham B., Weinberg M., Pepper J., Clapp C., Bousquet R., Hugh B., Kant R., Daly C., and Hauser E., Design, fabrication and vapor characterization of a microfabricated flexural plate resonance sensor and application to integrated sensor arrays, *Sens. Actuators B*, 73, 112 –123, 2001.

162. Land W. H., Jr., Leibensperger D., Wong L., Sadik O., Wanekaya A., Uematsu M., and Embrechts M. J., New results using multi array sensors and support vector machines for the detection and classification of organophosphate nerve agents, in *Systems, Man and Cybernetics*, IEEE International Conference, October, pp. 2883–2888, 2003.

163. Joshi K. A., Prouza M., Kum M., Wang J., Tang J., Haddon R., Chen W., and Mulchandani A., V-Type nerve agent detection using a carbon nanotube-based amperometric enzyme electrode, *Anal. Chem.*, 78, 331–336, 2006.

164. Pushkarsky M. B., Webber M. E., Macdonald T., Kumar C., and Patel N., High-sensitivity, high-selectivity detection of chemical warfare agents, *Appl. Phys. Lett.*, 88, 331–336, 2006.

165. Aifan C., Xiaodong H., Zhangfa T., Shouli B., Ruixian L., and Chiun L. C., Preparation, characterization and gas-sensing properties of SnO_2–In_2O_3 nanocomposite oxides, *Sens. Actuators B*, 115, 316–321, 2006.

166. Ponzoni A., Comini E., Sberveglieri G., Alessandri I., Bontempi E., and Depero L. E., Tin, niobium and vanadium mixed oxide thin films based gas sensors for chemical warfare agent attacks prevention, in IEEE Sensors Conference, pp. 1322–1325, 2007.

167. Seto Y., Kanamori-Kataoka M., Tsuge K., Ohsawa I., Maruko H., Sekiguchi H., Sano Y., Yamashiro S., Matsushita K., Sekiguchi H., Itoi T., and Iura K., Development of an on-site detection method for chemical and biological warfare agents, *Toxins Rev.*, 29, 299–312, 2007.

168. Pinnaduwage L. A., Gehl A. C., Allman S. L., Johansson A., and Boisen A., Miniature sensor suitable for electronic nose applications, *Rev. Sci. Instrum.*, 78, 1–3, 2007.

169. Ma X. F., Zhu T., Xu H. Z., et al., Rapid response behavior, at room temperature, of a nanofiber-structured TiO_2 sensor to selected simulant chemical-warfare agents, *Anal. Bioanal. Chem.*, 390(4), 1133–1137, 2008.

170. Chatterjee A. and Islam M. S., Fabrication and characterization of TiO_2–epoxy nanocomposite, *Mater. Sci. Eng,. A.*, 487(1–2), 574–585, 2008.

171. Houšková V., Štengl V., Bakardjieva S., and Murafa N., Photoactive materials prepared by homogeneous hydrolysis with thioacetamide: Part 2—TiO_2/ZnO nanocomposites, *J. Phys. Chem. Solids.*, 69(7), 1623–1631, 2008.

172. Sbaï M., Essis-Tome H., Gombert U., Breton T., and Pontié M., Electrochemical stripping analysis of methyl-parathion (MPT) using carbon fiber microelectrodes (CFME) modified with combinations of poly-NiTSPc and Nafion® films, *Sens. Actuators B*, 124(2), 368–375, 2007.

173. Rife J. C., Miller M. M., Sheehan P. E., Tamanaha C. R., Tondra M., and Whitman L. J., Design and performance of GMR sensors for the detection of magnetic microbeads in biosensors, *Sens. Actuators A*, 107(3), 209–218, November 1, 2003.

3 Detection of Volatile Organic Compounds
The Role of Tetrapyrrole Pigment-Oriented Thin Films

Hanming Ding
Department of Chemistry
East China Normal University
North Zhongshan Road, Shanghai,
People's Republic of China

CONTENTS

3.1 INTRODUCTION

The interaction between some ambient reactive compounds and organic or inorganic thin layers can cause variations in the physicochemical properties of the chemically interactive layers. Molecules in the gas phase, which are adsorbed onto the surface or absorbed in the bulk of the thin layer, generally modify the electrical, optical, or mass properties of the sensitive material, giving rise to a number of different kinds of chemical sensors based on different working principles. Metallophthalocyanines (MPcs) and metalloporphyrins (MPPs) represent a large family of functional π macrocycle materials with high chemical and thermal stability. Their general structures are shown in Figure 3.1. These compounds are the object of great interest to chemists,

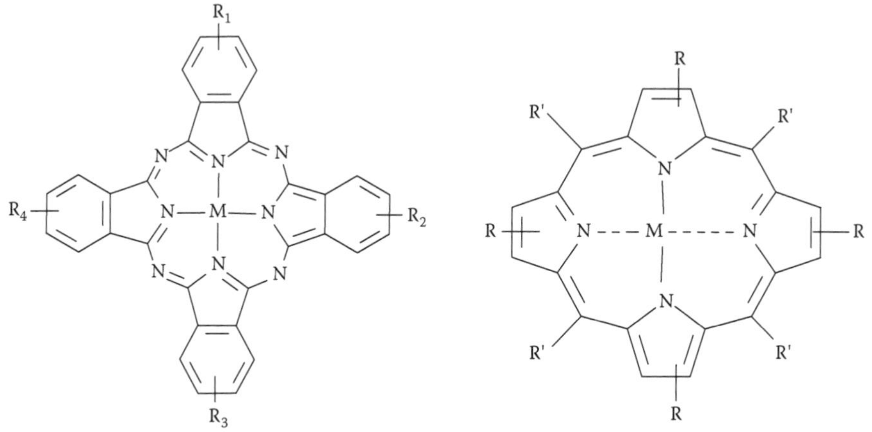

FIGURE 3.1 Chemical structures of phthalocyanines (left) and porphyrins (right), where M = H, or the metal atoms H and R, R′ = H, or the substituted organic groups.

physicists, and industrial scientists because of their potential role in emerging technologies, including photoconductors, organic light-emitting diode, photovoltaic cell, thin-film transistors, gas sensors, and biosensors [1]. These compounds, usually in their thin-film forms, interact with some inorganic gases such as H_2S, HCl, Cl_2, NH_3, or NO_x by absorption onto the sensing layer. Several review articles regarding these related sensing applications are available [2–4]. However, recent efforts have been made on these sensitive compounds to detect various volatile organic compounds (VOCs), such as aromatic compounds, alcohols, amines, and so forth [5–7]. In this chapter, we emphasize on the sensing properties of VOCs.

Volatile organic compounds are composed of several gases: hydrocarbons, solvents, and other organic compounds, which usually cause respiratory difficulties and irritation, and play important roles in the low atmospheric cycle, agriculture, medicines, foods, and so forth. MPcs and MPPs became a natural choice for the detection of VOCs because of their open coordination sites for axial ligation, their large spectral shifts upon ligand binding, and their intense coloration [8]. By varying the central atoms and the peripheral substituents, a wide range of volatile analytes could be detectable. Initially, the sensing properties of phthalocyanine and porphyrin derivatives are based on their π–π interaction with some aromatic compounds, which have π-system. Such interaction usually leads to the change in mass, conductivity, optical adsorption, and so forth of phthalocyanine and porphyrin derivatives. More recently, the analytes were extended to the vapors of amines, alcohols, and other volatile organic pollutes.

3.2 FABRICATION AND TRANSDUCTION

Sample preparation parameters, such as film structure, film morphology, and so forth, can determine the gas sensing properties. A smooth film with small grains generally presents a fast response and a high sensitivity. To be used in a gas sensor, the phthalocyanine and porphyrin materials are always prepared in thin-film form. Various growth techniques have been employed to prepare these thin films, including vacuum thermal evaporation, spin coating, plasma polymerization electrochemical deposition, Langmuir-Blodgett (LB) deposition, and self-assembly method, depending on the structures, solubility, and thermal/chemical stability of these materials. Thin films tend to behave in a different way by using different deposition techniques, thus influencing the performance of the sensors. Due to their insolubility in most of organic solvents, phthalocyanines are usually deposited onto various substrates by using vacuum evaporation. However, LB technique allows the fabrication of highly ordered ultrathin films, down to one single monolayer. Surface uniformity of LB films is useful to improve the performances of these sensors.

Different transduction methodologies are employed for the VOCs' detection, such as electrical conductivity [9], piezoelectric quartz crystal microbalance (QCM) [10–14], surface acoustic wave [15,16], field-effect transistor (FET) [17], surface plasmon resonance (SPR) [18,19], Kelvin probe [20], color variation [8,21,22], and UV-vis absorption [3,23,24], in the physical properties of the sensing elements.

Mukhopadhyay et al. first used LB films of specially substituted phthalocyanine molecules to sense toluene vapor based on the changes in the electrical conductivity [9]. Phthalocyanine and porphyrin derivatives are p-type semiconductors. The interaction with π-electron systems can lead to a cofacial orientation of the nucleus, resulting in a one-dimensional semiconducting system. The exposure to the VOCs may change the cofacial molecular orientation and, as a consequence, the conductivity. However, the interaction between the VOCs and the sensitive molecules is not very strong, as these VOCs are not strong electron donors or electron acceptors. A very low conductivity of 10^{-6} to 10^{-9} S/cm was usually measured when the sensitive layers were exposed to the VOCs, which is difficult to be detected. Therefore, in most cases, the mass transduction and UV-vis absorption method were adopted to detect the presence of organic vapors.

There are two types of piezoelectric sensors based on mass transduction, the surface acoustic wave (SAW) device and QCM. QCM has been widely used for mass measurement because its fundamental oscillating frequency is changed by adsorption of substances on the crystal surface, and the detection limit approaches as low as a few nanograms. SAW gas sensors operate on the same principle as QCM devices. The main drawback of these devices is associated with their high sensitivity. A response can be obtained not only for the specific adsorption but also for the nonspecific adsorption, even for the physical adsorption. To overcome this drawback, the sensor array coupled with the recognition pattern, namely electric nose, was usually used. In an electric nose, each sensor in the array behaves like a receptor by responding to different gases to varying degrees. These changes are identified by a pattern recognition system. The overall response pattern from the array is unique for a given gas in a family of gases [25].

The optical detection of vapors was based on the changes in optical properties of thin films, such as dielectric constant, refractive index, and so forth, when they were exposed to the VOCs. In the solid thin films, there are π–π interactions between the analytes and phthalocyanines/porphyrins. Interactions with VOCs can induce a change of these interactions, leading to broad, splitting, and shift of absorbance bands in their UV-vis spectra. UV-vis spectra of phthalocyanines are typically represented

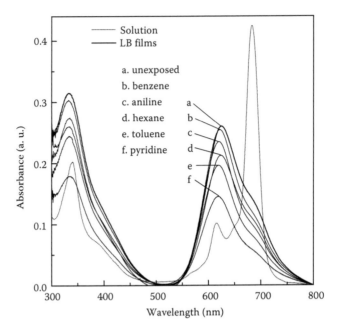

FIGURE 3.2 UV-vis spectra of CuPc A$_2$ LB films before (a) and after the exposure to the vapors of various organic solvents (b–f). Dashed line stands for UV-vis spectrum of chloroform solution of CuPc A$_2$ (10^{-6} M).

by two main absorption bands: the Q bands centered at about 680 (monomer) and 620 nm (dimer), and the Soret band at about 350 nm. Porphyrins, on the other hand, have an intense absorption band centered in the 400–420 nm spectral region [26].

Since Jiang found that the exposure to NH$_3$ has an effect on the structure of Pc LB films, further on their absorption spectra [27], the effect of various gases including the vapors of VOCs on the UV-vis adsorption of Pc and PP films has been widely investigated [23,24,28–31]. The effect of various VOC vapors on the absorption spectrum of tris-(2,4-di-*t*-amylpheoxy)-(8-quinolinoxy) CuPc (CuPc A$_2$) is shown in Figure 3.2.

In such Pc LB films, CuPc mainly exists as dimer. When the LB films were exposed to various organic vapors, the dimer absorbance band was blue-shifted from 625 nm initially to 624 (hexane), 622 (benzene), 620 (aniline), 620 (toluene), and 619 nm (pyridine), respectively (Figure 3.2). It is clear that the change becomes apparent when the films were exposed to the vapors of aromatic compounds. When such film was continuously exposed to benzene vapors, the dimer band was blue-shifted much more (Figure 3.3). However, the Soret bands have no change in their adsorption position. In all the cases, there is a hypochromic effect when exposed to the vapors. A same phenomenon was observed in LB films of bis(phthalocyaninate) rare earth compounds exposed to hexanal vapor [32]. These changes are associated with the formation of a charge-transfer (CT) complex between the phthalocyanine and the VOC [33,34].

A similar result was obtained for porphyrin derivatives. UV-vis spectroscopic measurements reveal a red shift and broadening of Soret and Q bands in LB films

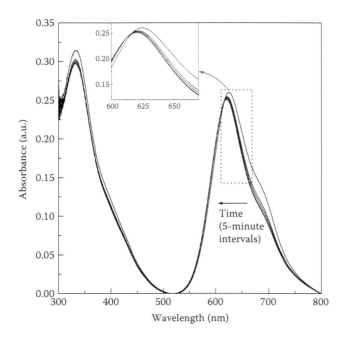

FIGURE 3.3 Time-dependent change in UV-vis spectra of CuPc A$_2$ LB films exposed to benzene vapor (5-minute interval).

of copper (II) tetra-(*tert*-butyl)-5,10,15,20-tetraazaporphyrin [CuPaz(t-Bu)$_4$], when exposed to the vapors of hexane, benzene, and toluene, as shown in Figure 3.4 [12].

Besides the formation of CT complex between phthalocyanines and the VOCs, the phase transition was observed after the exposure to the VOCs. The treatments of vanadyl-phthalocyanine and titanylphthalocyanine with thermal annealing and ethanol vapor exposure lead to the phase transition from amorphous to β-form in their evaporated thin films [35].

Although optical method is widely used to detect various VOCs, the selectivity is not satisfied, as many organic compounds, including aromatic and nonaromatic, can induce the changes in the optical adsorption of phthalocyanines and porphyrins. The large sensitivities and wide selectivities are particularly appealing for electronic nose applications [36,37]. Di Natale et al. designed an optical multisensor (optoelectronic nose) for the detection of VOCs [38]. The sensor sketch is shown in Figure 3.5. Four different porphyrins with different central metals and substituents were deposited onto one of the internal surfaces of the transparent walls of a Plexiglas chamber. Each porphyrin layer lies on a different optical path from the light-emitting diode (LED) to its relevant photodiode. Data were analyzed considering both stand-alone sensors and each sensor as a component of an optoelectronic nose. The capability to distinguish different volatile compounds and the contribution of each sensor was realized by means of a self-organizing map [39]. This sensor design shows a practicable way to fabricate broad-selectivity sensors. These sensors show a certain response not only to those analytes with specific interactions but also to those with

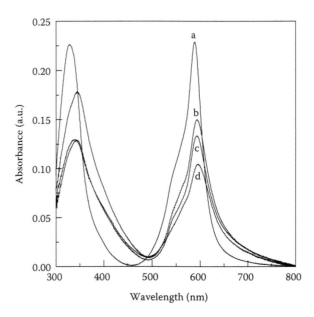

FIGURE 3.4 UV-vis spectra of 20-layer CuPaz(t-Bu)$_4$ LB films before (a) and after the exposure to the saturated vapors of benzene (b), toluene (c), and hexane (d).

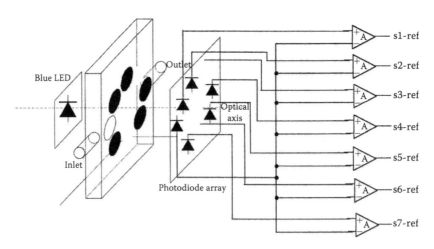

FIGURE 3.5 Schematic sketch of the optoelectronic nose. In the figure, eight different optical paths are displayed, seven of them are covered with sensitive layers, while the last is left uncoated, as reference. (Reprinted from Di Natale C., Salimbeni D., Paolesse R., Macagnano A., and D'Amico A., Porphyrins-based opto-electronic nose for volatile compounds detection, *Sens. Actuators B*, 65, 220–226, 2000, Copyright 2000, with permission from Elsevier Science.)

weak interactions, such as alkanes and aldehydes. They suggest that sapphyrin layer tends to be more effective in the detection of amines and acids, and the tetraphenylporphyrins in the detection of the alcohols and the corrole.

Suslick and coworkers have taken advantage of the large color changes induced in MPPs upon ligand binding and first reported the colorimetric array detection of a wide range of odorants using MPPs [8,21,22,40,41]. The design of the colorimetric sensor array is based on two fundamental requirements: (1) the chemo-responsive dye must contain a center to interact strongly with analytes and (2) this interaction center must be strongly coupled to an intense chromophore [21]. The array sensor is composed of four families of chemically responsive dyes: (1) a series of metalated tetraphenylporphyrins was used to differentiate analytes based on metal-selective coordination [40]; (2) bis-pocketed Zn porphyrins were used to differentiate on the basis of the size and shape of the analyte; (3) pH indicator dyes were used to differentiate on the basis of Brønsted basicity; and finally (4) highly solvatochromic dyes were used to indicate the polarity of the analyte.

The sensor array responses were determined for a series of different volatiles representing the common organic functionalities: amines, arenes, alcohols, aldehydes, carboxylic acids, esters, halocarbons, ketones, phosphines, sulfides, and thiols. Each analyte response is represented as the red, green, and blue values of each of the 24 dyes, that is, a 72-dimensional vector. These results suggest that the familial similarities among compounds of the same functionality are exceptional: amines, alcohols, aldehydes, esters, and so forth are all easily distinguished from each other.

3.3 SENSING APPLICATIONS

3.3.1 AROMATIC COMPOUNDS

Gas sensors based on phthalocyanine and porphyrin films were firstly used to detect the vapors of organic planar compounds with conjugated double bonds, due to their strong π–π interactions. Table 3.1 provides a list of phthalocyanines and porphyrins used in various sensors for the detection of aromatic compounds.

Mukhopadhyay et al. first used thin films of substituted phthalocyanine molecules to sense toluene vapor at levels of 5–9 ppm [9], and 2,4-toluene diisocyanate and 2,6-toluene diisocyanate down to 35 ppb [42]. The interaction of the gas with the LB films led to significant changes in the electrical conductivity. The response and sensitivity are dependent on the structure of the thin films. Fleischer et al. used CuPc as a sensitive layer based on Kelvin probe and QCM measurements [20,43]. The sensor has a sensitivity of 1 ppm for the detection of toluene and 10 ppm for acetone and ammonia. Granito et al. employed LB films of substituted phthalocyanines using SPR to detect toluene vapor, by varying the central metal ions (Cu or Ni). The sensor could detect concentrations of toluene vapor down to 50 ppm. However, the sensor has a poor reversibility, which may be associated with the interaction of toluene with the underlying silver substrate [18].

Fietzek et al. studied various MPcs as sensitive coatings on QCM for the detection of a variety of VOCs [44]. They also used thickness shear mode resonators (TSMRs) to study the interaction of substituted phthalocyanines (PCs) with benzene, toluene,

TABLE 3.1
Phthalocyanines and Porphyrins Used in Various Sensors for the Detection of Aromatic Compounds

Materials	Analytes	Sensitivity	Transduction	References
Substituted phthalocyanine	Toluene	5–9 ppm	Conductivity	[9]
Substituted phthalocyanines (copper or nickel)	Toluene	50 ppm	SPR	[18]
Copper (II) tetra-(*tert*-butyl)-5,10,15,20-tetraazaporphyrin	Benzene Toluene		QCM	[12]
Fe phthalocyanine (pp-FePc) / Fe tetraphenyl-porphyrin (pp-FeTPP)	Benzo[α]pyrene	20–76 μM	QCM	[48]
[Tetrakis-(3,3-dimethyl-1-butoxycarbonyl)] CuPc	Toluene and tetrachloroethene		Ellipsometric measurement	[46]
Octaethylporphyrins (OEP) and tetraphenylporphyrins (TPP): (OEP)InCl, (OEP)MnCl, (OEP)GaCl, (TPP)Pd, and (TPP)Rh	2,4-dinitrotrifluoro-methoxybenzene		QCM	[14]
Tert-butyl-substituted copper azaporphyrines	Benzene/hexane		Microgravimetry	[49]
Substituted phthalocyanines	Benzene Toluene	0.5 ppm 0.2 ppm	Thickness shear mode resonators	[45]
Copper phthalocyanine	Toluene	1 ppm	Kelvin probe	[20]

m-xylene, and *n*-octane [45]. The aromatic compounds revealed extremely nonlinear adsorption isotherms with high intensities at low concentrations. The calculated limit of detection for a phthalocyaninatoplatin(II) coating was 0.5 ppm for benzene and 0.2 ppm for toluene. Due to the limited number of available π-electron sites only a restricted amount of aromatics can interact sensitively via π-stacking and the rest of the aromatics, particularly at higher pressures, interact by weaker dispersion forces only. These behaviors can be understood quantitatively using a two-step sorption model [44]: the initial highly sensitive response at low concentrations by preferential absorption and the less sensitive response at high concentrations by nonpreferential absorption for different analyte/MPc combinations. This model allows evaluating the sensitivities of the phthalocyanines.

Ellipsometric measurements were carried out on LB films of [tetrakis-(3,3-dimethyl-1-butoxycarbonyl)] CuPc in the presence of dry air, toluene, and tetrachloroethene. The relative thickness, the normalized refractive index, and extinction coefficient at a fixed wavelength were changed, when these films are exposed to different concentrations of toluene and tetrachloroethene [46].

A vibration frequency of the oscillating piezoelectric SAW crystal coated with plasma-polymerized CuPc layer is decreased by adsorption of various VOCs, and the partition coefficients K of various VOCs were measured from the frequency decrease [10]. Analysis on the relationships between log K and the structure of the VOCs suggests the importance of the π–π interaction [47]. For example, the presence of a double bond can increase log K, and the aromatic ring, and further with polar substituents, would induce even greater increase. Mutagens were detected in solution by using a QCM electrode coated with a plasma-polymerized FePc or a tetraphenyl FePP film. These sensors have positive effect on benzo[α]pyrene in the concentration range 20–76 μM, but no response to n-alcohol, n-hexane, and benzene [48].

Five octaethylporphyrins (OEP) and tetraphenylporphyrins (TPP), (OEP)InCl, (OEP)MnCl, (OEP)GaCl, (TPP)Pd, and (TPP)Rh, were deposited on QCM to detect 2,4-dinitrotrifluoromethoxybenzene vapor [14]. When exposed to the nitroaromatic compound, a large and significant response was recorded for every porphyrin, the detection process being slightly reversible. Along with a good sensitivity, the sensors exhibit an excellent selectivity when common solvents are used as interfering vapors. Among all the studied derivatives, (OEP)MnCl appears as the most sensitive and selective coating.

The effect of LB films of *tert*-butyl-substituted copper azaporphyrines, namely binuclear phthalocyanine (Cu_2Pc_2) and porphyrazine (CuPaz), on sorption of benzene and hexane was studied microgravimetrically [49]. The greatest benzene/hexane selectivity was found for the homogeneous film of Cu_2Pc_2, as Cu_2Pc_2 is more π-rich than CuPaz molecule, and thus binds aromatic molecules more strongly than aliphatic ones.

We used LB films of copper (II) tetra-(*tert*-butyl)-5,10,15,20-tetraazaporphyrin [CuPaz(t-Bu)$_4$] to quantity vapors of hexane, benzene, and toluene [12]. Nanogravimetric measurements showed that the LB films have a good sensitivity to the vapors of benzene and toluene, suggesting aromatic compounds have a strong interaction with porphyrins.

Manganese corrolate LB films deposited onto QCM electrode was used to sense different VOCs [50]. The selectivity was specificated according to the partition coefficient, which was calculated at the limit of detection and at very high concentrations. The high gain was obtained for benzene, and no gain was shown for hexane and acetaldehyde, which have no specific interactions with the corrole ring. They also showed that it is possible to use Kelvin probe technique for the vapor sensing by employing manganese corrolate LB films or self-assembled monolayers of thiol-functionalized porphyrins [43,51].

QCM coated with thin films of metal complexes of protoporphyrin IX dimethyl ester (M-PPIX) deposited by electropolymerization was used to detect the vapors of triethylamine, acetic acid, ethanol, and toluene [52]. Poly-Ni(PPIX) shows larger sensitivity to toluene due to π–π interaction.

3.3.2 AMINES

Amines are a group of biologically active and even toxic compounds that cause environmental problems in agriculture, water, food, and medicines [19]. It was first showed by Saito et al. that CuPc layer has sensitivity to 100 ppm in the vapor of trithylamine and benzaldehyde by monitoring changes in the electric resistance [53]. The response time is short for triethylamine. The relationship between degree of response and concentration is shown in Figure 3.6.

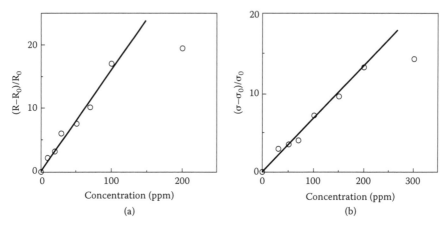

FIGURE 3.6 Relationship between degree of response and concentration of (a) teriethyla-mine and (b) benzaldehyde. (Reprinted from Saito M., Koyano T., Miyamoto Y., and Kaifu K., Electric responses of odor sensor using vapor-deposited copper-phthalocyanine film, *Jpn J. Appl. Phys.*, 34, 3271–3272, 1995, Copyright 1995, with permission from Institute of Pure and Applied Physics.)

The utilization of MPPs and related compounds as active materials to detect tri-ethylamine, based on the changes on mass and conductivity, was reported by Di Natale [54]. The selectivity depends mostly on the central metallic ion, which shows the possibility of changing the sensor selectivity with minor modification of the synthesis process. Conductivity measurements confirmed that the charge transport happens inside the MPPs, indicating the occurrence of a different conduction mecha-nism among these macrocycle compounds.

Delmarre et al. grafted cobalt (II) porphyrins by using a triethoxysilyl group and pure tetramethoxysilane as the precursor in a porous sol–gel matrix for the detection of amines [55]. The results on pyridine sensing show that the diffusion of pyridine occurs with a lower rate than that in organic matrices.

Spin-coated films of MPcs show good sensitivity and selectivity toward *tert*-bu-tylamine, diethylamine, dibutylamine, 2-butanonea, and acetic acid vapors, which depend on both the central metal ion and the peripheral substitutes of the macro-cycles [56,57].

By using principal component analysis (PCA), it is possible to evaluate the ana-lytical dispersion (i.e., selectivity) of an array toward some simple vapors. PCA is a powerful unsupervised linear data analysis technique widely used in gas-sensing area to extract the main relationships in the data matrix containing the sensor responses and to obtain qualitative results for pattern recognition [58]. Spadavecchia et al. measured the UV-vis spectrum of Langmuir-Schafer film of tris-(2,4-di-*t*-amylphenoxy)-(12-hydroxy-1,4,7,10- tetraoxadodecyl) CuPc in four spectral regions, 300–400, 550–600, 600–640, and 640–700 nm, and made a four-sensor array, where each selected spectral region generates an independent sensor [28]. The array sensor was successively exposed to a successive exposition of vapors of *tert*-butylamine, methanol, ethanol, hexane, and ethyl acetate. Based on the same principle, they also fabricated a four-sensor array using four MPc derivative LB films to detect terbutylamine [19].

LB films of tetra-4-*tert*-butyl- and tetra-(3-nitro-5-*tert*-butyl)-substituted CoPcs were used to detect pyridine, primary aliphatic amines, and benzylamine, by means of microgravimetry, UV-Vis spectroscopy, and optic microscopy [59]. The sorption occurs as stepwise intercalation of the sorbate molecules into the supramolecular 3D structure of the phthalocyanine assembly followed by formation of the donor-acceptor complexes. Both intercalation depth and stoichiometry of the complexes are determined by the molecular structure of amines. The supramolecular factor allows discrimination between amines in air but not in aqueous solutions because of concurrent intercalation of water.

Rella et al. used spin-coated films of three zinc phthalocyanines as optochemically interactive materials for the detection of amines in the UV-vis spectral range [30,31]. By adopting a multipeaks lorentzian deconvolution of the absorption spectra, the Q absorption band typical of MPc macromolecule was separated into the Q_I and Q_d bands. The dynamic optical responses toward VOCs were calculated taking into account the variation in the integral area of the selected Q_I or Q_d peak in dry air and in the presence of vapor, respectively. They found that the optical responses depend on the peripheral substituents of the macromolecules. The three systems present different responses toward same VOCs, which show the possibility to develop an array of independent and not redundant sensors for optoelectronic nose applications. The structure and morphology of the active sensing layer are determinant parameters of this vapor/surface interaction, especially the orientation of the molecular arrangement or their transitional dipole moments, and aggregation form in the surface.

A polymer film-based optical sensor responds reversibly to gaseous amines at sub-ppm levels [60]. The sensor is based on the equilibrium of an indium(III) octaethylporphyrin hydroxide ion-bridged dimer species with corresponding monomeric porphyrins within a thin poly(vinyl chloride) film. The presence of amines causes the dimeric species to be converted to monomer, which yields a significant change in the Soret band. Response to different amines is based on their relative partition coefficient into the polymer film and their strength of axial ligation reactions. With optimized film compositions, 1-butylamine can be detected to levels approaching 0.1 ppm, while less lipophilic ammonia can be monitored down to 10 ppm, with fully reversible responses to each species. The authors presented a simple mathematical model to explain the response of the amine sensor and to predict the optical behavior observed.

Suslick and coworkers reported the use of a colorimetric array sensor that is capable of the highly sensitive and highly selective discrimination of amines [8,22,40]. Amines span a wide range of molecular shapes, sizes, and electronic properties, and the selective discrimination within a larger family of amines is not easily achieved. Responses to 12 amines comprising linear, branched, and cyclic structures of similar molecular weight were recorded to provide a stringent test of the molecular recognition by a 24-dye sensor array. Each amine gives a unique color-difference map. Because the colorimetric dye array probes a wide range of intermolecular interactions, a very high level of dispersion was observed. When PCA is applied to this family of

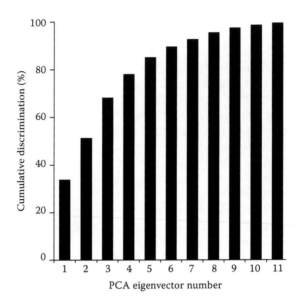

FIGURE 3.7 Principal component analysis from 12 very closely related amines show that the 24-dye sensor array has a very high level of dispersion. (Reprinted from Rakow N. A., Sen A., Janzen M. C., Ponder J. B., and Suslick K. S., Molecular recognition and discrimination of amines with a colorimetric array, *Angew. Chem. Int. Ed.*, 44, 4528–4532, 2005, Copyright 2005, with permission from Wiley-VCH Verlag GmbH & Co. KGaA, Weinheim.)

closely related analytes, there are 6 dimensions necessary for 90% discrimination, 8 for 95%, and 10 for 99%, as shown in Figure 3.7. A 36-dye array has an extraordinarily high level of dispersion: 11, 14, and 21 dimensions are required to define 90%, 95%, and 99% of the total variance, respectively. Moreover, such an array sensor was able to discriminate analytes in aqueous solutions. For these sensor arrays, every analyte at a different concentration may be considered a different analyte. Amines have both the lowest detection limits and recognition limits in aqueous solution. The lower detection limits range from 2 to 0.02 ppm in mole fraction for amines. These researches demonstrate the potential of the colorimetric sensor array to discriminate one analyte from another in a complex mixture.

3.3.3 ALCOHOLS

Although there is no π–π interaction between phthalocyanines/porphyrins and alcohols, there are many works carried out in sensing alcohols by using thin films of phthalocyanines and porphyrins. LB films of tetra-α-(2,2,4-trimethyl-3-pentyloxy) CuPc show good sensitivity to vapor of alcohol [61]. The response time of the LB film to alcohol is 2 minutes and recovery time is 1 minute. The response should be due to the hydrogen bonding between alcohol and the substituents.

Spadavecchia et al. [23] found that the response and selectivity depend on both the central metal ion of the molecules and the peripheral substituents. They used spin-coated films of tetra-4-(2,4-di-t-amylphenoxy) ZnPc and RuPc, tris-(2,4-di-t-amylphenoxy)-(12-hydroxy-1,4,7,10-tetraoxadodecyl) CuPc, and tetrakis-(*p-tert*-butylphenyl) CuPP for sensing the vapors of some alcohols, alkanes, esters, chetones, aldehydes, and pyridines, based on the variation of the UV-Vis optical absorption [23]. They employed a specific optical technique, which includes selection of four specific spectral regions taken in the UV-Vis spectral range corresponding to the typical Q and Soret bands of the phthalocyanine and porphyrin macromolecules and their corresponding blends [24,29]. This technique extracts the main relationships in the data matrix containing the sensor responses and provides qualitative useful results for pattern recognition [62]. The heterogeneous sensing layer (i.e., ZnPc/CuPP blend)-based sensor shows a different behavior from single homogeneous films.

Crone et al. investigated a kind of pattern produced by an array of 11 different organic FETs, monitoring in response to polar and nonpolar organic vapors, including alcohols, ketones, thiols, nitriles, esters, and ring compounds [17]. Responses were distinguishable for different classes of analytes among semiconductors with different lengths of side chains, for example, sexithiophene and dialkylsexithiophenes, and different carrier types, for example, CuPc and perfluorinated CuPc. For the alcohols, nonanol has a saturated vapor pressure of 10 ppm while hexanol has a vapor pressure of >100 ppm.

Lead phthalocyanine tetracarboxylic acid was used to sense humidity and alcohol vapors [63]. Remarkable improvement in the selectivity with respect to ethyl alcohol and reduction in the sensitivity for humidity was observed when the surface was treated with electron cyclotron resonance plasma. The response and recovery time were 50 and 30 seconds, respectively. The increased cross-linking of PbPc is responsible for the creation of new functional groups that have imparted the sensing of alcohol vapor through extrinsic doping.

LB films of copper (II) octa-*n*-butoxy-2,3-naphthalocyanine show good sensitivities to vapor of alcohols, with the following sequence of sensitivities: *i*-PrOH, EtOH, MeOH [64]. The response time and recovery time to vapor of MeOH, EtOH, and *i*-PrOH [volume fraction $(1–5) \times 10^{-5}$] were within 2 and 5 seconds, respectively, while those of the LB films to ammonia (1×10^{-4}) were 30–60 seconds and 4–5 minutes, respectively.

Nanocomposite LB films of CuPc A_2 and Fe_2O_3 were sensitive to ethanol vapor in the range of 2–8 ppm (the alternated film) or 100–200 ppm (the capped film) at room temperature and show better humidity-resist than the CuPc alone [65]. An monolayer of meso-tetra(4-sulfanatophenyl) CuPP deposited on a quartz substrate was used to detect the saturated vapor of ethanol, 2-propanol, and cyclohexane [66]. The thin film exhibited a good sensitivity and reproducibility toward all vapor samples.

Abrass and the coworkers used cyclic voltammetry to investigate the electrochemical reaction of propanethiol at a CoPc-modified screen-printed carbon electrode coated with a hydrogel [67]. Cyclic voltammograms exhibit one anodic peak at +0.45 V, resulting from the electrocatalytic oxidation of propanethiol to produce the corresponding disulfide. A linear response could be obtained between 6 and 22 vpm. The response time was 2 minutes.

Based on a thin-film bulk acoustic wave resonators (TFBAR) coated with tetraphenyl CoPP, an electroacoustic chemical sensor can detect ethanol vapor down to 500 ppm [68]. TFBAR operates on the same principle of the QCM, at an operation frequency extended up to several GHz. The larger output signal, associated with the higher operation frequency, is a condition to improve the device sensitivity. Sensor based on hybrid CoPP-SnO$_2$ thin films showed fast and reversible responses toward methanol vapors and highest responses at temperature of 250°C [69]. This experiment suggests that MPPs can be used to modify the selectivity of SnO$_2$ sensors.

Akrajas used optical technique to enrich the selectivity of four octaethyl metalloprophyrins (M = Mn, Fe, Co, and Ru) LB films toward four vapors of 2-propanol, ethanol, acetone, and cyclohexane [70]. An optical system was developed using these metalloprophyrins LB films as sensing elements and four LED's of different colors: red, yellow, green, and blue as light sources. The sensing sensitivity was based on the change on the light intensity at the peak wavelength of light sources after being reflected by the films, which depends on the wavelength of the light source and the central metal atoms. Each thin film produced 16 signals for a particular vapor, which constituted the pattern of the signature of the vapor. This work shows a possibility of using a sensor array to enhance the selectivity.

Spin-coated films of mesogenic octa-substituted phthalocyanine derivatives MPcR$_8$, where M = Ni(II), R = $-$S(CH$_2$)$_{11}$CH$_3$, $-$SCH(CH$_2$OC$_{12}$H$_{25}$)$_2$, $-$S(CH$_2$CH$_2$O)$_3$ CH$_3$, and 13,17-dioxanonacosane-15-sulfanyl, were used to detect organic solvent vapors by utilizing QCM and SPR [71,72]. These films show higher sensitivity and partition coefficient for ethanol, dichloromethane, chloroform, acetone, n-hexane, and benzene. The molecules with saturated C–C bonds such as ethanol interact with phthalocyanine films predominantly by formation of hydrogen bonds, and the sensor response to π-bond-containing compounds such as acetone is the result of their π–π interaction with conjugated Pc ring. However, the SPR response on exposures to benzene vapor is found to be independent of the type of substituents.

Thiol-functionalized MPP deposited as self-assembled monolayer onto the gold pad of QCM was used to detect VOC vapors [73]. The selectivities of these sensors depend on the nature of the metal coordinated to the thiol-functionalized porphyrin, which can be predicted by the hard–soft acid–base principle, for example, Mn-porphyrinates showed a higher sensitivity to hard ligands, such as alcohols, whereas cobalt complexes had a relative higher response to molecules with soft donor atoms, like dimethylsulfide. The selectivity is higher than that of the corresponding casting coated sensors, suggesting that ordered structure of thin films is a benefit to the selectivity.

3.3.4 AROMAS AND ODORS

Aromas and odors are mixture of different classes of chemical species often found in foods, beverages, and medicines. It is difficult to use a single chemosensor to discriminate these analytes. Sensor arrays, in which different sensor elements are used with data analysis, and subsequent pattern recognition will make it possible to characterize gas mixtures quantitatively or odors qualitatively. Sensors that incorporate pattern recognition software are usually referred to as "electronic noses" [74]. Recently, there has been much interest in the composition of the suites of organic vapors emitted by food products.

Rodriguez Mendez and coworkers did a series of researches in the development of new sensors able to detect the odors and aromas [32,58,75–78]. They first used lutetium bisphthalocyanine (LuPc$_2$) LB and evaporated films for the detection of the aroma of olive oil or wine (hexanol, hexanal, n-buthyl acetate, and acetic acid) based on the changes in their conductivity [76]. The kinetics and the intensity of the response of the films depend not only on the morphology and the thickness of the films but also on the nature of the reactant gas. They detected acetic acid down to 88 mmol/L, based on the refractive index changes in combination with an optical fiber [79]. They further employed LB films of bis(phthalocyaninate) rare earth compounds (LnPc$_2$) (including praseodymium, gadolinium, and lutetium bis(phthalocyanine)s and their octa-tertbutyl derivatives) as fiber optic sensors in near-infrared region [32]. This sensor was used to measure NO$_x$ and vapors of odors in foods and beverages such as alcohols, aldehydes, and esters. The gas-sensing properties depend on the central metallic ion and the substituents. For example, the response of chemiresistors based on LB films of bis[octakis(propyloxy)-phthalocyaninato] samarium(III) toward VOCs and HNO$_3$ was much more intense than that observed in unsubstituted derivatives [58]. In contrast, the lifetime of the sensors is considerably reduced.

Sensor arrays coupled with pattern recognition are useful in the discrimination of the aromas. The same group designed a five-sensor array based on LB films of LnPc$_2$ to discriminate among diverse virgin olive oils. They used unsubstituted bisphthalocyanines with different central metal atom (PrPc$_2$ and LuPc$_2$) and an octatertbutyl substituted bisphthalocyanine. The sensor array was used to discriminate four types of Spanish olive oils, based on the changes in the conductivity, coupled with a pattern recognition technique [34]. They also constructed a multichannel taste sensor based on electrochemical measurements to evaluate the five types of basic tastes (sweet, bitter, salty, acid, and umami), by using PCA [78]. The significant differences in the electrochemical responses obtained toward the five basic tastes lead to a fingerprint, which allows the different basic flavors to be distinguished.

LB films of n-tetraphenyl porphine iron (III) chloride, n-tetraphenyl porphine manganese (III) chloride, n-tetrakis (4-methoxyphenyl) porphine cobalt (II), and n-octaethyl porphine cobalt (II) mixed with arachidic acid were used as sensing layers for optical detection of capsicum (chili) aroma. The sensing sensitivity of aroma was based on changes of optical absorption of the films taken at four different wavelengths of 646, 615, 601, and 585 nm. The responses of the films upon aroma exposure were fast and recoverable. The patterns of the absorption changes were able to distinguish three capsicum samples: dried capsicum annum, fresh capsicum annum, and capsicum minimum [80].

A electronic nose based on an array of eight quartz microbalance-based (QCM) sensors coated with modified MPPs (5,10,15,20-tetraphenylporphyrin) was used for apple aroma measurements [13]. The response of each QCM sensor was modeled with Brunauer-Emmett-Teller (BET) adsorption isotherms. By means of multivariate analysis on all sensor responses, the different compounds could be discriminated well and quantified accurately. This calibration protocol can be used to characterize the sensors for the vapors of complex mixtures.

LB films of octa-(15-crown-5)-lutetium bisphthalocyanine (CR-Pc$_2$Lu) were sensitive to electron donor and electron acceptor gases as well as tobacco smoke [75]. The presence of crown ether groups on the phthalocyanine ring increased the sensitivity of the films to oxidizing gases.

3.3.5 ALKANES

There are few works involved in the detection of aliphatic hydrocarbons, due to their very week interactions between the analytes and Pcs. Urbanczyk and coworkers employed SAW technique to detect trichloroethylene [15]. The acoustic waveguide was fabricated on the y-cut of the $LiNbO_3$ piezoelectric substrate. The changes in the physical properties of the CuPc layer placed on a piezoelectric crystal surface can be recorded as a change in differential frequency in a dual delay-line oscillator system, under the exposure of the vapors of the VOCs. The sensitivity is normally quoted as differential response, that is, $\Delta fp/ppm$ of gas, and the greatest sensitivity (approximately 0.1 Hz/ppm) was obtained for trichloroethylene.

Monomeric soluble transition metal phthalocyanines $MPcR_4$ (R = *tert*-butyl or 2,2-dimethyl-3-phenyl-propoxy) and $MPcR_8$ (R = heptyl = C_7H_{15}) used as sensitive materials coated on QCM show reversible interaction and high sensitivity for tetra-chlorocthylene and n-propanol [81,82]. The variation in the sensitivities was associated with the difference in the boiling temperatures of the analytes. The solvent molecule/MPc interactions are mainly determined by Van der Waals bond energies between the polarizable aromatic rings of the phthalocyanine and the molecules, not metallic electronic states and π-bonds. The sensitivities have drastic differences between phthalocyanines without central metal atoms and with different central transition metal atoms. For example, the strong sensitivity of the metal-free $(t\text{-}Bu)_4H_2Pc$ toward alcohols with their hydroxyl group (OH) is strongly suppressed if the phthalocyanine ring contains a central metal atom. Such work suggests that the introduction of special peripheral substituents can improve sensor properties like better stabilities and faster response and recovery times because of an enhanced affinity of the VOCs and a higher mobility of the adsorbed analytes in the bulk.

A technique combining a sensor array with high-resolution gas chromatography (HRGC/SOMMSA) was used to study the responses of several chemosensors (metal oxides, surface acoustic wave devices, phthalocyanine sensors) to a number of selected food volatiles or aroma compounds [83]. It was shown that the temperatures and the dopants used significantly influence the sensitivity and/or selectivity of the sensors. For instance, a CuPc sensor selectively detected (E)-2-nonenal and (E,E)-2,4-decadienal in mixtures with several pyrazines. A combination of Headspace-HRGC with the SOMMSA technique is a useful approach to develop sensor arrays adapted to special targets in flavor control, based on quantitative correlations of key odorants with indicator volatiles.

3.4 SENSING PRINCIPLES

There are few principles for measuring VOCs by using phthalocyanines and porphyrins as the sensitive layers described in literature. Some effects of solvent vapors on organic films were suggested in literatures, such as effects on film thickness via swelling, effects on intermolecular interactions via intercalation, specific coordination effects, effects on the local electric field produced by the incident light beam at the individual molecular sites via modification of local dielectric constant, and so forth [28]. For example, in the detection of the aromatic VOCs, the sensing mechanism exploits the rearrangement of the electrical dipole in the thin film due to the interaction between macrocycle with the analytes, resulting in a change in absorption spectrum.

Normally, there is two-stage adsorption process: an initial fast change, followed by a slow shift to a steady value owing to the specific and nonspecific interactions. An example of this is shown in Figure 3.8. The first step is due to its strong contribution of the π-stacking analytes and the limited number of sites in the macrocyclic molecules, and it is ruled by a Langmuir isotherm. The second step is associated with its nonspecific weak contribution after the saturation of the specific sites, and the shape of the isotherm becomes linear (Henry-type behavior) [7,44,45]. The interactions are dependent on the nature of the sensitive materials and the analytes. MPcs and MPPs have ordinarily two possible sites for gas adsorption: (1) the central metal atom and (2) the conjugated π-electron system [72]. By varying the central ions and the peripheral substituents, MPcs and MPPs have different delocalized π-electron density; thus the interactions between them and the analytes will change.

Although attempts were made by means of spectroscopic methods to understand the nature of the interactions between VOCs and phthalocyanines [32,59], the sensing mechanism is not yet well understood. The interactions between phthalocyanines/porphyrins and the VOCs may be associated with bond formation, acid–base interactions, hydrogen bonding, dipolar, and multipolar interactions, π–π molecular

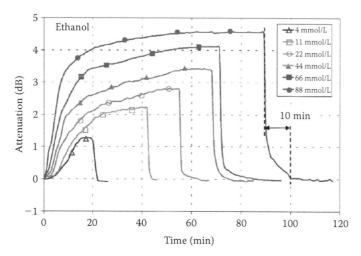

FIGURE 3.8 Response of an optical fiber sensor based on lutetium bisphthalocyanine to the presence of ethanol vapors at various concentrations. (Reprinted from Bariain C., Matias I. R., Fernandez-Valdivielso C., Arregui F. J., Rodriguez-Mendez M. L., and de Saja J. A., Optical fiber sensor based on lutetium bisphthalocyanine for the detection of gases using standard telecommunication wavelengths, *Sens. Actuators B*, 93, 153–158, 2003, Copyright 2003, with permission from Elsevier.)

complexation, Van der Waals interaction, and physical adsorption [21]. For the vapors of π-bond containing compounds or aromatic compounds such as acetone, tetrachloroethylene, benzene, toluene, and so forth, there are π–π interactions between conjugated macrocycles and the π-systems. The vapors with saturated C–C bonds such as ethanol, methanol, tetrachloromethane, dichloromethane, and so forth, interact with macrocycles predominantly by formation of hydrogen bonds between electron donor atoms (O, Cl) of these vapor molecules and hydrogen atoms of alkyl chains of the substituents [47,71]. For the metal-ligating vapors such as amines, the metal-coordination interactions exist between the central metals and the vapor molecules because phthalocyanines and porphyrins provide their open coordination sites for axial ligation.

3.5 CONCLUDING REMARKS

MPcs and MPPs are promising materials to be used for the fabrication of chemosensors in the detection of various volatile compounds, including aromatic compounds, amines, alcohols, alkanes, and so forth. The diversity in the coordinated metal ions, the peripheral substituents, and the conformations of the macrocyclic skeleton of MPcs and MPPs provide the possibility that these compounds can be employed to sense all kinds of VOCs. Precisely, this occurs because the interactions between phthalocyanines/porphyrins and the analytes are various and complicated. According to the current research results, the adsorption properties of phthalocyanines and porphyrins are characterized by large sensitivities and wide selectivities. Although many efforts have been taken for the detection of various VOCs, there is no possibility to detect one component exclusively. In a normal case, a chemosensor fabricated from the macrocycle material is sensitive to several VOCs in different degrees.

In many cases, the sensing goal is a comparison of identity between one complex mixture and another. In other cases, the goal is to monitor change in one or in a few components against a complex but constant background. To satisfy these goals, a sensor array, namely electronic nose technology, was usually used. In practice, array sensors based on QCM, UV-vis absorbance, and colorimetry have been fabricated. In the first two cases, data analysis techniques, such as PCA and pattern recognition, are included. In the latter case, a hierarchical cluster analysis (HCA) was performed to examine the multivariate distances between the analyte responses in this 72-D RGB color space [21]. Using metal centers that span a range of chemical hardness and ligand binding affinity and substituents that allow to adjust the accessibility of the metal ion to the ligand and further induce shape-selective ligation, a wide range of volatile analytes are differentiable [8]. By proper tuning of the structure of the sensitive materials, these array sensors show the possibility to distinguish several VOCs solely or in a mixture. This type of sensing array will be of practical importance for general-purpose vapor dosimeters and analyte-specific detectors.

ACKNOWLEDGMENTS

The financial supports from the SRF for ROCS, State Education Ministry and Shanghai Rising-Star Program (03QB14015) are gratefully acknowledged.

REFERENCES

1. Wang K., Xu J.-J. and Chen H.-Y., A novel glucose biosensor based on he nanoscaled cobalt phthalocyanine–glucose oxidase biocomposite, *Biosens. Bioelectron.*, 20, 1388–1396, 2005.
2. Zhou R., Jose F., Gopel W., Ozturk Z. Z., and Bekaroglu O., Review phthalocyanines as sensitive materials for chemical sensors, *Appl. Organomet. Chem.*, 10, 557–577, 1996.
3. Valli L., Phthalocyanine-based Langmuir-Blodgett films as chemical sensors, *Adv. Colloid Interface Sci.*, 116(1–3), 13–44, 2005.
4. Saja J. A. De and Rodriguez-Mendez M. L., Sensors based on double-decker rare earth phthalocyanines, *Adv. Colloid Interface Sci.*, 116, 1–11, 2005.
5. Josse F. J., Zhou R., Altindal A., Dabak S., and Bekaroglu Ö, Sensitive properties of soluble dodecylsulfanyl phthalocyanines for organic vapors using impedance spectroscopy and QCR, *Chemical Microsensors and Applications, Proceedings of SPIE–The International Society for Optical Engineering*, 3539, 74–84, Boston, MA, USA, 4–5 November 1998.
6. Altindal A., Patel R., Zhou R., Josse F., Ozturk Z. Z., and Bekaroglu O., Soluble dodecylsulfanylphthalocyanines as sensitive coatings for chemical sensors in gas phase, *Proceedings of the 1998 IEEE International Frequency Control Symposium*, pp. 676–684, Pasadena, CA, USA, 27–29 May 1998.
7. D'Amico A., Natale C. D., Paolesse R., Macagnano A., and Mantini A., Metalloporphyrins as basic material for volatile sensitive sensors, *Sens. Actuators B*, 65, 209–215, 2000.
8. Rakow N. A. and Suslick K. S., Colorimetric sensor array for odour visualization, *Nature*, 406, 710–713, 2000.
9. Mukhopadhyay S., Hogarth C. A., Thorpe S. C., and Cook M. J., Room temperature toluene sensing using phthalocyanine Langmuir-Blodgett films, *J. Mater. Sci. Mater. Electron.*, 5, 321–323, 1994.
10. Kurosawa S., Kamo N., Matsui D., and Kobatake Y., Gas sorption to plasma-polymerized copper phthalocyanine film formed on a piezoelectric crystal, *Anal. Chem.*, 62(4), 353–359, 1990.
11. Brunink J. A. J., Natale C. D., Bungaro F., Davide F. A. M., D'Amico A., Paolesse R., Boschi T., Faccio M., and Ferri G., The application of metalloporphyrins as coating material for quartz microbalance-based chemical sensors, *Anal. Chim. Acta*, 325, 53–64, 1996.
12. Ding H., Erokhin V., Ram M. K., Paddeu S., Valkova L., and Nicolini C., A physical insight into the gas-sensing properties of copper (II) tetra-(*tert*-butyl)-5,10,15,20-tetraazaporphyrin Langmuir-Blodgett films, *Thin Solid Films*, 379, 279–286, 2000.
13. Saevels S., Berna A. Z., Lammertyn J., Di Natale C., and Nicolai B. M., Characterisation of QMB sensors by means of the BET adsorption isotherm, *Sens. Actuators B*, 101, 242–251, 2004.
14. Montmeat P., Madonia S., Pasquinet E., Hairault L., Gros C. P., Barbe J.-M., and Guilard R., Metalloporphyrins as sensing material for quartz-crystal microbalance nitroaromatics sensors, *IEEE Sens. J.*, 5, 610–614, 2005.
15. Urbanczyk M., Jakubik W., and Kochowski S., Investigation of sensor properties of copper phthalocyanine with the use of surface acoustic waves, *Sens. Actuators B*, 22, 133–137, 1994.
16. Caliendo C., Verardi P., Verona E., D'Amico A., Di Natale C., Saggio G., Sarafini M., Paolesse R., and Huq S. E., Advances in SAW-based gas sensors, *Smart Mater. Struct.*, 6, 689–699, 1997.
17. Crone B., Dodabalapur A., Gelperin A., Torsi L., Katz H. E., Lovinger A. J., and Bao Z., Electronic sensing of vapors with organic transistors, *Appl. Phys. Lett.*, 78, 2229–2231, 2001.

18. Granito C., Wilde J. N., Petty M. C., Houghton S., and Iredale P. J., Toluene vapour sensing using copper and nickel phthalocyanine Langmuir-Blodgett films, *Thin Solid Films*, 284–285, 98–101, 1996.

19. Spadavecchia J., Ciccarella G., Rella R., Capone S., and Siciliano P., Metallophthalocyanines thin films in array configuration for electronic optical nose applications, *Sens. Actuators B*, 96, 489–497, 2003.

20. Fleischer M., Simon E., Rumpel E., Ulmer H., Harbeck M., Wandel M., Fietzek C., Weimar U., and Meixner H., Detection of volatile compounds correlated to human diseases through breath analysis with chemical sensors, *Sens. Actuators B*, 83, 245–249, 2002; Conf. Proc. Transducers 01 Munich (FRG) (2001).

21. Suslick K. S., An optoelectronic nose: Colorimetric sensor arrays, *MRS Bull.*, 29, 720–725, 2004.

22. Rakow N. A., Sen A., Janzen M. C., Ponder J. B., and Suslick K. S., Molecular recognition and discrimination of amines with a colorimetric array, *Angew. Chem. Int. Ed.*, 44, 4528–4532, 2005.

23. Spadavecchia J., Ciccarella G., Vasapollo G., Siciliano P., and Rella R., UV-Vis absorption optosensing materials based on metallophthalocyanines thin films, *Sens. Actuators B*, 100, 135–138, 2004.

24. Spadavecchia J., Ciccarella G., Siciliano P., Capone S., and Rella R., Spin-coated thin films of metal porphyrin–phthalocyanine blend for an optochemical sensor of alcohol vapours, *Sens. Actuators B*, 100, 88–93, 2004.

25. Arshak K., Moore E., Lyons G. M., Harris J., and Clifford S., A review of gas sensors employed in electronic nose applications, *Sens. Rev.*, 24, 181, 2004.

26. Supriyatno H., Yamashita M., Nakagawa K., and Sadaoka Y., Optochemical sensor for HCl gas based on tetraphenylporphyrin dispersed in styrene–acrylate copolymers: Effects of glass transition temperature of matrix on HCl detection, *Sens. Actuators B*, 85, 197–204, 2002.

27. Jiang D. P., Zhang L. G., Fan Y., Ren X. G., Guan Z. S., Li Y. J., and Lu A. D., The effects of detected gases on spectroscopic properties of phthalocyanine Langmuir-Blodgett films, *Thin Solid Films*, 293, 277–280, 1997.

28. Spadavecchia J., Ciccarella G., Valli L., and Rella R., A novel multisensing optical approach based on a single phthalocyanine thin films to monitoring volatile organic compounds, *Sens. Actuators B*, 113, 516–525, 2006.

29. Spadavecchia J., Ciccarella G., Stomeo T., Rella R., Capone S., and Siciliano P., Variation in the optical sensing responses toward vapors of a porphyrin/phthalocyanine hybrid thin film, *Chem. Mater.*, 16, 2083–2090, 2004.

30. Rella R., Siciliano P., Capone S., Spadavecchia J., Ciccarella G., and Vasapollo G., Optical sensing properties of phthalocyanines thin films in array configuration and their application in VOCS detection, pp. 115–120, Sensors and microsystems, Proceedings of the 8th Italian Conference, Trento, Italy, 12–14 February 2003.

31. Spadavecchia J., Ciccarella G., and Rella R., Optical characterization and analysis of the gas/surface adsorption phenomena on phthalocyanines thin films for gas sensing application, *Sens. Actuators B*, 106, 212–220, 2005.

32. Rodríguez-Méndez M. L., Gorbunova Y., and De Saja J. A., Spectroscopic properties of Langmuir-Blodgett films of lanthanide bis(phthalocyanine)s exposed to volatile organic compounds. Sensing applications, *Langmuir*, 18, 9560–9565, 2002.

33. Rodríguez-Méndez M. L., Aroca R., and De Saja J. A., Electrochromic and gas adsorption properties of Langmuir-Blodgett films of lutetium bisphthalocyanine complexes, *Chem. Mater.*, 5, 933–937, 1993.

34. Gutierrez N., Rodriguez-Mendez M. L., and de Saja J. A., Array of sensors based on lanthanide bisphtahlocyanine Langmuir-Blodgett films for the detection of olive oil aroma, *Sens. Actuators B*, 77, 437–442, 2001.

35. Del Caño T., Parra V., Rodríguez-Méndez M. L., Aroca R. F., and De Saja J. A., Characterization of evaporated trivalent and tetravalent phthalocyanines thin films: different degree of organization, *Appl. Surf. Sci.*, 246, 327–333, 2005.

36. Di Natale C., Macagnano A., Davide F., D'Amico A., Paolesse R., Boschi T., Faccio M., and Ferri G., An electronic nose for food analysis, *Sens. Actuators B*, 44, 521, 1997.

37. Gouma P. and Sberveglieri G., Novel materials and applications of electronic noses and tongues, *MRS Bull.*, 29, 697–702, 2004.

38. Di Natale C., Salimbeni D., Paolesse R., Macagnano A., and D'Amico A., Porphyrins-based opto-electronic nose for volatile compounds detection, *Sens. Actuators B*, 65, 220–226, 2000.

39. Di Natale C., Paolesse R., Macagnano A., Mantini A., Goletti C., and D'Amico A., Electronic nose and sensorial analysis: Comparison of performances in selected cases, *Sens. Actuators B*, B52, 162–168, 1998.

40. Sen A. and Suslick K. S., Shape-selective discrimination of small organic molecules, *J. Am. Chem. Soc.*, 122, 11565–11566, 2000.

41. Drain C. M., Hupp J. T., Suslick K. S., Wasielewski M. R., and Chen X., A perspective on four new porphyrin-based functional materials and devices, *J. Porph. Phthalocyanines*, 6, 243–258, 2002.

42. Agbabiaka A. A., Mukhopadhyay S., Mukherjee D., and Thorpe S. C., Molecules of synthesized phthalocyanine as a material for the detection of 2,4-toluene diisocyanate (TDI), *Supramol. Sci.*, 4, 185–190, 1997.

43. Di Natale C., Goletti C., Paolesse R., Drago M., Macagnano A., Mantini A., Troitsky V. I., Berzina T. S., Cocco M., and D'Amico A., Kelvin probe investigation of the thickness effects in Langmuir-Blodgett films of pyrrolic macrocycles sensitive to volatile compounds in gas phase, *Sens. Actuators B*, B57, 183–187, 1999.

44. Fietzek C., Bodenhoefer K., Haisch P., Hees M., Hanack M., Steinbrecher S., Zhou F., Plies E., and Göpel W., Soluble phthalocyanines as coatings for quartz-microbalances: Specific and unspecific sorption of volatile organic compounds, *Sens. Actuators B*, 57, 88–98, 1999.

45. Fietzek C., Bodenhoefer K., Hees M., Haisch P., Hanack M., and Göpel W., Soluble phthalocyanines as suitable coatings for highly sensitive gas phase VOC-detection, *Sens. Actuators B*, 65, 85–87, 2000.

46. Rella R., Siciliano P., Valli L., Spaeth K., and Gauglitz G., An ellipsometric study of LB films in a controlled atmosphere, *Sens. Actuators B*, B48, 328–332, 1998.

47. Kurosawa S. and Kamo N., Characteristics of sorption of various gases to plasma-polymerized copper phthalocyanine, *Langmuir*, 8, 254–256, 1992.

48. Kurosawa S., Tawara-Kondo E., Minoura N., and Kamo N., Detection of polycyclic compounds as mutagens using piezoelectric quartz crystal coated with plasma-polymerized phthalocyanine derivatives, *Sens. Actuators B*, B43, 175–179, 1997.

49. Valkova L., Borovkov N., Maccioni E., Pisani M., Rustichelli F., Erokhin V., Patternolli C., and Nicolini C., Influence of molecular and supramolecular factors on sensor properties of Langmuir-Blodgett films of tert-butl-substituted copper azaporphyrines towards hydrocarbons, *Colloids. Surf. A.*, 198–200, 891–896, 2002.

50. Paolesse R., Di Natale C., Macagnano A., Sagone F., Scarselli M. A., Chiaradia P., Troitsky V. I., Berzina T. S., and D'Amico A., Langmuir-Blodgett films of a manganese corrole derivative, *Langmuir*, 15, 1268–1274, 1999.

51. Di Natale C., Paolesse R., Mantini A., Macagnano A., Boschi T., and D'Amico A., Kelvin prove investigation of self-assembled-monolayers of thiol derivatized porphyrins interacting with volatile compounds, *Sens. Actuators B*, B48, 368–372, 1998.

52. Paolesse R., Natale C. D., Dall'Orto V. C., et al., Porphyrin thin films coated quartz crystal microbalances prepared by electropolymerization technique, *Thin Solid Films*, 354, 245–250, 1999.

53. Saito M., Koyano T., Miyamoto Y., and Kaifu K., Electric responses of odor sensor using vapor-deposited copper-phthalocyanine film, *Jpn J. Appl. Phys.*, 34, 3271–3272, 1995.

54. Di Natale C., Macagnano A., Repole G., Saggio G., D'Amico A., Paolesse R., and Boschi T., The exploitation of metalloporphyrins as chemically interactive material in chemical sensors, *Mater. Sci. Eng. C*, 5, 209–215, 1998.

55. Delmarre D. and Bied-Charreton C., Grafting of cobalt porphyrins in sol–gel matrices: Application to the detection of amines, *Sens. Actuators B*, 62, 136–142, 2000.

56. Rella R., Spadavecchia J., Ciccarella G., Siciliano P., Vasapollo G., and Valli L., Optochemical vapour detection using spin coated thin films of metal substituted phthalocyanines, *Sens. Actuators B*, 89, 86–91, 2003.

57. Spadavecchia J., Ciccarella G., Buccolieri A., Vasapollo G., and Rella R., Synthesis and structure of amorphous phase Cr(II) hemiporphyrazine using energy dispersive X-ray diffraction, *J. Porph. Phthalocyanines*, 7, 572–578, 2003.

58. Gorbunova Y., Rodríguez-Méndez M. L., Kalashnikova I. P., Tomilova L. G., and de Saja J. A., Langmuir-Blodgett films of bis(octakispropyloxy) samarium bisphthalocyanine. Spectroscopic and gas-sensing properties, *Langmuir*, 17, 5004–5010, 2001.

59. Valkova L., Borovkov N., Koifman O., Kutepov A., Berzina T., Fontana M., Rella R., and Valli L., Sorption of amines by the Langmuir-Blodgett films of soluble cobalt phthalocyanines: Evidence for the supramolecular mechanisms, *Biosens. Bioelectron.*, 20, 1177–1184, 2004.

60. Qin W., Parzuchowski P., Zhang W., and Meyerhoff M. E., Optical sensor for amine vapors based on dimer–monomer equilibrium of indium(iii) octaethylporphyrin in a polymeric film, *Anal. Chem.*, 75, 332–340, 2003.

61. Suwannet W., Jaisutti R., Chamlek O., Kerdcharoen T., Osotchan T., Jarubundit O., Kuen-asa P., Donprajam K., and Pratontep S., Homemade quartz crystal microbalance systems, for alcohol sensor, *Proceedings of the First National Symposium on Physics Graduate Research (1st NSPG)*, 29 June, 1 July, Chulabhorn Dam, Khon San, Chaiyaphum, Thailand, IAP02-1-IAP02-5, 2006.

62. Hierlemann A., Weimar U., Kraus G., Gauglitz G., and Göpel W., Environmental chemical sensing using quartz microbalance sensor arrays: Application of multicomponent analysis techniques, *Sens. Mater.*, 7, 179–189, 1995.

63. Naddaf M., Chakane S., Jain S., Bhoraskar S. V., and Mandale A. B., Modification of sensing properties of metallophthalocyanine by an ECR plasma, *Nucl. Instrum. Methods Phys. Res., Sect. B*, 194, 54–60, 2002.

64. Wang B., Zuo X., Wu Y. Q., and Chen Z. M., Preparation, characterization and gas sensing properties of lead tetra-(*tert*-butyl)-5,10,15,20-tetraazaporphyrin spin-coating films, *Sens. Actuators B*, 125(1), 268–273, 2007.

65. Huo L. H., Lia X. L., Li W. M., and Xi S. Q., Gas sensitivity of composite Langmuir-Blodgett films of Fe_2O_3 nanoparticle-copper phthalocyanine, *Sens. Actuators B*, 71, 77–81, 2000.

66. Umar A. A., Salleh M. M., and Yahaya M., Utilization of albumin-based sensor chips for the detection of metal content and characterization of metal–protein interaction by surface plasmon resonance, *Sens. Actuators B*, 101, 231–235, 2004.

67. Abass A. K. and Hart J. P., Electrocatalytic, diffusional and analytical characteristics of a cobalt phthalocyanine modified, screen-printed, amperometric gas sensor for propanethiol, *Sens. Actuators B*, 41, 169–175, 1997.

68. Benetti M., Cannat D., Di Pietrantonio F., Foglietti V., and Verona E., Microbalance chemical sensor based on thin-film bulk acoustic wave resonators, *Appl. Phys. Lett.*, 87, 173504-1–173504-3, 2005.

69. Nardis S., Monti D., Di Natale C., D'Amico A., Siciliano P., Forleo A., Epifani M., Taurino A., Rella R., and Paolesse R., Preparation and characterization of cobalt porphyrin modified tin dioxide films for sensor applications, *Sens. Actuators B*, 103, 339–343, 2004.

70. Muhamad Mat Salleh, dan Muhammad Yahaya A., Enriching the selectivity of metal-loporphyrins chemical sensors by means of optical technique, *Sens. Actuators B*, 85, 191–196, 2002.

71. Basova T. V., Tasaltin C., Gurek A. G., Ebeoglu M. A., Ozturk Z. Z., and Ahsen V., Mesomorphic phthalocyanine as chemically sensitive coatings for chemical sensors, *Sens. Actuators B*, 96, 70–75, 2003.

72. Basova T., Kol'tsov E., Ray A. K., Hassan A. K., Gurek A. G., and Ahsen V., Liquid crystalline phthalocyanine spun films for organic vapour sensing, *Sens. Actuators B*, 113, 127–134, 2006.

73. Paolesse R., Di Natale C., Macagnano A., Davide F., Boschi T., and D'Amico A., Self-assembled monolayers of mercaptoporphyrins as sensing material for quartz crystal microbalance chemical sensors, *Sens. Actuators B*, B47, 70–76, 1998.

74. Honeybourne C. L., Organic vapor sensors for food quality assessment, *J. Chem. Educ.*, 77, 338–344, 2000.

75. Rodríguez-Méndez M. L., Souto J., J. Gonzalez J., and de Saja J. A., Crown-ether lutetium bisphthalocyanine Langmuir-Blodgett films as gas sensors, *Sens. Actuators B*, B31, 51–55, 1996.

76. Rodríguez-Méndez M. L., Souto J., de Saja R., Martinez J., and de Saja J. A., Lutetium bisphathalocyanine thin films as sensors for volatile organic components (VOCs) of aromas, *Sens. Actuators B*, B58, 544–551, 1999.

77. Rodríguez-Méndez M. L., Langmuir-Blodgett films of rare-earth lanthanide bisphthalo-cyanines. Applications as sensors of gases and volatile organic compounds, *Comments. Inorg. Chem.*, 22, 227–239, 2000.

78. Arrieta A., Rodriguez-Mendez M. L., and De Saja J. A., Langmuir–Blodgett film and carbon paste electrodes based on phthalocyanines as sensing units for taste, *Sens. Actuators B*, 95, 357–365, 2003.

79. Bariain C., Matias I. R., Fernandez-Valdivielso C., Arregui F. J., Rodriguez-Mendez M. L., and de Saja J. A., Optical fiber sensor based on lutetium bisphthalocyanine for the detection of gases using standard telecommunication wavelengths, *Sens. Actuators B*, 93, 153–158, 2003.

80. Mat Salleh M. and Akrajas M. Yahaya, Optical sensing of capsicum aroma using four porphyrins derivatives thin films, *Thin Solid Films*, 416, 162–165, 2002.

81. Schierbaum K. D., Zhou R., Knecht S., Dieing R., Hanack M., and Göpel W., The interaction of transition metal phthalocyanines with organic molecules: A quartz-microbalance study, *Sens. Actuators B*, 24, 69–71, 1995.

82. Öztürk Z. Z., Zhou R., Weimar U., Ahsen V., Bekaroğlu O., and Göpel W., Soluble phthalocyanines for the detection of organic solvents: Thin film structures with quartz microbalance and capacitance transducers, *Sens. Actuators B*, 26–27, 208–212, 1995.

83. Schieberle P., Hofmann T., Kohl D., Krummel C., Heinert L., Bock J., and Traxler M., *ACS. Symp. Ser.*, 705, 359–374, 1998.

4 Modeling of Surface Acoustic Wave Sensor Response

Subramanian K. R. S. Sankaranarayanan,
Venkat R. Bhethanabotla, Babu Joseph
Sensors Research Laboratory
Department of Chemical Engineering
University of South Florida
Tampa, Florida

CONTENTS

4.1   INTRODUCTION

A sensor allows the transduction of chemical and/or physical properties at an interface into usable information. A chemical sensor generates an output signal, which is a function of the chemical entity and/or concentration (Figure 4.1). The transduction methods are usually classified as electrochemical, acoustic, optical, and thermal.

Acoustic wave sensors are devices that allow transduction between electrical and acoustical energies and are so named because they utilize a mechanical or acoustic wave as sensing mechanism [1]. As the acoustic wave propagates through or on the surface of a material, changes in the propagation path affect the amplitude and/or velocity of the wave. By monitoring the changes in wave velocity and amplitude, the frequency or phase characteristics of a sensor can be correlated to the corresponding physical quantity being measured. Virtually all acoustic wave devices and sensors use a piezoelectric material to generate the acoustic wave. Piezoelectricity refers to production of electrical charges on imposition of mechanical stress and occurs in crystals that lack center of symmetry [2]. The converse is also true, which means that application of electrical field results in mechanical displacement.

Piezoelectric acoustic wave sensors apply an oscillating electric field to create a mechanical wave, which propagates through the substrate and is then converted back to an electric field for measurement. There are several piezoelectric substrates that can be used for acoustic wave generation. The most common are quartz (SiO$_2$), lithium niobate (LiNbO$_3$), and lithium titanate (LiTaO$_3$). Other materials that have commercial applications include gallium arsenide (GaAs), silicon carbide (SiC), langasite (LGS), zinc oxide (ZnO), aluminum nitride (AlN), lead zirconium titanate (PZT), and polyvinylidene fluoride (PVF). Each of these materials has its own

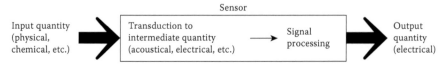

FIGURE 4.1   Schematic diagram of a sensor producing electrical output in response to analyte presence.

advantages and disadvantages, which include cost, temperature dependence, attenuation, and propagation velocity. The properties of these materials are influenced by the crystal cut and orientation [3,4].

The devices commonly used for sensor applications and material characterization include thickness shear mode (TSM) resonator [5], the surface acoustic wave (SAW) device [6], the acoustic plate mode (APM) device, and the flexural plate wave (FPW) device. In one-port acoustic devices such as TSM, a single port serves as both the input and the output port, whereas in two-port devices such as SAW, APM, and FPW, one port is used for input and the other serves as an output port. The input signal generates an acoustic wave that propagates to a receiving transducer, which regenerates a signal at the output port. The sensor response is determined based on the relative signal levels and phase delay between the input and the output ports.

Chemical sensitivity is imparted to the device by attaching a thin film to the acoustically active region [7,8]. The film serves as a chemical-to-physical transducer wherein one or more of its properties change in response to the presence of the species to be detected. Sensor response commonly relies on increased mass density of the film arising from the species accumulation. However, changes in other film parameters such as elastic and electrical properties also contribute to the response. Acoustic devices such as SAW with substantial surface normal displacement components are suitable for gas-sensing applications [9]. On the other hand, acoustic devices generating shear motion in the liquid, for example, TSM and shear horizontal acoustic plate mode (SH-APM) can operate with excessive damping when in contact with liquid and hence find applications in liquid sensing [10–13]. The interaction mechanism of SAW and SH-SAW devices [14] with their immediate environment (thin film, liquid, or both) as well as the resulting response forms the focus of this chapter.

4.2 SURFACE ACOUSTIC WAVES: SAW AND SH-SAW

SAW are elastic waves that propagate along the surface of an elastic body, with most of the energy density confined to a depth of about one wavelength below the surface. There exist two main categories of surface waves, each with varying propagation characteristics.

4.2.1 RAYLEIGH WAVE (GAS SENSING)

In 1887, Lord Rayleigh discovered the SAW mode of propagation and predicted the properties of these waves [15]. The Rayleigh waves have a longitudinal component and a vertical component that can couple with the medium placed in contact with the device's surface. The Rayleigh mode SAW has predominantly two particle displacement components in the sagittal plane [16–18]. The surface particles move in elliptical paths characterized by a surface normal and a surface parallel component. The surface parallel component is parallel to the wave propagation direction. The generated electromechanical field travels in the same direction. The velocity of the wave depends on the substrate material and the cut of the crystal. Typically, the energies of the SAW are confined to a zone close to the surface a few wavelengths

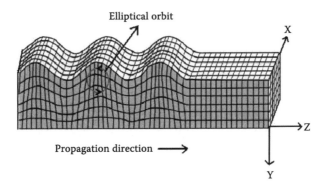

FIGURE 4.2   Rayleigh wave propagation in Y-Z LiNbO$_3$.

thick. An example of piezoelectric substrate is lithium niobate (LiNbO$_3$), where the dominant acoustic mode propagating on a Y-cut Z propagating LiNbO$_3$ is the Rayleigh mode (Figure 4.2). The use of Rayleigh SAW sensors is applicable only to the gas media as the Rayleigh wave is severely attenuated in liquid media.

4.2.2   SHEAR HORIZONTAL WAVE (BIOSENSING)

The SH-SAW devices are very similar to the SAW devices described earlier [19,20]. However, the selection of a different piezoelectric material and appropriate crystal cut yields shear horizontal waves instead of Rayleigh waves. An example of piezoelectric substrate is lithium titanate (LiTaO$_3$), where the dominant acoustic mode propagating on a 36°-rotated Y-cut X propagating LiTaO$_3$ is the SH mode [21] (Figure 4.3). The particle displacements in this type of wave are transverse to the wave propagation direction and parallel to the plane of the surface. This makes SH-SAW devices suitable for operation in liquid media, where propagation at the solid-liquid media can be attained with minimal energy losses [12]. The appearances of these devices are very similar to that of Rayleigh mode devices, but a thin solid film or grating is added to prevent wave diffraction into the bulk.

The mechanical and electrical displacements for metalized and free surfaces at liquid-36 YX LiTaO$_3$ interface are shown in Figure 4.4a and 4.4b [22]. Most of the acoustic energy is confined to within one wavelength from the surface of the substrate. When the surface is metalized and electrically shorted, the potential on the surface is zero. In this case, only the normalized displacement (u_2) interacts with the liquid loading, and the phenomenon is called mechanical perturbation. If the surface is free and electrically open, then both u_2 and normalized electric potential (Φ) interact with the adjacent liquid medium (Figure 4.4b). Interactions of Φ and the electrical properties of the liquid constitute the acoustoelectric interaction. The influence of both the mechanical and acoustoelectrical interactions on sensor response and material characterization is discussed in the subsequent sections.

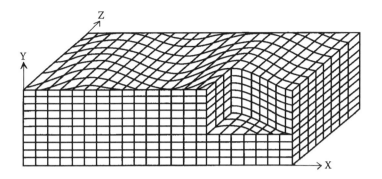

FIGURE 4.3   Shear horizontal wave propagation in 36°-rotated Y-X LiTaO$_3$.

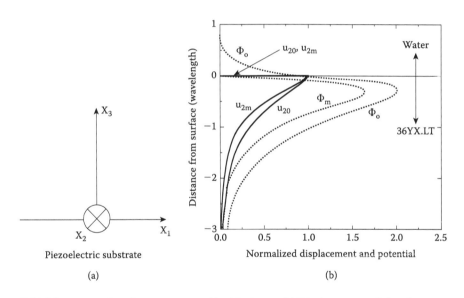

FIGURE 4.4   (a) Coordinate system used in this chapter (b) Displacement and electric potential profiles in 36 YX LiTaO$_3$ with liquid loading. Kondoh J. and Shiokawa S: Shear Horizontal Surface Acoustic Wave Sensors. *Sensors Update*. 2001. 6. Copyright Wiley-VCH Verlag GmbH & Co. KGaA. Reproduced with permission.

4.3   SENSOR RESPONSE

Acoustic wave devices use piezoelectric materials for excitation and detection of acoustic waves (Figure 4.5). The nature of all of the parameters involved with sensor applications concerns either mechanical or electrical perturbations [23,24]. An acoustic device is thus sensitive mainly to the physical parameters, which may interact with the mechanical properties of the wave and/or its associated electrical field. For chemical or biosensing applications, a transduction (sensing) layer is used to convert

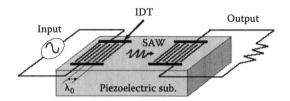

FIGURE 4.5 SAW device used in sensing applications.

the value of desired parameter (e.g., analyte concentration etc.) into mechanical and electrical perturbation that can disturb the acoustic wave properties.

The acoustic wave velocity is affected by several factors, each of which possesses a potential sensor response [9,25–27].

$$\frac{\Delta V}{V_0} \cong \frac{1}{V_0}\left(\frac{\partial V}{\partial \text{mass}}\Delta\text{mass} + \frac{\partial V}{\partial \text{elec}}\Delta\text{elec} + \frac{\partial V}{\partial \text{mech}}\Delta\text{mech} + \frac{\partial V}{\partial \text{envir}}\Delta\text{envir}\right) \quad (4.1)$$

Equation 4.1 illustrates the perturbation of acoustic velocity due to various factors. A sensor response may be due to a combination of these factors. Understanding the acoustic wave perturbation due to each of the above factors would help gain insights into the sensing mechanism as well as in designing efficient sensors.

4.4   MATERIALS CHARACTERIZATION

The recent progress in the area of materials science has resulted in newer materials being synthesized and used/developed for applications such as paints and coatings, corrosion protection, lubrication, electronics, chemical separations, and so forth [28–30]. The properties of these materials are often complex and very different from simple ideal substances. The ability of the material to meet the stringent specifications required for a specific application depends on its chemical and physical properties. Thus, characterizing the material properties plays a vital role in materials science.

Thin films form an important category of materials that find applications in a wide variety of industrial applications [31]. Optimization of thin film properties requires techniques, which can directly characterize the same. SAW devices are ideally suited to thin-film characterization due to their extreme sensitivity to thin-film properties (Equation 4.1). The sensitivity of SAW devices to a variety of film properties such as mass density, viscoelasticity, and conductivity makes them versatile characterization tools. The ability of SAW devices to rapidly respond to changes in thin-film properties allows for monitoring dynamic processes such as film deposition, chemical modification, and diffusion of species in and out of the film. The thin-film focus should not be viewed as a limitation of SAW devices. Bulk material properties can be derived from thin-film data, although such extrapolations should be performed with care [1]. In this chapter, approximate expressions showing applicability of SAW devices to characterize physical and chemical properties of thin-film materials are derived.

4.5   MODELING OF SAW DEVICES

Many models have been proposed for the analysis of SAW sensor response to the various mechanical and electrical perturbations [12,13,32–34]. The simplest of these rely on perturbation theory, or use analytical solutions based on approximations such as isotropic media with negligible piezoelectricity [35–37]. These techniques provide valuable insight into the effects of changes in parameters such as layer height, liquid viscosity, or mass loading, but are limited by the assumptions made [38]. The response of SAW sensors have also been studied by Green's function methods using the quasistatic approximation [39]. This is appropriate for delay line devices where reflection, regeneration, and bulk wave effects are negligible. Most existing studies assume that electrodes are located on the upper surface of the SAW, whereas in actual sensing applications they are often placed between the substrate and the guiding layer. Periodic Green's function yields a great deal of information for SAW signal processing components [40]. The models described above as well as other simple models (Mason's model [41], equivalent circuit models [42]) either introduce simplifying assumptions or else handle only small segments of the SAW devices. For an accurate calculation of piezoelectric devices operating in the sonic and ultrasonic range, numerical methods such as finite element and/or boundary element methods are the preferred choice [33,43,44].

In the following sections, perturbation theory approach as well as precise numerical technique such as finite elements are used to study the effect of various mechanical and electrical perturbations on SAW sensor response and are shown to yield useful information for efficient sensor design and material characterization.

4.6   PERTURBATION THEORY

Perturbation theory is concerned with small changes in solution caused by small changes in the physical parameters of the problem. In acoustic wave problems, this theory is used for calculating effects of small parameter changes on a numerically computed solution [35,36]. For example, it can be used to find the attenuation of a surface wave solution that has been numerically solved for a lossless case. Other applications include evaluating temperature coefficient of propagation velocity for a numerically computed surface wave solution. Perturbation theories generally serve as guidelines for computation. Knowledge of various perturbations of numerical solutions shows trends that are useful in selecting cases to be carried through a full-scale computation.

4.6.1   PERTURBATION APPROACH TO CALCULATE SAW SENSOR RESPONSE

The perturbation approach is one of the commonly used theories to calculate acoustic sensor response in gas and liquid sensing [45–47]. Perturbation theory assumes that changes in system parameters are small enough to allow the use of exact numerical solution to be a starting point for derivation of the perturbed model. Thus, the perturbation theory is not accurate for large variations. The perturbed terms are replaced by exact, unperturbed solution to leave only the unknown quantities of interest in the final approximation. The exact perturbation expression derived from the complex reciprocity relation assuming no source terms are present and that the solution varies exponentially with time is given by [36]

$$\Delta\beta_n = \beta_n' - \beta_n = \frac{-i\left\{-v_n^*.T_n' - v_n'.T_n^* + \Phi_n'(i\omega D_n') + \Phi_n'(i\omega D_n)^*\right\}.\hat{y}\bigg]_0^h}{\int_0^h\left\{-v_n^*.T_n' - v_n'.T_n^* + \Phi_n'(i\omega D_n') + \Phi_n'(i\omega D_n)^*\right\}.\hat{z}dy} \tag{4.2}$$

In the above expression, ' and * indicate the perturbed quantities and complex conjugates. Also, v_n represents the particle velocity in nth direction, T_n is the surface stress, Φ is the electric potential, D is the electrical displacement, ω is the angular velocity, h is the crystal thickness, and β is the propagation constant. In order to use the above exact expression, the perturbed quantities must be known. The perturbation calculation is unnecessary for cases where exact calculation of the perturbed problem is available. Therefore, an approximate solution of the perturbed quantities must be sought.

Since, the perturbation in Equation 4.2 is assumed to be small, the perturbed fields in the denominator may be replaced by the unperturbed fields

$$4P_n = 2\,\mathrm{Re}\int_0^h\left[-v_n^*.T_n + \Phi_n(i\omega D_n)^*\right].\hat{z}dy \tag{4.3}$$

where P_n is the average unperturbed power flow per unit width along x. The same approximation if applied to the numerator results in $\Delta\beta = 0$. Therefore, the resulting boundary perturbation formula is given as

$$\Delta\beta_n = \frac{-i\left\{-v_n^*.T_n' - v_n'.T_n^* + \Phi_n'(i\omega D_n') + \Phi_n'(i\omega D_n)^*\right\}.\hat{y}\bigg]_0^h}{4P_n} \tag{4.4}$$

Assuming small perturbations to be directly additive, the mechanical and electrical perturbation can be treated independently. In most of the SAW sensors, a metalized conducting layer is employed, resulting in negligible electroacoustic effect. Under these conditions, the third and fourth terms in Equation 4.4 are omitted and the equation is simplified as

$$\Delta\beta_n = \frac{-i\left\{-v_n^*.T_n' - v_n'.T_n^*\right\}.\hat{y}\bigg|_0}{4P_n} \tag{4.5}$$

In case of SAW sensors, only the upper surface (y) incorporating the sensing layer is assumed to be perturbed due to the shallow penetration depth of the surface acoustic mode into the bulk of the crystal. To evaluate $\Delta\beta$, it is necessary to relate the perturbed surface stresses $(T.\hat{y})$ in terms of the unperturbed fields. In acoustic wave problems, the stress is delineated as the surface acoustic impedance (Z) seen by the polarized particles [37],

$$-T.\hat{y}\big|_{y=0} = Z.v\big|_{y=0} \text{ (unperturbed case)} \tag{4.6}$$

$$-T.\hat{y}\big|_{y=0} = Z.v'\big|_{y=0} \text{ (perturbed case)} \tag{4.7}$$

Taking the complex conjugate of both sides for the unperturbed case,

$$-T^*.\hat{y}\big|_{y=0} = Z^*.v^*\big|_{y=0} \tag{4.8}$$

In the above equations, it has been assumed that the particle velocity field is unchanged by the perturbation. By substituting Equations 4.7 and 4.8 into Equation 4.5, the normalized perturbation equation at the boundary is given by

$$\Delta\beta_n = \frac{-i\left\{v_n^*.Z'.v + v_n'.Z^*.v^*\right\}}{4P_n} \tag{4.9}$$

The amplitude of the SAW is dependent upon time t and propagation path x. The dependency is assumed as $\exp(i(\omega t - \beta x))$. A complex propagating factor β is defined by the wave number k and attenuation α as

$$\beta = k - i\alpha = \frac{\omega}{V} - i\alpha \tag{4.10}$$

The variation of the complex propagating factor for constant frequency is derived as

$$\frac{\Delta\beta}{k} = -\frac{\omega\Delta V}{V^2}.\frac{V}{\omega} - i\frac{\Delta\alpha}{k} = -\frac{\Delta V}{V} - i\frac{\Delta\alpha}{k} \tag{4.11}$$

The interaction on the surface caused by any gas/liquid analyte can be derived from the fractional phase velocity change $(\Delta V/V)$ and the normalized attenuation $(\Delta\alpha/k)$. Comparing Equations 4.9 and 4.11, the fractional velocity change and the normalized attenuation change are given by the real part and imaginary part, respectively, as [47]

$$\frac{\Delta V}{V} = \text{Re}\left[\frac{-iV\left\{v_n^*.Z'.v + v_n'.Z^*.v^*\right\}}{4\omega P_n}\right] \tag{4.12}$$

$$\frac{\Delta\alpha}{k} = \text{Im}\left[\frac{-iV\left\{v_n^*.Z'.v + v_n'.Z^*.v^*\right\}}{4\omega P_n}\right] \tag{4.13}$$

The above theory can be used to obtain explicit, although approximate, expressions for film-induced velocity and attenuation changes by relating them to surface mechanical impedances (a measure of the difficulty in displacing the film). The mechanical impedances depend on the property of the film and the detailed manner in which the film is translated and deformed by the passing wave.

4.6.1.1   Perturbation of Rayleigh and SH Wave due to Viscoelastic Layer

The changes in complex propagation factor β are related to surface mechanical impedances Z_i experienced by surface displacement components in translating and deforming the film overlay. By considering both the in-plane and cross-plane displacement gradients, the final results are applicable to both thin films as well as acoustically thick films [37] (Figure 4.6). In calculating the SAW displacement components, the SAW surface displacement is resolved into surface normal and in-plane displacement components. Assuming the SAW to be a plane wave, the continuing equation is rewritten as follows:

$$\frac{\partial T_{i1}^f}{\partial x_1} + \frac{\partial T_{i3}^f}{\partial x_3} = \rho \dot{v}_i, \quad i = 1, 2, 3 \tag{4.14}$$

T_{ij} is the stress tensor that indicates the force per unit area in the ith direction in planes normal to the jth direction and ρ is the film density.

The in-plane displacement gradients give rise to $T_{i3}^f = E^{(i)} \dfrac{\partial u_i}{\partial z}$, with the various moduli given by

$$E^{(1)} = \frac{\partial u_{x1}^f}{\partial x_1} = \frac{T_{11}}{S_{11}} = \frac{4G_f(3K_f + G_f)}{3K_f + 4G_f} \tag{4.15}$$

$$E^{(2)} = \frac{\partial u_{x2}^f}{\partial x_1} = \frac{T_{21}}{S_{21}} = G_f \tag{4.16}$$

$$E^{(3)} = \frac{\partial u_{x3}^f}{\partial x_1} = \frac{T_{31}}{2S_{31}} = 0 \tag{4.17}$$

E_f^i refers to the Young's modulus for the film layer generated by the ith displacement component. G_f and K_f are the shear and bulk modulus of the film, respectively.

Similarly, the cross-film gradients give rise to

$$T_{i3}^f = M_f^{(i)} \frac{\partial u_i}{\partial z} \tag{4.18}$$

$M_f^{(i)}$ is the generalized modulus for the film layer and given as $M_f^{(1)} = M_f^{(2)} = G_f$, $M_f^{(3)} = K_f$.

By substituting T_{i1}^f and T_{i3}^f into the continuity equation, a two-dimensional wave equation for the displacements in the film is obtained:

$$M_f^{(i)} \frac{\partial^2 u_{x3}^f}{\partial x_1^2} + E_f^{(i)} \frac{\partial^2 u_i^f}{\partial x_1^2} = \rho_f \dot{v} \tag{4.19}$$

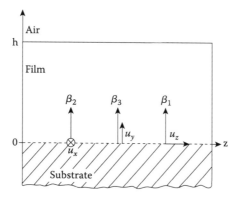

FIGURE 4.6 The film displacement results from superposition of waves generated by the surface displacements u_{i0} at the film-surface interface and radiated across the film with propagation factors β_i. The surface normal components u_{y0} generates the compressional wave while the in-plane components (u_{x0} and u_{z0}) generate the shear waves. (Reprinted with permission from Martin S. J., Frye G. C., and Senturia S. D., Dynamics and response of polymer-coated surface acoustic wave devices: Effect of viscoelastic properties and film resonance, *Anal. Chem.*, 66, 2201–2219, 1994. Copyright (1994) American Chemical Society.)

Substituting the harmonic solution $\vec{u}_i = \hat{u}_i(x_3)e^{j(\omega t - \beta x_1)}$ into Equation 4.19 yields a homogeneous differential equation for the displacement profile $u_i(x_3)$ in the film,

$$\frac{\partial^2 u_i^f}{\partial x_3^2} + \beta_i^2 u_i^f = 0 \tag{4.20}$$

and

$$\beta_i = \omega \left(\frac{\rho_f - \dfrac{E_f^{(i)}}{V^2}}{M_f^{(i)}} \right)^{1/2} \tag{4.21}$$

β_i is the complex propagation factor for displacement u_i ($i = 1, 2, 3$) propagating across the film. In the regime of the film thickness ($0 \leq x_3 \leq h$), the solution is given as

$$u_i^f(x_3) = Ae^{j\beta_i x_3} + Be^{-j\beta_i x_3} \tag{4.22}$$

In order to determine the constants A and B, we require two boundary conditions. The first boundary condition stipulates that the displacement at the film/substrate interface be continuous, that is, $u_i(0^+) = u_{i0}$, where $u_i(0^+)$ is the displacement at the lower film surface, u_{i0}.

Equation 4.22 gives

$$A + B = u_{i0} \tag{4.23}$$

The second boundary condition corresponds to stress free boundary condition at the upper surface (film/air interface).

$$T_{i3}\big|_{x_3 = h} = 0 \tag{4.24}$$

As a result we have

$$Ae^{j\beta_i x_3} - Be^{-j\beta_i x_3} = 0 \tag{4.25}$$

Solving Equations 4.23 and 4.25 simultaneously,

$$A = \frac{u_{io}e^{-j\beta_i h}}{e^{j\beta_i h} + e^{-j\beta_i h}} \tag{4.26}$$

$$B = \frac{u_{io}e^{j\beta_i h}}{e^{j\beta_i h} + e^{-j\beta_i h}} \tag{4.27}$$

For an isotropic film, the displacement components can be considered independent of each other, while calculating the surface mechanical impedance (Z_i). Therefore, the surface mechanical impedance associated with each displacement component u_i is given by

$$Z_i = -\frac{T_{i3}}{v_i}\bigg|_{x_3 = 0} \tag{4.28}$$

The interfacial stress used to evaluate the above impedance is found using Equation 4.18:

$$T_{i3}(0) = M_f^{(i)} \frac{\partial u_i}{\partial x_3}\bigg|_{x_3 = 0} = j\beta_i M_f^{(i)}(A - B) \tag{4.29}$$

and the interfacial velocity is evaluated as

$$v_i(0) = \dot{u}_i(0) = j\omega u_i(0) = j\omega(A + B) \tag{4.30}$$

Substituting Equations 4.29 and 4.30 into 4.28 gives the following:

$$Z_i = -\frac{\beta_i M_f^{(i)}}{\omega}\left(\frac{A - B}{A + B}\right) = \frac{\beta_i M_f^{(i)}}{\omega}\tanh(j\beta_i h) = j\frac{\beta_i M_f^{(i)}}{\omega}\tan(\beta_i h) \tag{4.31}$$

Therefore,

$$Z_i = \begin{bmatrix} j\dfrac{\beta_1 M_f^{(1)}}{\omega}\tan(\beta_1 h) & 0 & 0 \\[2ex] 0 & j\dfrac{\beta_2 M_f^{(2)}}{\omega}\tan(\beta_2 h) & 0 \\[2ex] 0 & 0 & j\dfrac{\beta_3 M_f^{(3)}}{\omega}\tan(\beta_3 h) \end{bmatrix} \qquad (4.32)$$

where

$$\beta_1 = \omega\left(\frac{\rho_f - \dfrac{E_f^{(1)}}{V^2}}{M_f^{(1)}}\right)^{1/2} = \omega\sqrt{\frac{\rho_f - \dfrac{4G_f\left(\dfrac{3K_f + G_f}{3K_f + 4G_f}\right)}{V^2}}{G_f}} \qquad (4.33)$$

$$\beta_2 = \omega\left(\frac{\rho_f - \dfrac{E_f^{(2)}}{V^2}}{M_f^{(2)}}\right)^{1/2} = \omega\sqrt{\frac{\rho_f - \dfrac{G_f}{V^2}}{G_f}} \qquad (4.34)$$

$$\beta_3 = \omega\left(\frac{\rho_f - \dfrac{E_f^{(3)}}{V^2}}{M_f^{(3)}}\right)^{1/2} = \omega\sqrt{\frac{\rho_f}{K_f}} \qquad (4.35)$$

Substituting Equation 4.32 into Equation 4.9 gives the change in the complex propagating factor:

$$\frac{\Delta\beta}{k} = \frac{V}{4\omega P}\left[\frac{\beta_1 G_f}{\omega}\tan(\beta_1 h).v_1^2 + \frac{\beta_2 G_f}{\omega}\tan(\beta_2 h).v_2^2 + \frac{\beta_3 K_f}{\omega}\tan(\beta_3 h).v_3^2\right] \qquad (4.36)$$

For Rayleigh wave propagation, Equation 4.36 reduces to

$$\frac{\Delta\beta}{k} = \frac{V}{4\omega P}\left[\frac{\beta_1 G_f}{\omega}\tan(\beta_1 h).v_1^2 + \frac{\beta_3 K_f}{\omega}\tan(\beta_3 h).v_3^2\right] \qquad (4.37)$$

For SH-SAW propagation [48], Equation 4.36 reduces to

$$\frac{\Delta\beta}{k} = \frac{V}{4\omega P}\left[\frac{\beta_2 G_f}{\omega}\tan(\beta_2 h).v_2^2\right] \qquad (4.38)$$

The velocity shift and attenuation changes are obtained by substituting Equation 4.37 or Equation 4.38 into Equation 4.11.

4.6.1.2 SAW Response from Viscoelastic Films

Viscoelastic polymeric film coatings are commonly used as sensing layers in SAW chemical and biological sensors [49]. For such coatings, the response in terms of velocity and attenuation changes can be predicted from Equations 4.37 and 4.38 using phase shifts $\Phi_i = \text{Re}(\beta_i h)$. For most of the polymeric films, high K_f values (10^{10} dynes/cm^2) result in negligible contributions from compressional wave shifts ($\Phi_3 = \text{Re}(\beta_3 h) < \pi/2$). The main contributions arise from the shear displacement terms or Φ_1 and Φ_3. If a piezoelectric substrate such as ST-quartz is used, then the shear component (Φ_1) dominates.

The attenuation and velocity changes calculated from Equation 4.36 versus the shear wave phase shift for various values of film loss parameter $r_i = -\text{Im}(\beta_i)/\text{Re}(\beta_i)$ are shown in Figure 4.7. The film loss parameter is a ratio of power dissipation to energy storage.

It can be seen from Figure 4.7 that the velocity decreases linearly with Φ_1 for $\Phi_1 \leq \pi/2$. Above $\pi/2$, the velocity sees an upward transition whereas the attenuation goes through a maximum. This behavior is mainly attributed to combination of responses arising from interference between waves generated at the lower film surface and

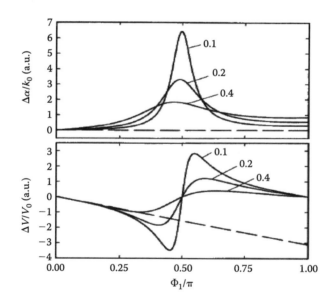

FIGURE 4.7 SAW velocity and attenuation changes vs. shear wave phase shift (Φ_1) for several values of film loss parameter. The dashed line corresponds to prediction using Tiersten formula. (Reprinted with permission from Martin S. J., Frye G. C., and Senturia S. D., Dynamics and response of polymer-coated surface acoustic wave devices: Effect of viscoelastic properties and film resonance, *Anal. Chem.*, 66, 2201–2219, 1994. Copyright (1994) American Chemical Society.)

those reflected from upper (film/air) interface. The resonant response at $\pi/2$ recurs at higher harmonic values such as $3\pi/2$, $5\pi/2$, $7\pi/2$, and so forth.

4.6.1.3 Material Characterization Using Rayleigh Mode: Perturbation of Elastic Properties of Thin-Palladium Film

The SAW hydrogen responses measured for four different crystal structures involving a Pd sorbent film [50] is given in Table 4.1. Find the relative variations in the elastic constant and the density of the film using perturbation theory approach. The values of film density and elastic constants are $\rho = 11000$ kg/m^3, $C_{11} = 190$ GPa, and $C_{44} = 40$ GPa.

The perturbation expansion derived [50,51] on the basis of Equation 4.37 is given by

$$\frac{\Delta V}{V} = \frac{\pi h}{2\lambda} \left[-\frac{\Delta\hat{\rho}}{\hat{\rho}} \left\{ (A_x^2 + A_y^2 + A_z^2)\hat{\rho}V^2 \right\} + \frac{\Delta\hat{C}_{44}}{\hat{C}_{44}} \left\{ (4A_z^2 + A_x^2)\hat{C}_{44} \right\} + \left\{ \frac{\left(1 - \frac{\Delta\hat{C}_{44}}{\hat{C}_{44}} \right)}{\left(1 - \frac{\Delta\hat{C}_{11}}{\hat{C}_{11}} \right)} \left(4A_z^2 \frac{\hat{C}_{44}^2}{\hat{C}_{11}} \right) \right\} \right]$$

(4.39)

where $A_i = \left(\frac{|v_i|}{P^{1/2}} \right)_{y=0} \omega^{1/2}$ represents normalized mechanical displacement components.

To find the variations in film density and elastic constants, the measured responses (Table 4.1) would be substituted in Equation 4.39. This yields a set of four simultaneous equations with the unknowns being $\Delta\rho/\rho$, $\Delta C_{11}/C_{11}$, and $\Delta C_{44}/C_{44}$. Solving the above equations using best-fit procedure, the results obtained within error limits of $\pm20\%$ are

$$\Delta\rho/\rho = +0.11 \%, \quad \Delta C_{11}/C_{11} = +29\% \quad \text{and} \quad \Delta C_{44}/C_{44} = -28\%.$$

As per the notations used, the density and elastic constant C_{11} of the Pd film decrease, while constant C_{44} is increased upon interaction with hydrogen.

Using the above result, the response of Pd film in terms of velocity shift on the ZnO/(001)Si and ZnO/(111)Si substrates has been calculated. The ratios of film thickness (h) to wavelength (λ) for the two substrates are 0.043 and 0.1, respectively. Substituting known values of density, elastic constants and their relative variations in Equation 4.39, the theoretical response ($\Delta V/V$) for ZnO/(001)Si and ZnO/(111)Si are obtained as -110 ppm and -375 ppm, respectively.

TABLE 4.1
SAW Hydrogen Responses for Different Test Structures

Test Structure	$\Delta V/V$ (ppm)
Pd/YZ-LiNbO$_3$	-22
Pd/(001), <110>-BGO	-53
Pd/ZX-CdS	-67
Pd/ST, X-SiO$_2$	-104

4.6.2 Acoustoelectric Perturbation of SAW

The propagation of surface acoustic wave is associated with generation of a layer of bound charges at the surface that accompanies the mechanical wave. This bound charge is the source of the evanescent electric field and is the source of the electric potential (Φ). The coupling of an acoustic wave propagating in a piezoelectric substrate with charge carriers in the adjacent medium provides a mechanism for studying the changes in electrical conductivity in thin solid films and solution. This acoustoelectric effect is utilized to construct chemical sensors as well as to study conductivity effects.

4.6.2.1 Acoustoelectric Effect Associated with Rayleigh Wave Propagation

The deposition of a conductive film onto the SAW medium results in redistribution of the charge carriers to compensate for the layer of bound charge generated by the passing wave. For a SAW device employing the Rayleigh mode, a surface film having sheet conductivity (σ_s) perturbs the wave velocity by an amount ($\Delta v/v$) and changes the attenuation ($\Delta\alpha/k$) by an amount given by [52]

$$\frac{\Delta V}{V} = -\frac{K^2}{2} \frac{\sigma_s^2}{\sigma_s^2 + (V(\varepsilon_0 + \varepsilon_s))^2} \tag{4.40}$$

$$\frac{\Delta\alpha}{k} = \frac{K^2}{2} \frac{V(\varepsilon_0 + \varepsilon_s)\sigma_s}{\sigma_s^2 + (V(\varepsilon_0 + \varepsilon_s))^2} \tag{4.41}$$

K^2 is the electromechanical coupling coefficient squared. ε_0 and ε_s are the air and substrate dielectric permittivities, respectively. The above expressions have also been derived by Ballantine et al. [1] on the basis of an equivalent circuit model. The velocity and attenuation changes measured as a function of the nickel film deposited onto ST-cut quartz SAW device are shown in Figure 4.8. The SAW velocity undergoes a monotonic decrease, whereas attenuation goes through a maximum (in the 10–30 Å thickness range) with increasing metal thickness. The magnitude of the acoustoelectric response is proportional to the electromechanical coupling factor (K^2) and therefore depends on the substrate properties.

4.6.2.2 Acoustoelectric Effect Associated with Shear Horizontal Wave Propagation

The sensor sensitivity equation for the acoustoelectric interaction can be derived from an extension of the perturbation method of Auld (Section 4.6.1). The electrical properties of liquid are represented by the relative permittivity ε_r, and conductivity, σ. An approximate theory for the acoustoelectric interaction has been derived by Kondoh et al. [21,22,38,48,54–56] assuming a nonconductive liquid as reference.

$$\varepsilon_l = \varepsilon_r \varepsilon_0 \tag{4.42}$$

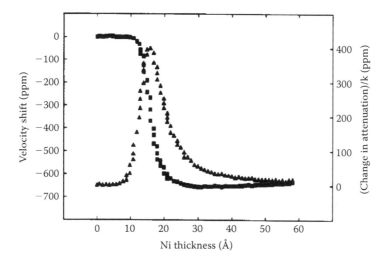

FIGURE 4.8 Attenuation and velocity responses arising from acoustoelectric effect for a Ni film deposited on a 97MHz ST Quartz device. The fractional velocity change has been corrected for the effect of mass loading. (Reprinted with permission from Ricco A. J. and Martin S. J., *Multiple frequency surface acoustic wave devices as sensors*, presented at IEEE Ultrasonics Symposium, June 4–7, 1990. Hilton Head Island, SC, USA. (© 1990 IEEE).)

ε_l is the permittivity of the reference liquid. The electrical properties of the sample liquid are expressed as a complex permittivity:

$$\varepsilon_l' = \varepsilon_r'\varepsilon_0 - j\frac{\sigma'}{\omega} \tag{4.43}$$

Here the ′ represents the perturbed quantity corresponding to the sample liquid. The change from Equation 4.42 to Equation 4.43 results in velocity and attenuation change of the SH-SAW as per the following equations [55]:

$$\frac{\Delta V}{V} = -\frac{K_s^2}{2}\frac{(\sigma'/\omega)^2 + \varepsilon_0(\varepsilon_r' - \varepsilon_r)(\varepsilon_r'\varepsilon_0 - \varepsilon_p^T)}{(\sigma'/\omega)^2 + (\varepsilon_r'\varepsilon_0 - \varepsilon_p^T)^2} \tag{4.44}$$

$$\frac{\Delta\alpha}{k} = \frac{K_s^2}{2}\frac{(\sigma'/\omega)^2(\varepsilon_r\varepsilon_0 - \varepsilon_p^T)}{(\sigma'/\omega)^2 + (\varepsilon_r'\varepsilon_0 - \varepsilon_p^T)^2} \tag{4.45}$$

K_s is the electromechanical coupling coefficient when the liquid is loaded at the free surface and ε_p is the permittivity of the crystal. A highly sensitive sensor is realized for a material with high electromechanical coupling coefficient. For a 36 YX LiTaO$_3$, $K_s = 0.1643$ and $\varepsilon_p = 4.58 \times 10^{-10}$ F/m.

4.6.2.3 Material Characterization Using SH-SAW Device: Perturbation of Electrical Properties to Calculate Conductivity and Permittivity

If the velocity shift and the attenuation responses are known from experiments, the unknown parameters in Equations 4.44 and 4.45 are the electrical properties of the liquid sample [21,22,38,48,54,55]. Therefore, simultaneous evaluation of ε_r' and σ' is possible. A permittivity-conductivity chart proposed by Kondoh et al. can be used to derive the electrical properties of the sample liquid. The chart is formed by eliminating the conductivity and permittivity from Equations 4.44 and 4.45 and solving the following equations of circle:

$$\left(\frac{\Delta V}{V} + \frac{K_s^2}{4} \frac{\varepsilon_0 (2\varepsilon_r' - \varepsilon_r) + \varepsilon_p^T}{\varepsilon_r' \varepsilon_0 + \varepsilon_p^T} \right)^2 + \left(\frac{\Delta \alpha}{k} \right)^2 = \left(\frac{K_s^2}{4} \frac{\varepsilon_r \varepsilon_0 + \varepsilon_p^T}{\varepsilon_r' \varepsilon_0 + \varepsilon_p^T} \right)^2 \tag{4.46}$$

$$\left(\frac{\Delta V}{V} + \frac{K_s^2}{4} \right)^2 + \left(\frac{\Delta \alpha}{k} - \frac{K_s^2}{4} \frac{\varepsilon_0 \varepsilon_r + \varepsilon_p^T}{(\sigma'/\omega)} \right)^2 = \left(\frac{K_s^2}{4} \frac{\varepsilon_r \varepsilon_0 + \varepsilon_p^T}{(\sigma'/\omega)} \right)^2 \tag{4.47}$$

The results of permittivity-conductivity chart by plotting Equations 4.46 and 4.47 are shown in Figure 4.9 with distilled water as the reference liquid ($\varepsilon_r = 80$) [38,56] in the $\Delta V/V$–$\Delta \alpha/k$ plane. The experimental result of sample liquid loaded on a 100 MHz SH-SAW is shown at location A. The relative permittivity and conductivity are 40 and 0.6 S/m, respectively.

4.7 FINITE ELEMENT MODELS

The models commonly used to simulate the mechanical and electrical behavior of piezoelectric transducers generally introduce simplifying assumptions that are often invalid for actual designs [57]. The geometries of practical transducers are often two- (2-D) or three-dimensional (3-D) [58]. Simulations of piezoelectric media require the

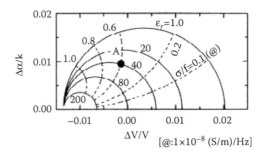

FIGURE 4.9 Permittivity-conductivity chart to derive electrical properties from SH-SAW responses. Kondoh J. and Shiokawa S: Shear Horizontal Surface Acoustic Wave Sensors. *Sensors Update.* 2001. 6. Copyright Wiley-VCH Verlag GmbH & Co. KGaA. Reproduced with permission.

complete set of fundamental equations relating mechanical and electrical quantities to be solved. The finite difference or finite element scheme is sufficient to handle the differential equations [59–65]. The finite element method has been a preferred method for modeling acoustic wave sensors such as SAW because of its ability to handle complicated geometries.

Finite element was applied by Lerch [58] to calculate the natural frequencies with related Eigen modes of the piezoelectric sensors and actuators as well as their responses to various dependent mechanical and electrical perturbations. A direct finite element analysis was carried out by Xu to study the electromechanical phenomena in SAW devices [59]. The influence of the number of electrodes on the frequency response was analyzed. The finite element calculations were able to evaluate the influence of the bulk waves at higher frequencies. Ippolito et al. have investigated the effect of electromagnetic feed through as wave propagation in layered SAW devices [60]. The same model was extended to study electrical interactions occurring during gas sensing [62]. Recently, a 3-D finite element model (FEM) was developed for a SAW palladium thin film hydrogen sensor [66]. The effect of the palladium thin film on the propagation characteristics of the SAW was studied in the absence and presence of hydrogen. The variations in mass loadings, elastic constants, and conductivity were the factors used in evaluating the velocity change of the wave. All the above demonstrate the feasibility of FEMs to adequately model SAW sensor response under varying conditions.

4.7.1 FEM Formulation for Piezoelectric Materials

The constitutive equations of piezoelectric media in linear range coupling the two are given by

$$\mathbf{T} = c^E \mathbf{S} - e^t \mathbf{E} \tag{4.48}$$

$$\mathbf{D} = e S + \varepsilon^S \mathbf{E} \tag{4.49}$$

where $\mathbf{T}$ is the vector of mechanical stresses, $\mathbf{S}$ is the vector of mechanical strains, $\mathbf{E}$ is the vector of electric field, $\mathbf{D}$ is the vector of dielectric displacement, c^E is the mechanical stiffness matrix for constant electric field E, ε^S is the permittivity matrix for constant mechanical strain, and e is the piezoelectric matrix.

These matrix equations relating the various electrical and mechanical quantities in piezoelectric media form the basis for the derivation of the FEM. The electric field $\mathbf{E}$ and mechanical strain $\mathbf{S}$ are related to the electrical potential and mechanical displacement, respectively, as follows:

$$\mathbf{E} = -\,\mathrm{grad}\;\Phi \tag{4.50}$$

$$\mathbf{S} = Bu \tag{4.51}$$

$$\text{where } B = \begin{bmatrix} \partial/\partial x & 0 & 0 \\ 0 & \partial/\partial y & 0 \\ 0 & 0 & \partial/\partial z \\ \partial/\partial y & \partial/\partial x & 0 \\ 0 & \partial/\partial z & \partial/\partial y \\ \partial/\partial z & 0 & \partial/\partial x \end{bmatrix} \quad \text{in Cartesian coordinates.}$$

The propagation of acoustic waves in piezoelectric materials is governed by the mechanical equations of motion and Maxwell's equations for electrical behavior. The elastic behavior of piezoelectric media is governed by the Newton's law:

$$\text{DIV} T = \rho \partial^2 u / \partial t^2 \tag{4.52}$$

where DIV is the divergence of dyadic and ρ is the density of the piezoelectric medium.

The electrical behavior is described by the Maxwell's equation, considering that piezoelectric media are insulating and have no free volume charge:

$$\textbf{div D} = 0 \tag{4.53}$$

Equations 4.48 through 4.53 constitute a complete set of differential equation, which can be solved with appropriate mechanical and electrical boundary conditions. The mechanical boundary conditions are mainly in the form of displacements and forces, whereas electrical boundary conditions are represented in terms of potential and charges.

Extending Hamilton's variational principle to piezoelectric media gives an equivalent description of the above boundary value problem (BVP):

$$\delta \int E \text{dt} = 0 \tag{4.54}$$

Here the operator δ denotes first-order variation, and the Lagrangian term E is determined by the energies available in the piezoelectric medium

$$E = E_{\text{kinetic}} - E_{\text{elastic}} + E_{\text{dielectric}} + W \tag{4.55}$$

The respective energies used in the above equation are listed below:

Kinetic energy:

$$E_{\text{kinetic}} = \frac{1}{2} \iiint \rho u^2 dV \tag{4.56}$$

Elastic energy:

$$E_{\text{kinetic}} = \frac{1}{2} \iiint S^t T dV \tag{4.57}$$

Dielectric energy:

$$E_{\text{kinetic}} = \frac{1}{2} \iiint D' E dV \tag{4.58}$$

External mechanical and electrical excitation generates an energy W given by

$$W = \iiint_V u' F_b dV + \iint_A u' F_s dA - \iint_A \Phi q_s dA + \sum F_p - \sum \Phi Q_p \tag{4.59}$$

Here, F_b is the vector of mechanical body forces, F_s is the vector of mechanical surface forces, F_p is the vector of mechanical point forces, q_s is the vector of surface charge, and Q_p is the vector of point charges.

Finite element formulation involves subdivision of the body to be modeled into small discrete elements (called finite elements). The system of equations represented from 4.48 to 4.59 are solved for at the nodes of these elements and the values of mechanical displacements u and forces F as well as the electrical potential Φ and charge Q. The values of these mechanical and electrical quantities at an arbitrary position on the element are given by a linear combination of polynomial interpolation functions $N(x, y, z)$ and the nodal point values of these quantities as coefficients. For an element with n nodes (nodal coordinates: (x_i, y_i, z_i); $i = 1, 2, \ldots, n$) the continuous displacement function $u(x, y, z)$ (vector of order three), for example, can be evaluated from its discrete nodal point vectors as follows (the quantities with the sign "^" are the nodal point values of one element):

$$u(x, y, z) = N_u(x, y, z)\hat{u}(x_i, y_i, z_i) \tag{4.60}$$

Here, $\hat{u}$ is a vector of nodal point displacements (order $3n$) and N_u is the shape function or interpolation function for displacement. Similarly, other mechanical and electrical quantities are interpolated using appropriate shape functions. With the shape functions for displacement (N_u) and the electric potential (N_Φ), Equations 4.50 and 4.51 can be written as

$$E = -\text{grad}\Phi = \text{grad}(N_\Phi \hat{\Phi}) \tag{4.61}$$

$$S = Bu = BN_u \hat{u} \tag{4.62}$$

A set of linear differential equations describing one single piezoelectric finite element by substituting polynomial interpolation functions (N_x) into Equation 4.54:

$$m\hat{\ddot{u}} + d_{uu}\hat{\dot{u}} + k_{uu}\hat{u} + k_{u\Phi}\hat{\Phi} = \hat{F}_B + \hat{F}_S + \hat{F}_P \tag{4.63}$$

$$k_{u\Phi}^t \hat{u} + k_{\Phi\Phi}\hat{\Phi} = \hat{Q}_S + \hat{Q}_P \tag{4.64}$$

where $\hat{\dot{u}}$, $\hat{\ddot{u}}$ represent vectors of nodal velocities and accelerations. The various matrices in the above coupled equations are listed below [58]:

Mechanical stiffness matrix:

$$k_{uu} = \iiint B_u^t c^E B_u dV \tag{4.65}$$

Mechanical damping matrix:

$$d_{uu} = \alpha^{(e)} \iiint \rho N_u^t N_u dV + \beta^{(e)} \iiint B_u^t c^E B_u dV \tag{4.66}$$

Piezoelectric coupling matrix:

$$k_{u\Phi} = \iiint B_u^t e^t B_\Phi dV \tag{4.67}$$

Dielectric stiffness matrix:

$$k_{\Phi\Phi} = \iiint B_\Phi^t \varepsilon^S B_\Phi dV \tag{4.68}$$

Mass matrix:

$$m = \iiint \rho N_u^t N_u dV \tag{4.69}$$

Mechanical body forces:

$$\hat{F}_B = \iiint N_u^t N_{FB} f_B^{(e)} dV \tag{4.70}$$

Mechanical surface forces:

$$\hat{F}_S = \iiint N_u^t N_{FS} f_S^{(e)} dV \tag{4.71}$$

Mechanical point forces:

$$\hat{F}_P = N_u^t F_P^{(e)} \tag{4.72}$$

Electrical surface charges:

$$\hat{Q}_S = -\iint N_\Phi^t N_{QS} q_S^{(e)} dA \tag{4.73}$$

Electrical point charges:

$$\hat{Q}_P = -N_\Phi^t Q_P^{(e)} \tag{4.74}$$

where the following are the forces acting at any element (e),

$f_B^{(e)}$ — External body force
$f_S^{(e)}$ — External surface force
$F_P^{(e)}$ — External point force
$q_S^{(e)}$ — External surface charge
$Q_P^{(e)}$ — External point charge

$\alpha^{(e)}$ and $\beta^{(e)}$ are the damping coefficients of element (e). The magnitudes of the damping matrices depend on the energy dissipation characteristics of the modeled structure.

Subdividing the body to be computed into finite elements results in a mesh composed of numerous single elements. A set of linear differential equation represents the complete finite element mesh of the modeled piezoelectric substrate.

$$M\ddot{u} + D_{uu}\dot{u} + K_{uu}\dot{u} + K_{u\Phi}\Phi = F_B + F_S + F_P \qquad (4.75)$$

$$K'_{u\Phi}u + K_{\Phi\Phi}\Phi = Q_S + Q_P \qquad (4.76)$$

The quantities in these sets of equation u, Φ, F_B, F_S, Q_S, and Q_P are globally assembled field quantities. If the entire mesh is composed of a total of n nodes and the model is solved for four degrees of freedom (three displacements and potential), then matrix Equation 4.75 will consist of $3*n$ and 4.76 will consist of n linear differential equation, thus resulting in a total of $4*n$ linear equations.

The solution to Equations 4.75 and 4.76 yields the mechanical displacements and the electrical potential in piezoelectric medium. The above mechanical and electrical equations are coupled by matrix $\mathbf{K_{u\Phi}}$ that is represented in terms of the piezoelectric stress tensor $\mathbf{e}$. As $\mathbf{e} \rightarrow 0$, $\mathbf{K_{u\Phi}} \rightarrow 0$ and the two sets (Equations 4.75 and 4.76) then represent pure mechanical finite element and electrostatic field models, respectively. The sets of equation represented by Equation 4.75 and 4.76 can be solved using various commercially available packages such as ANSYS [67], PZFLEX, ABAQUS, and so forth (Section 4.7.6), all of which offer excellent postprocessing capabilities.

4.7.2 FEM MODELING OF SAW DEVICES

The computation of a complete SAW device with FEM is at present impossible. For example, a conventional two-port SAW device (Figure 4.5) consisting of interdigital transducers (IDT) and end reflectors may have—especially on substrate materials with low piezoelectric coupling constants—a length of thousands of wavelengths and a lateral overlap (aperture) of hundred wavelengths. Depending on the working frequency, the substrate, which carries the electrodes, has also a depth of up to hundred wavelengths. Taking into account that FEM requires a spatial discretization with at least 10 first-order finite elements per wavelength and that an arbitrary piezoelectric material has at least four degrees of freedom, this leads to 4×10^{-8} unknowns in the three-dimensional case [68].

Typically, the size of MHz frequency SAW devices simulated using 2-D and 3-D FE models are currently restricted to ~20–30 wavelengths along the propagation direction, 4–5 wavelengths in depth and 4–5 wavelengths wide. The number of IDT finger pairs at each port is also limited to ~5–10. The mesh generated by spatial discretization for one such SAW device geometry is shown in Figure 4.10. Both the 2-D and 3-D FE models are shown. The transmitting and receiving IDTs are shown in green. By applying a known input signal at the transmitting IDT, the corresponding sensor response can be obtained at the receiving or output IDT nodes. Discussion on SAW sensor response to various input signals is presented in the subsequent sections.

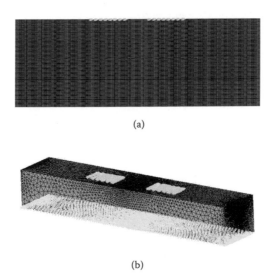

(a)

(b)

FIGURE 4.10 Meshed structure showing SAW device (a) 2-D model and (b) 3-D model. The IDT fingers are represented by coupled set of nodes.

4.7.3 FEM SIMULATIONS OF SAW SENSOR RESPONSE

The frequency response of a SAW device $H(f)$ can be obtained from its impulse response $h(t)$ by taking the Fourier transform.

$$H(f) = \int_{-\infty}^{\infty} h(t)e^{-2\pi ft}\,dt \qquad (4.77)$$

To obtain the impulse response, a signal of the following form is input at the transmitting electrodes [59]:

$$V_i = \begin{cases} \pi f_i \sin(2\pi f_i t) & 0 < t \le 1/2 f_i \\ 0 & t > 1/2 f_i \end{cases} \qquad (4.78)$$

The above signal becomes an impulse as f_i approaches infinity, which certainly cannot be handled practically. Several numerical studies have shown that the above signal can adequately represent an impulse by taking $f_i = 2f_0$, where f_0 is the center frequency of the filter. The impulse response detected by the output transducer is obtained directly from the finite element simulation. An example of the input voltage waveform and the corresponding frequency response obtained using Equation 4.78 is shown in Figure 4.11a and 4.11b.

Simulations involving SAW sensor typically generate responses of SAW device to various mechanical and electrical perturbations. Variations in the mass loading, elastic constants, and conductivity are some of the factors, which contribute to the velocity change of the surface acoustic wave. The change in velocities results in a

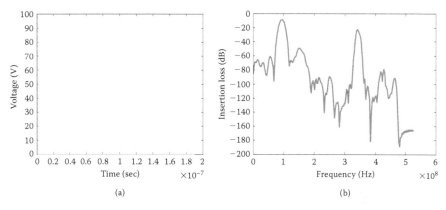

FIGURE 4.11   (a) Input voltage waveform and (b) frequency response.

corresponding shift in frequency. A series of simulations can be carried out and the impulse response of the SAW device to various perturbations can be studied.

4.7.4 MODELING OF A TYPICAL SAW H_2 GAS SENSOR RESPONSE

FEM models have been used to gain insights into the SAW gas sensor response. Atashbar et al. [66,69] used a 3-D FEM model with an AC analysis to study the response of SAW gas sensor resulting from perturbations of a thin film of palladium due to absorption of hydrogen gas. The absorption of hydrogen by palladium thin film changes the SAW propagation velocity. As brought out in Section 4.6, variations in mass loading, elastic constants, and conductivity are some of the factors that contribute to the velocity change. In case of hydrogen gas sensors, the absorption of H_2 causes a decrease in density and Young's modulus of elasticity of the palladium film.

By simulating for known changes in material properties (Table 4.2), the variations in the voltage and displacement waveforms (obtained by solving Equations 4.75 and 4.76) at the output IDT resulting from gas absorption can be determined. The response can then be calculated using Equation 4.77. The voltage and the displacement waveforms in the presence and absence of hydrogen for a SAW device based on a YZ- $LiNbO_3$ is shown in Figures 4.12 and 4.13.

From the voltage (Figure 4.12) and displacement (Figure 4.13) waveforms, it can be seen that the variation in the material properties of Pd film in accordance with H_2 absorption leads to a time delay in the waveforms at the nodes in the output IDT. These are a result of the velocity change of the SAW. The delay in the displacement and voltage waveforms is approximately 2 ns (Figure 4.14).

An impulse response analysis (Equation 4.78) can be performed on the SAW sensor model to gain insights into the sensor response. The frequency response is calculated using Equation 4.77. Figure 4.15 shows the insertion loss computed for the device over a frequency span of 500 MHz. The insertion loss of the sensor resulting from H_2 absorption ~1.5 dB.

TABLE 4.2
Property Changes on H₂ Absorption

Properties	Changes on 3% H₂ Absorption	References
Density	2%	[70]
Young's modulus	15%	[50]
Volume	10%	[66]

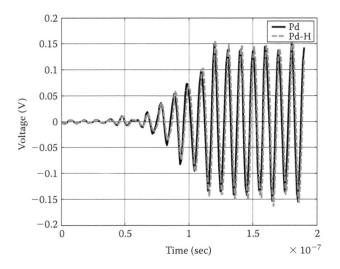

FIGURE 4.12 Voltage waveform in the presence and absence of hydrogen. The input voltage is alternating with 5-V (peak-peak) signal.

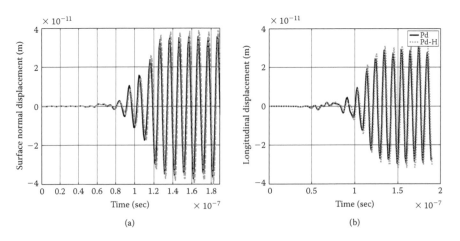

FIGURE 4.13 The displacement waveforms at the output IDT fingers with and without hydrogen along the (a) surface normal direction and (b) longitudinal direction.

4.7.5 MODELING OF COMPLICATED TRANSDUCER GEOMETRIES: HEXAGONAL SAW DEVICE

The main advantage of numerical methods like finite elements lie in their ability to model SAW devices involving complicated transducer geometries. One such geometry is the hexagonal SAW device proposed by Cular et al. [71] shown in Figure 4.16. The three different delay paths could be used for simultaneous

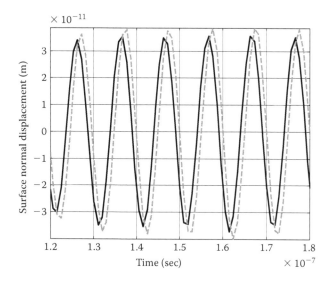

FIGURE 4.14 Time delay in the displacement waveform in the surface normal direction. A 2 ns time delay results from the absorption of H_2.

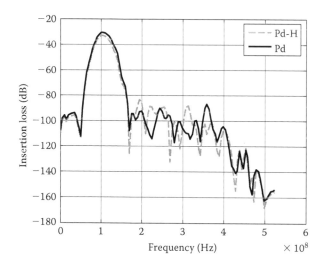

FIGURE 4.15 Frequency response showing attenuation (transmission losses) resulting from gas absorption.

detection and the data collected across the three delay paths allows for better characterization of the sensing (thin film) material. This design allows for the simultaneous extraction of multiple properties (film material density or thickness, Lamé and shear moduli, and sheet conductivity) of a thin film material to achieve a more complete characterization than when a single SAW device is utilized. In sensor applications, this capability translates to better discrimination of the analyte and possibly more accurate determination of the concentration. Preliminary experimental results have shown increased sensitivity for these devices when used as a chemical sensor. Other application of the hexagonal SAW in biosensing involves the ease of detection as well as removal of nonspecifically bound proteins (acoustic streaming) enabling the repeated use of sensor device. One of the delay paths is used for detection, whereas the other delay paths are used to remove the nonspecifically bound proteins using acoustic streaming phenomenon. The fabrication of a hexagonal SAW device can be carried out on any piezoelectric substrate such as lithium tantalate and lithium niobate. However, prior to the device fabrication, it is important to establish the type of waves that are generated along the various delay paths. The choice of a delay path for any specific application depends on the propagation characteristics of the wave generated along the crystal cut and orientation corresponding to that delay path.

4.7.5.1 FEM Simulation of Hexagonal SAW Device

In this section, a hexagonal SAW device based on $LiNbO_3$ substrate is modeled using finite element technique. The wave propagation characteristics along the three different delay paths corresponding to crystal orientation with Euler angles (0, 90, 90),

FIGURE 4.16 Hexagonal SAW device used for chemical and biosensing applications as well as materials characterization.

(0, 90, 30), and (0, 90, 150), respectively, are evaluated. Using FEM, the types of waves that exist along the three delay lines are calculated.

The 3-D FE model describes three two-port delay line structures along each of the Euler direction and consists of three finger pairs in each port. The interdigital transducer (IDT) fingers are defined on the surface of a lithium niobate substrate and the fingers are considered as massless electrodes to ignore the second-order effects arising from electrode mass, thereby simplifying computation. The periodicity of the finger pairs is 40 microns and the aperture width is 200 microns. The transmitting and receiving IDTs are spaced 130 microns or 3.25λ apart.

The substrate for (0, 90, 90) or YZ-$LiNbO_3$ was defined as 800 microns in propagation length, 300 microns wide, and 150 microns deep. For simulating the other two directions, the geometry of substrate is kept the same, whereas the crystal coordinates are rotated. To achieve this, the material properties, that is, stiffness, piezoelectric, and permittivity matrices are rotated by 60° and −60° along the x–z plane to model Euler directions (0, 90,150) and (0, 90, 30), respectively. The simulated models have a total of, approximately, 25,0000 nodes and are solved for four degrees of freedom (three displacements and voltage). The model was created to have the highest densities throughout the surface and middle of the substrate. Two kinds of analysis are carried out along each of the three delay lines:

1. An impulse input of 10 V over 1 ns is applied to study the frequency response of the device.
2. AC analysis with a 5 V peak-peak input and 100 MHz frequency to study the wave propagation characteristics.

4.7.5.2 Impulse Response

By applying an impulse function as an input, the frequency response can be calculated (Section 4.7.2). The velocities corresponding to wave propagation along the three Euler angles are given in Table 4.3. Since frequency is directly proportional to velocity, it is expected that the frequency response would follow the order (0, 90, 90) < (0, 90, 30) < (0, 90, 150). The calculated frequency response for an input impulse (1 ns) of 10 V is shown in Figure 4.17. The calculated device frequency along the three directions follows (0, 90, 90) < (0, 90, 30) < (0, 90, 150). The least attenuation occurs along the (0, 90, 90) direction, whereas the maximum is observed for (0, 90, 150).

4.7.5.3 AC Analysis

In order to gain insights into the types of wave that are generated and propagating along the three directions, an AC analysis can be carried out. This is done by applying a 5-V peak-peak signal input for 200 ns at a frequency of 100 MHz. The response was obtained at the output IDT node located 210 microns away from the input. The generated voltage and displacement waveforms are shown in Figures 4.18 through 4.20.

Figures 4.18 through 4.20a depict the generated voltage at the output node of the hexagonal SAW device in response to a signal with 5 V amplitude at a frequency

TABLE 4.3
Theoretical and Measured Wave Velocities along the Different Shorted Delay Paths of the Hexagonal SAW Device on Lithium Niobate

Orientation Euler Angle (φ, θ, ψ)	Theoretical (m/s)	Experimental (m/s)
(0, 90, 91)	3542.06	3593.30
(0, 90, 151)	3646.81	3721.85
(0, 90, 31)	3622.59	3620.73

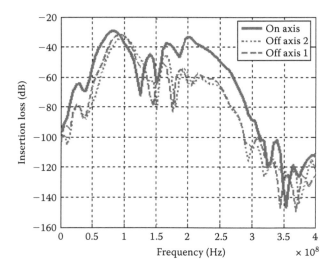

FIGURE 4.17 Calculated frequency response along the three Euler directions. On axis, Off Axis 1, and Off Axis 2 corresponds to (0, 90, 90), (0, 90, 30), and (0, 90, 150), respectively.

of 100 MHz. Voltage profiles obtained for the three Euler directions from the AC analysis corroborate the findings of the impulse response analysis. The stabilized value of the voltage obtained at the output IDT shows higher peak value for (0, 90, 90) direction followed by (0, 90, 30) and (0, 90, 150). This indicates lesser attenuation along (0, 90, 90) direction when compared to the other two. The anisotropic nature of the substrate results in varying amplitude values for displacement profiles along the surface normal, longitudinal, and shear horizontal direction as shown in Figures 4.17 through 4.19b–c. For the (0, 90, 90) direction, the surface normal and longitudinal components are an order of magnitude higher than the shear horizontal component indicative of wave motion, which is more or less ellipsoidal. This type of wave motion corresponds to that of the Rayleigh mode. The displacement profiles of the off-axis components signified by (0, 90, 30) and (0, 90, 150) directions show lesser amplitude variations among the three directions indicative of mixed wave modes, which are a combination of more than one wave type such as pure Rayleigh or shear horizontal modes. Analysis of displacement versus depth profiles as shown in Figure 4.4b can also be carried out and is likely to yield more insights into the propagation characteristics of the waves along the three directions (Figure 4.21).

The wave propagation along (0, 90, 30) direction in lithium niobate substrate is shown in Figure 4.21. At approximately 85 ns, the wave has reached the end of the substrate. Wave reflections along the edges are observed at higher simulation times. A 3-D view of the acoustic wave propagation can thus be useful to understand the acoustic wave propagation in piezoelectric substrates.

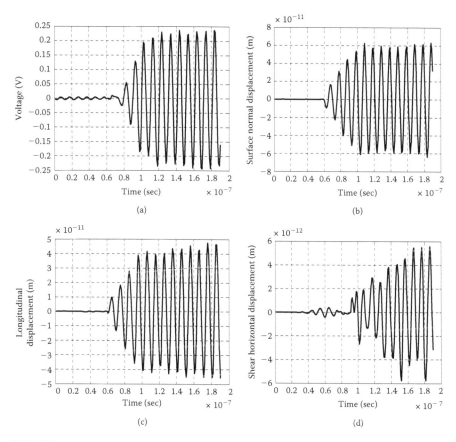

FIGURE 4.18 Voltage and displacement waveforms at the output IDT node along (0, 90, 90) Euler direction.

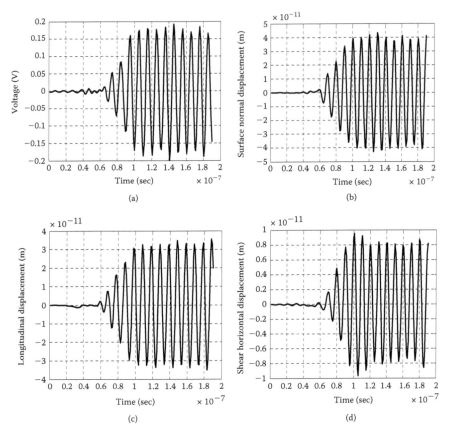

FIGURE 4.19 Voltage and displacement waveforms at the output IDT node along (0, 90, 30) Euler direction.

4.7.6 AVAILABLE FINITE ELEMENT PACKAGES

The main advantage in using commercially available packages lies in their postprocessing ability. Some of the commercially used finite element analysis packages with piezoelectric capability are listed as follows:

- ANSYS/Multiphysics—ANSYS Inc. [67]
- Abaqus—Hibbitt, Karlsson & Sorensen
- Pafec—SER Systems Ltd.
- Pzflex—Weidlinger Associates
- Femlab 3—Comsol
- Atila—Magsoft Corp.
- Adina-T—Adina R & D, Inc.

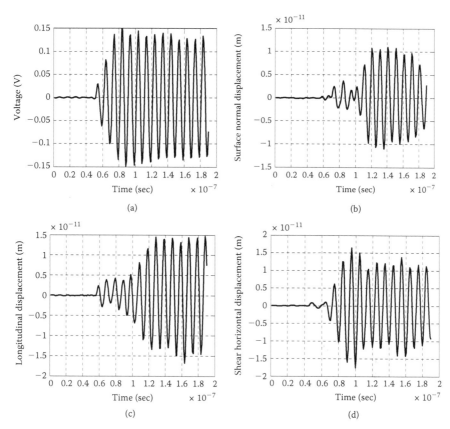

FIGURE 4.20   Voltage and displacement waveforms at the output IDT node along (0, 90, 150) Euler direction.

4.8   LIMITATIONS OF THE TWO APPROACHES

Perturbation theory is concerned mainly with the changes in the wave propagation characteristics arising from variations in various physical factors. Wave generation and identification of wave modes, however, require the use of other analytical techniques based on Green's function approach or numerical techniques such as FEM or a combination of both. Perturbational formulas are often valid only when the deviations from the starting solution are small (10%–15%). In most of the cases, there is a prior assumption of the wave type, which limits applicability of the derived perturbational formulas to the specific wave types. On the other hand, although numerical techniques such as finite element are more accurate than perturbation theory, several limitations exist. A fine mesh generation is required for accurate modeling of SAW sensors. In addition, the simulations are computationally intensive and therefore time consuming. Simulation times scale linearly with mesh size. Another major setback arises from acoustic wave reflection from the edges if the simulations are carried out

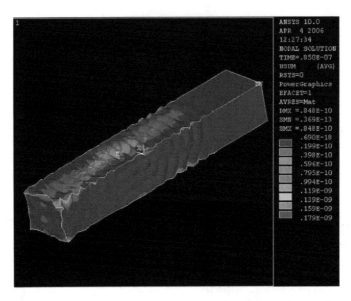

FIGURE 4.21   Surface acoustic wave propagation in lithium niobate substrate along (0, 90, 30) direction.

for sufficiently longer times. One of the ways to overcome this limitation is to employ damping elements at the ends of the substrate. In most of the experimental studies, the frequency shifts reported for sensor response are in the order of a few hundred KHz. To obtain this level of accuracy, longer simulation times with smaller time steps are required. While longer simulation times are also necessary to attain a stable state, too long a simulation time results in wave reflections causing instabilities to set in. A simulation time of 100–200 ns was found to be optimum for the substrate dimensions considered in the present study.

ACKNOWLEDGMENTS

One of the authors Subramanian K. R. S. Sankaranarayanan wishes to thank Stefan Cular, Samuel Ippolito, and Dr. Jun Kondoh for their help in getting this chapter done.

REFERENCES

1. Ballantine D. S., White R. M., Martin S. J., Ricco A. J., Frye G. C., Zellers E. T., and Wohltjen H., *Acoustic Wave Sensors: Theory, Design, and Physico-Chemical Applications* (New York: Academic Press, 1997).
2. Cady W. F., *Piezoelectricity*, Vol. 1 and 2 (New York: Dover, 1964).
3. Slobodnik A. J., Conway E. D., and Delmonico R. T., *Microwave Acoustics Handbook*, *Surface Wave Velocities*, Vol. 1A, Report, November 1971–April 1980 (Griffiss AFB, NY: Rome Air Development Center, 1974).

4. Slobodnik A. J., Conway E. D., and Delmonico R. T., *Microwave Acoustics Handbook, Surface Wave Velocities*, Vol. 2, Report, November 1971–April 1980 (Griffiss AFB, NY: Rome Air Development Center, 1974).

5. Nomura T. and Minemura, A., Behavior of a piezoelectric quartz crystal in an aqueous solution and the application to the determination of minute amount of cyanide, *Jpn. Chem. Soc.*, 10, 1621–1625, 1980.

6. White R. M. and Volltmer F. W., Direct piezoelectric coupling to surface elastic waves, *Appl. Phys. Lett.*, 7, 314, 1965.

7. Wohltjen H., Mechanism of operation and design considerations for surface acoustic wave device vapor sensors, *Sens. and Actuators B*, 5, 307–325, 1984.

8. Wohltjen H., 4,312,228. US patent, January 26, 1982.

9. Hommady M., Campitelli A., and Wlodarski W., Acoustic wave sensors: Design, sensing mechanisms and applications, *Smart Mater. Struct.*, 6, 647–657, 1997.

10. Josse F., Bender F., and Cernosek R. W., Guided shear horizontal surface acoustic wave sensors for chemical and biochemical detection in liquids, *Anal. Chem.*, 73, 5937–5944, 2001.

11. Josse F., Bender F., Cernosek R. W., and Zinszer K., *Guided SH-SAW sensors for liquid-phase detection*, presented at IEEE International Frequency Control Symposium and PDA Exhibition, June 6–8, 2001, Seattle, Washington, USA.

12. Shiokawa S. and Kondoh J., *Surface acoustic wave sensor for liquid-phase application*, presented at IEEE Ultrasonics Symposium, October 17–20, 1999, Caesars Tahoe, NV, USA.

13. Shiokawa S. and Kondoh J., Surface acoustic wave sensors, *Jpn. J. Appl. Phys.*, 43, 2799–2802, 2004.

14. Nakamura K., Kazumi M., and Shimizu H., *SH-type and Rayleigh-type surface waves on rotated Y-cut LiTaO₃*, presented at IEEE Symposium, October 26–28, 1977, Phoenix Arizona.

15. Rayleigh, L., On waves propagating along the plane surface of an elastic solid, *Proceedings of the London Mathematical Society* 17, 4–11, 1885.

16. Rosenbaum J. F., *Bulk Acoustic Wave Acoustic Theory and Devices* (London: Artech House Boston, 1988).

17. Royer D. and Dieulesaint E., *Elastic Waves in Solids*, Vol. I (New York: Springer, 1999).

18. Royer D. and Dieulesaint E., *Elastic Waves in Solids*, Vol. II (New York: Springer, 1999).

19. Martin F., Newton M. I., McHale G., Melzak K. A., and Gizeli E., Pulse mode shear horizontal-surface acoustic wave (SH-SAW) system for liquid based sensing applications, *Biosensors and Bioelectronics*, 19, 627–632, 2004.

20. Martin S. J., Ricco A. J., Niemczyk T. M., and Frye G. C., Characterisation of SH acoustic plate mode liquid sensors, *Sens. Actuators*, 20, 253–268, 1989.

21. Kondoh J. and Shiokawa S., A liquid sensor based on a shear horizontal SAW device, *Electron. Comm. Jpn II*, 76, 69–82, 1993.

22. Kondoh J. and Shiokawa S., Shear horizontal surface acoustic wave sensors, *Sens. Update*, 6, 2001.

23. Grate J. W., Martin S. J., and White R. M., Acoutic wave microsensors. Part I, *Anal. Chem.*, 65, 941A–948A, 1993.

24. Grate J. W., Martin S. J., and White R. M., Acoutic wave microsensors. Part II, *Anal. Chem.*, 65, 987A–996A, 1993.

25. Amico A. D' and Verona E., SAW sensors, *Sens. Actuators*, 17, 55–66, 1989.

26. Khlebarov Z. P., Stoyanova A. I., and Topalova D. I., Surface acoustic wave sensors, *Sens. Actuators B*, 8, 33–40, 1992.

27. McCallum J. J., Piezoelectric devices for mass and chemical measurements: An update, *Analyst*, 114, 1173–1189, 1989.

28. Marrion A. R., *The Chemistry and Physics of Coatings* (Boca Ratton, FL: Lewis, 1993).

29. Pesek J. J., *Chemically Modified Surfaces* (Boca Ratton, FL: Lewis Publishers, 1994).

30. Sibilia J. P., *A Guide to Materials Characterization and Chemical Analysis* (New York: VCH Publishers, 1988).

31. Thompson M. and Stone D. C., *Surface-launched Acoustic Wave Sensors (Chemical Sensing and Thin-Film Characterization)* (New York: John Wiley and Sons, 1997).

32. Shen C.-Y., Hsu C.-L., Hsu K.-C., and Jeng J.-S., Analysis of shear horizontal surface acoustic wave sensors with the coupling of modes theory, *Jpn J. Appl. Phys.*, 44, 1510–1513, 2005.

33. Ventura P., Hode J. M., Desbois J., and Solal H., Combined FEM and Green's function analysis of periodic SAW structure, application to the calculation of reflection and scattering parameters, *IEEE Trans. Ultrason. Ferroelectrics Freq. Contr.*, 48, 1259–1274, 2001.

34. Yoon S., Yu J.-D., Kanna S., Oshio M., and Tanaka M., *Finite element analysis of the substrate thickness effect on traveling leakey surface acoustic waves*, presented at IEEE Ultrasonics Symposium, October 5–8, 2003, Honolulu Hawaii Publication.

35. Auld B. A., *Acoustic Fields and Waves in Solids*, Vol. 1 (New York: John Wiley and Sons, 1973).

36. Auld B. A., *Acoustic Fields and Waves in Solids*, Vol. 2 (New York: John Wiley and Sons, 1973).

37. Martin S. J., Frye G. C., and Senturia S. D., Dynamics and response of polymer-coated surface acoustic wave devices: Effect of viscoelastic properties and film resonance, *Anal. Chem.*, 66, 2201–2219, 1994.

38. Kondoh J. and Shiokawa S., Measurements of conductivity and pH of liquid using surface acoustic wave devices, *Jpn J. Appl. Phys.*, 31(Suppl. 31–1), 82–84, 1992.

39. Peach R. C., *A general Green function analysis for SAW devices*, presented at IEEE Ultrasonics Symposium, November 7–10, 1995, Seattle, WA, USA.

40. Plessky V. P. and Thorvaldsson T., Rayleigh waves and leaky SAW's in periodic systems of electrodes: Periodic Green functions analysis, *Electron. Lett.*, 28, 1317–1319, 1992.

41. Mason W. P., *Electromechanical Transducers and Wave Filters*, 3rd ed. (Princeton, NJ: D. van Nostrand, 1948).

42. Datta S., *Surface Acoustic Wave Devices* (Englewood Cliffs: Prentice-Hall, 1986).

43. Kenny T. D., Pollard T. B., Berkenpas E. J., and Pereira da Cunha M., *FEM/BEM impedance and power analysis for measured LGS SH-SAW devices*, presented at IEEE Ultrasonics Symposium, August 24–27, 2004, Montreal, Canada.

44. Finger N., Kovacs G., Schoberl J., and Langer U., *Accurate FEM/BEM-simulation of surface acoustic wave filters*, presented at IEEE Ultrasonics, October 5–8, 2003, Honolulu, Hawaii.

45. Caliendo C., *The perturbative approach applied to the characterization of TiO_2 films for SAW sensor applications*, presented at Sensors and Microsystems, Proceedings of the Italian Conference, Roma, February 3–5, 1999.

46. Ogilvy J. A., Predicting mass loading sensitivity for acoustic wave sensors operating in air, *Sens. Actuators B*, B42, 109–117, 1997.

47. Li Z., Guided shear horizontal surface acoustic wave (SH-wave) chemical sensors for detection of organic contaminants in aqueous environments, PhD: Marquette University, 2005.

48. Kondoh J., Shiokawa S., Rapp M., and Stier S., Simulation of viscoelastic effects of polymer coatings on surface acoustic wave gas sensor under consideration of film thickness, *Jpn J. Appl. Phys.*, 37, 2842–2848, 1998.

49. Li Z., Jones Y., Hossenlopp J., Cernosek R., and Josse F., Analysis of liquid-phase chemical detection using guided shear horizontal-surface acoustic wave sensors, *Anal. Chem.*, 77, 4595–4603, 2005.

50. Anisimkin V. I., Kotelyanskii I. M., Verardi P., and Verona E., Elastic properties of thin-film palladium for surface acoustic wave (SAW) sensors, *Sens. Actuators B*, B23, 203–208, November 1–4, 1994, Hotel Martinez Cannes, France.

51. Anisimkin V. I., Kotelyanskii I. M., Fedosov V. I., Caliendo C., Verardi P., and Verona E., *Analysis of the different contributions to the response of SAW gas sensors*, presented at IEEE Ultrasonics Symposium, November 1–4, 1994, Hotel Martinez Cannes, France.

52. Ricco A. J., Martin S. J., and Zipperian T. E., Surface acoustic wave gas sensors based on film conductivity changes, *Sens. Actuators*, 8, 1985.

53. Ricco A. J. and Martin S. J., *Multiple frequency surface acoustic wave devices as sensors*, presented at IEEE Ultrasonics Symposium, Decembert 4–7, 1990, Honolulu, Hawaii.

54. Kondoh J. and Shiokawa S., Surface acoustic wave sensor based on film conductivity changes, *Sens. Actuators*, 8, 319, 1985.

55. Kondoh J. and Shiokawa S., Measurements of conductivity and pH of liquid using surface acoustic wave devices, *Jpn J. Appl. Phys.*, 31, 82–84, 1992.

56. Kondoh J., Saito K., Shiokawa S., and Suzuki H., Simultaneous measurements of liquid properties using multichannel shear horizontal surface acoustic wave microsensor, *Jpn J. Appl. Phys.* 35, 3093–3096, 1996.

57. Morgan D. P., *History of SAW devices*, presented at IEEE International Frequency control symposium, May 27–29, 1998, Ritz-Carlton Hotel, Pasadena, California.

58. Lerch R., Simulation of piezoelectric devices by two- and three-dimensional finite elements, *IEEE Trans. Ultrason. Ferroelectrics Freq. Contr.*, 37, 233–247, 1990.

59. Xu G., *Finite element analysis of second order effects on the frequency response of a SAW device*, presented at IEEE Ultrasonics Symposium, October 22–25, 2000, San Juan, Puerto Rico.

60. Ippolito S. J., Kalantar-Zadeh K., Powell D. A., and Wlodarski W., *A 3-dimensional finite element approach for simulating acoustic wave propagation in layered SAW devices*, presented at Ultrasonics, IEEE Symposium, October 5–8, Honolulu Hawaii Publication.

61. Ippolito S. J., Kalantar-zadeh K., Powell D. A., and Wlodarski W., *A finite element approach for 3-dimensional simulation of layered acoustic wave transducers*, presented at Optoelectronic and Microelectronic Materials and Devices, December 11–13, 2002, Sydney, Australia.

62. Ippolito S. J., Kalantar-zadeh K., Wlodarski W., and Matthews G. I., *The study of ZnO/XY LiNbO₃/sub 3/ layered SAW devices for sensing applications*, presented at IEEE Sensors, October 22–24, 2003, Sheraton Centre Hotel, Toronto, Canada.

63. Xu G., Direct finite element analysis of the frequency response of a Y-Z LiNbO$_3$ SAW filter, *Smart Mater. Struct.*, 9, 973–980, 2000.

64. Hasegawa K. and Koshiba M., Finite-element solution of Rayleigh-wave scattering from reflective gratings on a piezoelectric substrate, *IEEE Trans. Ultrason. Ferroelectrics Freq. Contr.*, 37, 99–105, 1990.

65. Hashimoto K.-Y., Omori T., and Yamaguchi M., *Recent progress in modelling and simulation technologies of shear horizontal type surface acoustic wave devices*, presented at International symposium on acoustic wave devices for future mobile communication systems, March 5–7, 2001, Chiba, Japan.

66. Atashbar M. Z., Bazuin B. J., Simpeh M., and Krishnamurthy S., *3-D Finite element simulation model of saw palladium thin film hydrogen sensor*, presented at IEEE International Ultrasonics, Ferroelectrics and Frequency control Joint 50th Anniversary Conference, August 24–27, 2004, Montréal, Canada.

67. ANSYS, "Trademark of ANSYS, Inc.," 10 eds, 2005.

68. Hofer M., Finger N., Kovacs G., Schoberl J., Langer U., and Lerch R., *Finite element simulation of bulk- and surface acoustic wave (SAW) interaction in SAW devices*, presented at IEEE Ultrasonics Symposium, October 8–11, 2002, Munich, Germany.

69. Atashbar M. Z., Bazuin B. J., Simpeh M., and Krishnamurthy S., 3D FE simulation of H_2 SAW gas sensor, *Sens. Actuators B*, B111–B112, 213–218, 2005.

70. Fabre A., Finot E., Demoment J., and Contreras S., In situ measurement of elastic properties of PdHx, PdDx, and PdTx, *J. Alloys. Compd.*, 356–357, 372–376, 2003.

71. Cular S., Branch D. W., and Bhethanabotla V. R., Hexagonal saw devices for enhanced sensing, presented at AICHE annual meeting, Cincinnati, OH, October 2005.

5 Recent Advances in the Development of Sensors for Toxicity Monitoring

Ibtisam E. Tothill
Associate Professor in Analytical Biochemistry
Cranfield Health
Cranfield University
Cranfield, Bedfordshire, U.K.

CONTENTS

5.1 INTRODUCTION

The increase in the presence of toxic substances in the environment has been high-lighted by global reports, regulators, and the scientific literature. Increasing environmental legislation, in the Unites States, Europe, and worldwide to control the release and levels of toxic chemicals in the environment, has created a need for rapid and reliable monitoring systems for air, soil, water, and food analysis. To assess the impact of toxins and devise management options, rapid sensing methods must be applied to monitor their concentration and assess their toxicity. This is to reduce and eliminate their effect on humans and the environment. Toxins are usually present and also harmful at very low concentrations ($<\mu g\,L^{-1}$) with many group compounds having similar structures. Therefore, rapid, sensitive and specific tests, and sensors need to be developed to achieve accurate assessment results. A range of chemical/

physical or biological methods are currently employed for monitoring toxic chemical compounds, but the effects of substances in the environment such as synergistic and antagonistic toxic effects, and the potential effects of unknown or very low traces of compounds may not be detected. Good manufacturing practice (GMP) and Hazard Analysis Critical Control Point (HACCP) for food, quality assurance, and quality control (QA/QC) procedures, in general, are usually applied through the production chain, but incidents can occur, for example, in food manufacturing processes or in water purification and distribution systems and also air pollution; therefore rapid monitoring technologies are essential to manage incidents that may prejudice safety or quality of life. Deliberate contamination and bioterrorism are also a new reality and can occur anywhere at anytime, and hence early warning systems (EWS) for toxins detection are essential for risk management. EWS are important tools to avoid or mitigate the impacts of an intentional contamination event in time to allow an effective local response that reduces or eliminates adverse impacts [1].

A range of bioassay methods has been developed based mainly on the use of whole cells or intact organisms for the assessment of toxicity. Such tests can provide a more direct measure of relevant toxicity than chemical and physical analyses, alone, which are not sufficient to assess the potential effects on human health and the ecosystems. The general principle is to expose the test organisms to various doses of the pollutant and to monitor their biological integrity as a measure of toxicity. Such tests are important in providing useful information critical in completing an overall risk assessment since it is not feasible to determine the specific toxicity of each of the toxic compounds in an environmental sample if total toxicity is required. Whole sample toxicity testing is a simple, cost-effective and a relevant means of determining sample toxicity. Biological assays can also monitor toxins not detected by the restricted range of specific sensors that can feasibly be operated in a system [2]. However, many of the old bioassays methods have now been supplemented by a range of chemical sensors and biosensors or developed into sensor systems to enable a more rapid, online/on-site signal transduction and analysis. The sensor devices can be based on a variety of technology platforms. Biosensors should also be distinguished from bioassays, where the transducer is not an integral part of the analytical system [3]. In recent years, however, many of the microorganisms used in bioassay tests have been referred to as "biosensor" by either the manufacturers or the tests developers, and, hence, many bioassays are referred to as biosensors. This makes separating the two technologies more difficult in some instances, and careful consideration should be taken to ensure that confusion is eliminated in future products.

The development of chemical and biological sensors for the rapid, reliable, and low-cost determination of toxic compounds is receiving great attention from research community. A sensor is a device capable of deriving qualitative or quantitative information regarding the chemical, biological, or physical state of a material. Biosensors consist of a recognition layer and a transducer as an integrated system. A range of selective and specific recognition layers is interfaced onto the surface of a transduction system. The interaction of the toxin being measured with the recognition layer induces a change in the latter, which is transduced (converted) into a measurable electronic signal. For chemical sensors, the recognition layer is a chemical (synthetic) recognition layer, such as a polymer film, a supramolecular host system, or a self-assembled arrangement that is designed for selective interaction with the substance to be sensed (analyte, target species) [4]. In biosensors, the recognition material is usually a natural biologically derived system such as microorganisms, enzyme, antibody, nucleic acid, or other natural molecular binding system [5]. The sensing principle is achieved through a whole spectrum of transduction principles, which can be easily measured, and miniaturized physical principle linked with a chemical/biochemical reaction. Thus, electrochemical (potentiometric, amperometric, conductimetric), optical (absorbance, fluorescence, chemiluminescence), thermal, and piezoelectric transduction systems can be employed. Today a wide range of antibodies, enzymes, and cell-based sensor systems have been developed for the detection and quantification of toxic compounds, and these are suitable for laboratory-based or field-based analysis. Sensors for toxins analysis are also very important for military and homeland security application. The sensors need to be an integrated system for monitoring, analyzing, interpreting, and communicating the data, which can then be used to make decisions that are protective of public health and minimize unnecessary concern and inconvenience to the public [6,7]. Toxicity monitoring can be categorized into the following three areas:

1. Broad spectrum: monitors acute toxicity and employs microbial cells and whole organisms assays. Examples, are bacterial bioluminescent assay and fish and daphnia tests.
2. Selective behavior: monitors specific toxicity and employs enzyme inhibition tests and reporter genes tests. Examples, are use of cholinesterase for organophosphates and urease for heavy metal ions detection and also transgenic microorganism for specific analyte assessment.
3. Analyte specific: monitors chronic toxicity and employs tests specific for a single toxin or for analysis of a class of toxins. This covers a wide spectrum of tests including antibody-based assays.

Analytical technology used for toxicity assessment covers a broad field. This chapter provides an overview of the state-of-the-art technologies used in the development of sensors for toxicity assessment. The chapter, however, will focus on broad-spectrum toxicity and selective behavior toxicity assessment. Analyte specific will only be introduced with further readings since this will be covered elsewhere. The chapter will not cover toxicity assays or methods presented as kits but mainly sensors. For more detailed review of toxicity bioassays please refer to References 2 and 8.

5.2 SIGNATURE INDICATION OF CONTAMINATION

The threat of environmental pollution and biological warfare agents has increased the need for high-throughput toxin detection systems, which can identify unknown toxic chemicals in continuous systems, such as clean water distribution systems. Assessment of toxicity can be carried out using several approaches, which include assessing the general states of the sample or assessing the sample for specific toxic compounds. Before reviewing the specific methods used to sense toxicity/toxic compounds, there are emerging technologies and some are commercially available sensors for the Water Industry and Regulatory Authority that can provide a "signature

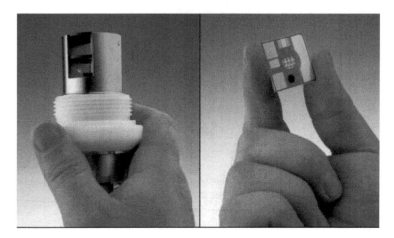

FIGURE 5.1   The color and turbidity measuring head and multiparameter sensor chip (Photo courtesy of Censar Technologies Inc.).

indication" of toxicity in water systems. For accidental or intentional contamination event for surveillance and monitoring of, for example, water as EWS for drinking water security and risk management, two-stage systems are being implemented to assess quality and toxicity. The first stage can employ continuous real-time in situ sensors that can provide generic states of the water sample. If the sample was designated as contaminated then second-stage assessment using specific and sensitive sensor technology will be used to confirm and identify the toxic compounds or toxicity. Sensors used for first-stage systems usually measure simple physiochemical parameters and contain array of sensors for water quality parameters, such as pH, conductivity, turbidity and color, dissolved oxygen, temperature, ORP/redox potential. These sensors are linked to data handling and processing and can provide fingerprint indication of contamination events by toxic compounds. Several sensor products are commercially available, for example, the electrochemical multiparameter sensor chip and the color and turbidity sensor (Figure 5.1) (Censar Technologies, Inc., Dorset, U.K.), the YSI Muiltiparameter Sonde (YSI, Hydrodata Ltd., Hertfordshire, U.K.), and the Six-Cense™ system (Dascore, Inc. South Carolina). Details of some of these sensors are available in the EPA report on EWS [9]. The sensor systems can give a first-step indication of toxicity assessment, as any change in the chemical/physical parameters of the sample will be recorded and taken as possible contamination, where samples will be removed and analyzed for toxicity confirmation. Further details on EWS are also covered by Grayman et al. [6] and Hasan et al. [7]. Toxicity systems based on sensing the toxic compounds are reviewed in the following sections.

5.3   WHOLE-CELL BIOSENSORS (MICROBIAL SENSORS)

Sensors based on the use of microorganisms or cellular response coupled with a transducer have been applied in toxicity assessment and in the detection of toxic compounds in diverse fields as medicine, environmental monitoring, defense, food processing, and safety [2,10]. Different types of microorganisms have been applied as sensing element, and these include prokaryotic (bacteria) as well as eukaryotic cells (fungi, yeast, plants, and animal cells). These cells are combined with a method of interrogating their general metabolic status, which involves detecting respiratory activity or metabolic functions of the cells to detect the compounds that are either a substrate or an inhibitor of these processes. This usually detects oxygen or substrate consumption, the production of carbon dioxide or metabolites, and detection of luminescence or direct electrochemical sampling of the electron transport chain. Transducers utilized in whole-cell sensors are varied but mostly are either optical or electrochemical [11]. Most of these sensors are unable to identify the compounds, but are sensitive to low level of toxin detection. However, specific microorganisms that may be very sensitive to specific toxic compounds or have been modified (transgenic) to acquire or to induce specific response have also been used in the development of toxicity sensors. Genetically engineered microorganisms based on fusing of the *lux* gene (from *Vibrio fischeri*), *gfp* gene (from Aequorea Victoria), or *lacZ* gene (from *Escherichia coli*), reporters to an inducible gene promoter, have been applied to toxicity assay and sensors [12]. There are several review papers and book chapters addressing microbial bioassays and biosensors developments [2,10,13–17].

For specific aspects of toxicity, whole-cell-based sensors such as BOD (biochemical oxygen demand) have been developed since BOD is one of the most widely used tests in the measurement of organic pollution in wastewaters, effluents, and polluted waters. Conventional methodology for assessing the amount of biodegradable substances in water (BOD_5) is slow (a five day test), expensive to perform, and is subject to a high degree of fluctuation. Hence, a range of BOD biosensors have been developed to provide fast and accurate method of BOD measurement of wastewater. BOD biosensors are usually based on the use of microorganisms immobilized behind a membrane at a Clark oxygen electrode. On introducing a sample containing readily biodegradable substances (such as wastewater) into the sensor, the respiratory activity of the microorganisms, and therefore their oxygen consumption, increases immediately. The decrease in oxygen is then correlated with the organic material present in the water sample [18]. The function of a microbial respiratory sensor is shown in Figure 5.2 [15]. This shows the two possibilities of measurement (1) endpoint measurement, in which the difference in current reflects the respiration rate of the substrates, and (2) the kinetic measurement, first derivative of the current-time response corresponding to the acceleration of respiration. Various BOD sensors using microorganisms such as *Trichosporon cutaneum*, *Hansenula anomala*, *Pseudomonas* sp., *Escherichia coli*, and *Bacillus subtilis*, have been developed. Since employing one strain may give a narrow substrate spectrum, a mixed culture of two or more strains have been used in BOD sensor to broaden the substrate spectrum and achieve a more representative signal. The "BOD module" biosensor produced by Medigen (Prufgerate-Werk Medigen GmbH, Germany) has a BOD measuring range of 2–22 mg L^{-1} BOD with a fast response time of less than 1 minute. Another sensor such as the "ARAS BOD Biosensor System" from Dr. Lange (Dr. Bruno Lange GmbH, Willstatterstr, Dusseldorf, Germany) is based on the same principle as above. This sensor has a measuring range of 6–600 mg L^{-1} BOD with a response time of 60 seconds.

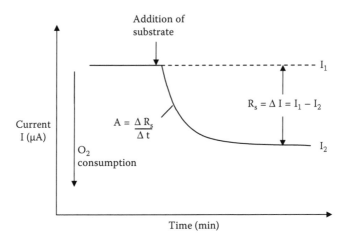

FIGURE 5.2 Typical response curve of a respiratory microbial sensor. Principles of measurement: (1) endpoint measurement (substrate respiration rate R_s), and (2) kinetic measurement (acceleration of respiration rate A) [figure taken from 15].

Another BOD sensor is from Nissan Electric Co. Ltd. (Tokyo), which has been incorporated into industrial standard methods in Japan. A range of other BOD sensors are also commercially available and marketed by a range of companies such as DKK Corporation, Japan; Autoteam FmbH, Germany; STIP Isco GmbH, Germany; Kelma, Belgium; Bioscience, Inc., U.S.; and USFilter, U.S. POLYTOX (InterLab Supply, Ltd, The Woodland, TX) is also a whole-cell system that uses a portable dissolved oxygen probe to monitor bacterial respiration to indicate toxicity of a water or wastewater stream including the presence of chemical and biological contaminants [5]. The BOD sensors are the most extensively used microbial biosensor today. The move to the use of miniaturized oxygen electrodes using screen-printed sensors is the new advances in this area. Electrodes based on platinum and gold working electrodes are being fabricated and used to replace the Clark oxygen electrode. The microbial cells are usually immobilized on the electrode surface using membranes (e.g., cellulose acetate) or polymers (e.g., poly-carbamoyl-sulfonate).

Microbial biosensors have also been developed to detect specific compounds rather than general toxicity detection. These types of sensors either use natural microorganisms, which have the capability of degrading the compound of interest, or use microorganisms, which have been genetically modified to acquire this capability. The literature contains a range of biosensors based on both principles. Whole-cell biosensors have been applied for the detection of toxic volatile organic compounds (VOCs) such as benzene in air analysis. The bacteria *Pseudomonas putida* ML2 is a naturally occurring microorganism originally isolated from refinery soil, and it is able to aerobically biodegrade benzene and more importantly utilize it as its sole carbon and energy source [19]. Figure 5.3 shows the biochemical pathway utilized by the microorganism in the catabolism of benzene. The increase in the respiration rate and corresponding oxygen consumption of *P. putida* ML2 during the aerobic assimilation of benzene can be used as the basis of a biosensor

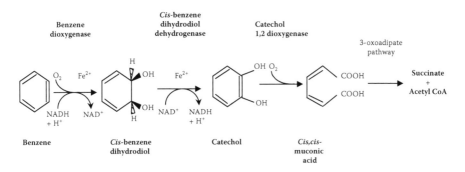

FIGURE 5.3   The biochemical pathway utilized by *P. putida* ML2 in the catabolism of benzene [figure taken from 19].

for the detection of this hydrocarbon in solution [20]. The bacterial cells were immobilized between two cellulose acetate membranes and fixed onto an ampero-metric Clark-type dissolved oxygen electrode (Figure 5.4). The sensor was based on flow injection analysis (FIA) and used to analyze air samples collected using charcoal adsorption tubes. Benzene was then extracted using dimethylformamide (DMF) [21]. The biosensor displayed a linear detection range between 0.025 and 0.15 mM benzene with a response time of 6 minutes. This linear detection range allows the analysis of air samples containing between 3 and 16 ppm benzene based on a 60-minute sampling period. The biosensor displayed no interference to other benzene-related compounds in the BTEX (Benzene Toluene, Ethylbenzene, Xylene) range. Lanyon et al. [22] have now transferred the assay principle to the surface of a gold screen-printed electrode by immobilizing *P. putida* bacteria on the electrode surface. This was to enable mass production, ease of use, and the fabrication of a more portable sensor system. The screen-printed sensor displayed comparable responses to measurements based on the oxygen probe with a linear detection range between 0.01–0.1 mM benzene and similar response characteristics.

Riley et al. [23] developed a cell-based device to evaluate toxicity of inhaled mate-rials. This is to assess health risks associated with inhalation of materials such as anthrax and indoor air pollution hazards. The sensor platform utilized cultured lung cells as biological recognition element, and the health and activity of the cells were characterized through a colorimetric metabolic assay that is monitored using light-emitting diodes and photodetectors as complete sensor system. The sensor showed good sensitivity and was able to detect 10^{-4}% by volume of volatizable pesticides and was also sensitive to particulate matter and metals. Another pesticide toxicity sensor was developed using confluent monolayer of cardiac cell cultured on microelectrode array (laminin modified surface) composed of 60 substrate-integrated electrodes. This sensor was developed to detect and evaluate toxicity, specifically for acute and eventually chronic insecticide (pyrethroids) interactions with cardiac cells [24]. Toxicity was recorded as changes in the electrophysiological properties of the car-diac myocytes, namely, reduction in beating frequency. Results showed that not only can be the toxins detected, but it might also be possible to classify them.

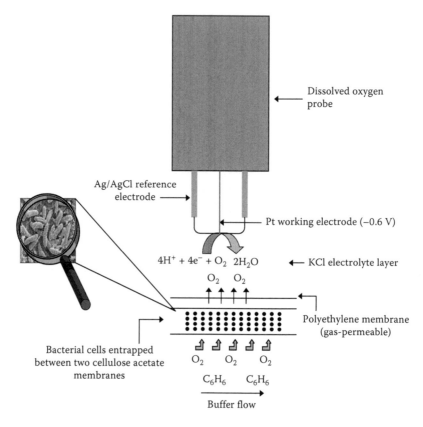

FIGURE 5.4 Scheme of the bacterial biosensor showing the passage of oxygen through the bacterial membranes to the surface of the working electrode, as the basis of measurement [figure taken from 22].

Sensors for phenol and substituted phenols have received attention due to the high toxicity of these compounds to the ecosystems. The bacteria *Arthrobacter* JS 443 and *Moraxella* sp., which degrade *p*-Nitrophenol (PNP), have been used with oxygen electrode and carbon electrodes to develop sensors capable of detecting PNP [25,26]. Other phenol-degrading microorganisms such as *Rodococcus erthropolis* immobilized on Clark oxygen electrode for 2,4-dinitrophenol detection and *P. putida* DSM 50026 immobilized on a screen-printed electrode for phenol detection [27,28]. Sensors for cyanide detection have been developed using a range of microorganisms. Most are based on respiration inhibition and consist of immobilized microorganisms such as *Nitrosomonas europaea*, *Thiobacillus ferrooxidans*, *Saccharomyces cerevisiae*, and *Pseudomonas fluorescence* on the surface of an oxygen electrode [13].

Whole-cell biosensors based on the inhibition of respiratory activity by the analyte of interest are widespread in the literature. Inhibition of microbial respiration by various toxic compounds has been measured using a range of transducers, although optical and electrochemical devices are the dominant. A whole-cell biosensor for the online detection of herbicides and pesticides in drinking water has been described.

The biosensor uses redox mediators to monitor the photosynthetic activity of the cyanobacterium *Synechococcus* electrochemically by transporting electrons from the cells to an electrode, resulting in the flow of current, which is measured in an external circuit. This sensor is reported to be capable of detecting herbicides from the nitriles, urease, anilides, and triazine families at 1–3 ppm concentrations. A mediated amperometric biosensor system for pollution monitoring and surface water intake protection has also been reported. This biosensor uses a chemical mediator to divert electrons from the respiratory systems of the physically immobilized *Escherichia coli* bacteria to an amperometric carbon electrode poised at 550 mV with respect to a silver/silver chloride reference/counter electrode. The sensor was reported to be capable of detecting a range of aquatic pollutants down to ppb levels within minutes of exposure [18]. CellSense™ (Euroclone, U.K.) is a commercial whole-cell biosensor system used for the detection of specific pollutants in water samples, for example, detection of the anionic surfactants linear alkylbenzene sulphonates, 2,4-dichlorophenoxyacetic acid, formaldehyde, sulphide, and chlorinated phenols. Other systems monitor bacterial metabolism based on color changes or on oxygen consumption. Examples are the commercially available systems as the ToxTrak (Hach Company, Loveland, CO), which is based on bacterial metabolism as indicated by a color change. A range of potentiometric transducers have been applied with whole-cell biosensors, such as the use of ion-selective electrode (pH electrode, NH_4^+ electrode, chloride electrode) or gas-sensing electrode (CO_2, NH_3) [13]. In addition to cells, plant tissues have also shown potential as sensing layers due to the enzyme system they contain. Examples are the use of sugar beet slices, peel of cucumber or squash, and potato and mushroom slices [29,30].

Optical biosensors based on UV-visible absorption, luminescence, and fluorescence as a result of interaction between the microbial cells and the target analytes are widely used in the development of whole-cell biosensors. Microorganisms that are natural or engineered (transgenic) have been applied for toxicity assessment [31]. Some systems are designed for total toxicity and others recognize and report the presence of specific toxic compounds. Bioluminescence bacteria usually emit light when they are healthy, but their metabolism is disrupted as a result of toxins exposure and therefore light emitting is reduced. The reduction in light can be correlated to toxins concentration, and hence this can play an important role in real-time process monitoring. Many test kits have been developed based on the use of natural luminescence bacteria using a companying monitoring instrument. Commercially available systems include ToxScreen II (Check Light, Ltd., Qiryat Tivon, Israel), which uses *Photobacterium leiognathi*; BioTox (Hidex Oy, Turku, Finland) which uses *Vibrio fischeri*; MicroTox/Delta Tox (Strategic Diagnostics Inc., Newark, DE) which uses *Vibrio fischeri* [32,33]. These systems are used to measure mixed toxicants (nonspecific) [34]. Coupling these cells with luminescence transducer has resulted in the development of whole-cell bioluminescence sensor. The literature contain a large number of sensors based on the use of fluorescence and luminescence transducers, but most are still in development stages [13]. These type of systems may, however, suffer from false positive since any reduction in the metabolic activity will reduce the signal output. Also a range of ions such as sodium, potassium, calcium, and magnesium ions all have been shown to have an influence on the light emission by *V. fischeri* [12].

Transgenic microorganisms are typically produced with a constructed plasmid in which genes that code for, for example, luciferase or β-galactosidase are placed under the control of a promoter that recognizes the toxins of interest. Because the organism's biological recognition system is linked to the reporting system, the presence of the toxin in the sample will result in the synthesis of inducible enzymes, which then catalyze reactions resulting in the production of detectable products. Recently, a range of transgenic microorganisms have been developed for either general toxicity monitoring or to detect specific compound. Lee et al. [35] have developed a cell-based array technology (chip array and plate array) that uses recombinant bioluminescent bacteria to detect and classify environmental toxicity. The sensor uses 20 recombinant bioluminescent bacteria having different promoters fused with *lux* genes and a CCD camera that measures and quantifies the signal density using image analysis software. Three different chemicals were used to test the sensor chip, and these can either cause superoxide damage (paraquat), DNA damage (mitomycin C), or protein/membrane damage (salicylic acid). The sensor system gave good results for the tested compounds.

Heavy metal mediated toxicity is dependant on bioavailability of the metals, and therefore a range of transgenic whole-cell biosensors have been developed to detect specific heavy metals in liquid samples and also soils. Biosensors for Ni^{2+} and Co^{2+} (*Ralstonia eutropha* AE 2515), Hg^{2+} (*E. coli* HMS174) and Cu^{2+} (*S. cerevisiae*) have been developed using transgenic microorganisms [36–38]. The Cu^{2+} biosensor was developed using transgenic *S. cerevisiae* containing plasmids with copper inducible promoter fused to the *LacZ* gene. When Cu^{2+} is present, the cells are able to utilize lactose as the carbon source and lead to an increase in oxygen consumption (Figure 5.5) [38]. Another example is the use of genetically modified *Moraxella* sp., and *P. Putida* with surface-expressed organophosphorus hydrolase was used to develop whole-cell biosensors for organophosphate compounds detection [39]. Other reporters such as the *gfp* gene coding for the green fluorescence protein have also been widely applied in a range of cells. Biosensors based on this type of transgenic cell have various applications such as assessing heterogeneity of iron bioavailability and toluene bioavailability [40]. A review of the different type of whole-cell biosensors using transgenic microorganism is available by Lei et al. [13] and Sørensen et al. [12].

Microorganism-based biosensors are relatively inexpensive to construct and can operate over a wide range of pH and temperature. Biosensors with immobilized active cells can show only short-term viability and stability. However, methods for improving cell stability in these types of sensors are in progress. Numerous reports have described the genetic engineering of bacterial sensor strains, but issues addressing the shelf life of these microorganisms are largely unresolved. Methods of preserving the activity of these types of sensors have been reviewed recently by Bjerketorp et al. [41]. General limitations involve the long assay time including the initial response and return to baseline. These characteristics are primarily determined by the cellular diffusion characteristics. But, the broad specificity of these biosensors to environmental toxins can be an advantage if toxicity of the sample needs to be assessed (qualitative analysis). However, the broad specificity (poor selectivity) of some of

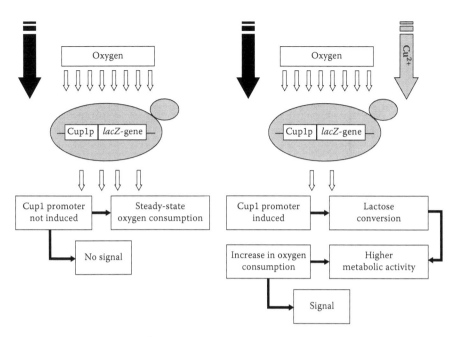

FIGURE 5.5   Principle of Cu^{2+} measurement using recombinant *Saccharomyces cerevisiae* [38].

these sensors may make them less desirable for specific compound sensing. These biosensor devices are suitable for mass production for rapid field tests or for online monitoring due to simple manufacturing procedures and the uncomplicated measuring principle they employ. Therefore, interest in these types of devices is increasing. They also have many advantages over the use of pure enzyme biosensors because of the following reasons: they are cheaper to fabricate; multienzyme reactions occur in the cells, and these can be used; coenzymes and activators are often present and cultivation of cells is easier and cheaper compared with pure enzyme preparations. Progress in recombinant DNA technologies has created tailored microorganisms with improved specificity and sensitivity. However, the use of transgenic microorganisms for field testing may be a problem due to legislations imposed on their field use in European countries. Different types of microbial sensors have been developed as portable system and these can be applied to monitor grab samples. The biosensors have also showed promise for use in early warning screening for drinking water because they can rapidly react to toxins [9]. However, the substances in the water systems that interfere with these types of sensors include chlorine, chloramine, and copper [42,43]. The use of sodium thiosuphate can be applied to remove chlorine residues from water samples before analysis. A new range of cells are being exploited in the construction of biosensors for toxicity monitoring such as B-cell, cardiac-cell, and, fish-cell systems, and some have also been applied to microelectrode arrays. Some examples of whole-cell biosensors for toxicity analysis are listed in Table 5.1.

TABLE 5.1
Whole-Cell Biosensors for Toxicity Assessment

Analyte	Microorganism	Transducer	Detection Limit	Reference
BOD	A. adeninivorans LS3	Oxygen electrode	2.61–524 mg l⁻¹	[44]
BOD	Activated sludge (mixed microbial consortium)	Oxygen electrode/flow injection system	>3.5 mg l⁻¹	[45]
BOD	T. cutaneum and B. subtilis	Oxygen electrode	0.5 mg l⁻¹	[46]
BOD	Microbial consortium	Oxygen electrode with wall-Jet flow cell	5.0 mg l⁻¹	[47,48]
BOD	P. putida SG10	Photocatalytic biosensor	1 mg l⁻¹	[49]
Acrylamide: acrylic acid	Brevibacterium sp.	Oxygen electrode	0.01–0.1 g l⁻¹	[50]
Cyanide	S. cerevisiae	Oxygen electrode	0.15–15 nM	[51]
Cyanide	T. ferrooxidans	Oxygen electrode	0.5 μM	[52]
Herbicides[a] (diuron and atrazine)	Chlroplast/thylakoid membranes	Pt-electrode in microelectrochemical cell	2×10⁻⁵ mM	[53]
Mono and polyphenols[a] (atrazine)	Potato (S. tuberosum) slices (polyphenol oxidase inhibition)	Oxygen electrode	20–130 μM	[29]
Phenolic compounds	P. putida	Oxygen electrode	100 μM	[54]
p-Nitrophenol	Arthrobacter sp. JS443	Oxygen electrode	0.2 μM	[25,39]
p-Nitrophenol	Moraxella sp.	Carbon past electrode	20 nM	[26]
Chlorophenols	Rhodococcus sp. Trichosporon beigelii	Oxygen electrode	0.004–0.04 and 0.002–0.04 mM	[55,56]
Chlorinated and brominated hydrocarbons (1-chlorobutane and ethylenebromide)	Rhodococcus sp. DSM6344	Ion-selective electrodes	0.22 and 0.04 mg l⁻¹	[57]

The detection limit values and other superscript markings above are reproduced below using LaTeX where applicable: detection limits such as $2.61\text{–}524\ \text{mg l}^{-1}$, $>3.5\ \text{mg l}^{-1}$, $0.5\ \text{mg l}^{-1}$, $5.0\ \text{mg l}^{-1}$, $1\ \text{mg l}^{-1}$, $0.01\text{–}0.1\ \text{g l}^{-1}$, $0.15\text{–}15\ \text{nM}$, $0.5\ \mu\text{M}$, $2\times10^{-5}\ \text{mM}$, $20\text{–}130\ \mu\text{M}$, $100\ \mu\text{M}$, $0.2\ \mu\text{M}$, $20\ \text{nM}$, $0.004\text{–}0.04$ and $0.002\text{–}0.04\ \text{mM}$, 0.22 and $0.04\ \text{mg l}^{-1}$.

Polycyclic aromatic hydrocarbons (Naphthalene)	Sphingomonas yanoikuyae B1 or Ps. Fluorescens WW4	Oxygen electrode	0.01–3.0 mg l^{-1}	[58,59]
Organophosphate nerve agents (paraxon, methyl parathion, diazinon)	E. coli^b (organophosphorous hydrolase)	Potentiometric	0.055–1.8, 0.06–0.91 and 0.46–8.5 mM	[60]
Organophosphate nerve agents (paraxon, parathion, coumaphos)	E. coli^b (organophosphorous hydrolase)	Fiber-optic	0.0–0.6, 0.0–0.03 and 0.0–0.075 mM	[61]
Organophosphorus nerve agents	P. putida^b JS 444	Oxygen electrode	0.058 mg l^{-1}	[39]
Genotoxicants	E. coli^b DPD1718	Luminescence	0.1mg l^{-1} mitomycin	[62]
Benzene	P. putida ML 2	Oxygen electrode	0.025–0.15mM	[20,21]
		Gold screen-printed electrode	0.01–0.1 mM	[22]
Pollutants such as diuron and mercuric chloride	Synechococcus sp. PCC7942	Photoelectrochemical	0.2 and 0.06 μM	[63]
Copper	S. cerevisiae^b	Oxygen electrode	0.5–2 mM	[38]
Cadmium	E. coli^b	Electrochemical cell	25 nM	[64]
Ni^{2+} and Co^{2+}	Ralstonia eutropha AE2515	Luminescence	0.1 μM Ni^{2+} 9 μM Co^{2+}	[36]
Hg^{2+}	E. coli^b HMS174	Luminescence	0.2 ng g^{-1}	[37]
Arsenic	E. coli^b DH5α	Fluorescence	-	[40]

^{a} Tissues or cellular organelle based.
^{b} Transgenic microorganisms.

5.4 ORGANISMS SENSOR SYSTEMS

Different types of organisms such as daphnia, mussels, algae, and fish have been extensively incorporated in toxicity tests for water assessment systems [65]. Most of these assays are developed as test systems with few as laboratory-based sensor systems. Membranes with their active enzyme system have also been implemented in the development of toxicity kits and sensors. An example is the MitoScan Kit (Harvard BioScience, Inc., Holliston, MA), which uses fragmented inner mitochondrial membrane vesicles isolated from beef heart (EPA, 2005 [9]). The submitochondrial particles contain complexes of enzymes responsible for electron transport and oxidative phosphorylation. When specific toxins are in the sample, the enzyme reactions are slowed or inhibited, and these are monitored spectophotometrically at 340 nm. This is still in a bioassay test kit format but may be developed to optical sensor system.

The use of algae in the development of biosensors is widely reported. Algae-modified carbon paste electrode has been used for copper measurement using differential pulse voltammetric quantitation of copper Gardea-Torresdey et al. [66]. The limit of detection of this sensor was 2×10^{-6} M. The detection of photosynthetic activity of algae to detect toxic compounds has been used in the system called the LuminoTox (Lab Bell Inc., Shawinigan, Canada). This allows for the specific detection of herbicides and organic solvents (e.g., gasoline, hydrocarbons). Toxicity can be measured in 10–15 minutes using a handheld luminometer [9]. A compact amperometric algal biosensor based on the use of the unicellular microalga *Chlorella vulgaris* has been developed by immobilizing the cells on the surface of transparent indium tin oxide electrode and used for the evaluation of water toxicity [67].

A range of commercially available toxicity test/sensor as an integrated complete systems based on the use of mussels such as the MosselMonitor (Delta Consult, the Netherlands), fish as the Bio-Sensor (Biological Monitoring, Inc., Blacksburg, VA), Daphnia as the Daphnia Toximeter (bbe moldaenke, Keil-Kronshagen, Germany), and Algae as the Algae Toximeter (bbe moldaenke, Keil-Kronshagen, Germany), are today referred to as biosensor systems. Many of these systems monitor the physical and/or chemical behavior of these organisms and are known to be affected by the chlorine content of the water sample and thus are more suitable for grab sample analysis. A full review of these systems is available in the EPA report on water security [9].

5.5 ENZYME-BASED SENSORS

Enzymes-based biosensors are well reported in the literature for chemical toxicity screening. The sensor devices produced using enzymes are usually simple and easy to fabricate, inexpensive, and sensitive to low levels of toxicants. Immobilization of enzymes on the electrode surface can include adsorption, covalent attachment, or film deposition using a range of procedures [68–70]. The sensor system relies primarily on two enzyme mechanisms: catalytic transformation of a pollutant and detection of pollutants that inhibit or mediate the enzyme's activity. In catalytic enzyme biosensor, the enzyme specific for the substrate of interest (toxin in this case) is immobilized on the sensor surface. The substrate is catalyzed as it enters the enzyme layer, and the resulting signal is proportional to the concentration of the toxin as the enzyme substrate. Sensors based on catalytic transformation can be configured to operate continuously and reversibly and includes, for example, the use of the enzyme tyrosinase for the detection of phenolic compounds, and the use of organophosphate hydrolase and choline oxidase for the detection of organophosphorus pesticides. Mechanisms for the tyrosinase biosensors involve the detection of phenols either through the electrochemical reduction of quinone intermediates or through oxygen consumption (O_2 is a cosubstrate) using a Clark electrode. New advances using nanomaterials such as gold nanoparticles have been applied in the development of tyrosinase sensor for phenol index monitoring in wine [71]. However, sensors based on catalytic enzymes activity are limited since there are only few enzymes that use toxic compounds as substrates with easily detectable system and also since the detection limit seems to be high for these types of sensors when compared to other sensing systems.

Enzyme inhibition sensors are the most commonly reported enzyme-based biosensors for the detection of toxic compounds and heavy metal ions. The sensors are based on the selective inhibition of specific enzymes by classes of compounds or by the more general inhibition of enzyme activity. Most of the research carried out has been directed toward the detection of organophosphorus and carbamate insecticides and the triazine herbicides and metal ions analysis [72,73]. Several enzymes have been used in inhibition sensors for pesticides and heavy metal analysis using water, soil, and food samples including choline esterase, horseradish peroxidase, polyphenol oxidase, urease, and aldehyde dehydrogenase.

Inherent advantages for these formats involve the larger number of toxic compounds, usually of a particular chemical class, that inhibit the enzyme and the low concentrations needed to affect the enzyme activity. Examples for these biosensors are the sensors that use the enzyme acetyl cholinesterase (AChE) and butyrylcolinesterase for organophosphorus compounds and carbamate pesticide analysis. These sensors are known to be very sensitive, with detection limits for cholinesterase biosensors being reported to be in the $\mu g\,l^{-1}$ to $ng\,l^{-1}$ range for compounds such as aldicarb, carbaryl, carbofuran, and dichlovos. Cholinesterase (ChE) biosensors have emerged as reliable and rapid techniques for toxicity analysis mainly for the detection of pesticides (organophosphates and carbamate), nerve gas, and heavy metals. Two types of ChEs are used that have different substrate specificity: acetylcholinesterase (AChH), which preferentially hydrolyses acetyl esters such as acetylcholine, and butyrylcolinesterase (BuChE), which hydrolyzes butyrylcholine. A new paper has been published recently that gives complete review of ChE biosensors, their fabrication, and application in toxicity monitoring [73]. ChE biosensors have been fabricated using a range of transducers including electrochemical, optical, conductimetric, and piezoelectric using mono or bi enzyme system. ChE inhibition has been used to detect and quantify nerve agents as a commercial test kit such as the Severn Trent Field Enzyme Test (Severn Trent Services, U.K.). The test kit is based on the use of a membrane disk saturated with the enzyme, which is dipped into the liquid sample for 1 minute. In the absence of pesticide/nerve agent the enzyme will react with the esters, resulting in a blue color formation. If a detectable quantity of the pesticide/nerve agent is present, the color will not be formed due to the inhibition of the enzyme (remains white). This test has a detection limit of 0.1–5 mg l^{-1} for carbamates, 0.5–5 mg l^{-1} for thiophosphates, and 1–5 mg l^{-1} for organophosphates [9]. Moving from the test kits into sensors development, several sensor formats have been developed recently. Many incorporate bienzyme system to enable transduction of the reaction catalysis to the sensor. Many biosensors use ChE coupled to choline oxidase (ChO) configuration, where ChE hydrolyzes acetylcholine or butyrylcholine to choline and acetate. Choline is then used by the enzyme ChO to produce electrochemically active products (H$_2$O$_2$) [74]. Other sensors use oxygen electrode to measure the consumption of oxygen in the ChO-catalyzed reaction [75]. Single enzyme sensors (ChE sensors) have also been applied mainly based on the measurement of the change in pH and/or redox potential. Many biosensors have been developed using this transduction system, with recent potentiometric sensor using pH-sensitive sensors as ion-selective field-effect transducers (ISFET) or light-addressable potentiometric sensors (LAPS) fabricated using microelectronic technology [76]. New developments for single enzyme sensor use nonspecific substrate such as acetylthiocholine (ACTh) and butyrylthiocholine (BuTCH) or incorporation of electronic mediators as tetracyanoquinodimethane (TCNQ) and cobalt phtalocyanine (CoPC) [73].

New advances in enzyme inhibition biosensors apply nanostructured materials and nanocomposites to obtain increased sensitivity and also use transgenic enzyme systems [77,78].

Horseradish peroxidase (HRP) is also used for the detection of toxic compounds. A chemiluminescence test based on the reaction of luminol and an oxidant in the presence of the enzyme HRP has been developed to indicate the presence of toxins in a sample. The HRP-catalyzed reaction produces light that is measured by a luminometer or a luminescence transducer. This enzyme has been used to detect a range of compounds such as phenols, amines, heavy metals, or compounds that interact with the enzyme, reduce light output, and indicate contamination. Test kits such as the Eclox Water Test Kit (Seven Trent Services, U.K.) is based on the use of HRP in the test format described earlier. This type of test is designed for the qualitative assessment of water samples for a range to compounds that inhibit the HRP activity.

The use of photosynthetic enzymes isolated from plants has been implemented in a toxicity monitor (LuminoTox, Lab_Bell Inc., Shawinigan, Canada). This system can detect a range of compounds such as hydrocarbons, herbicides, phenols, polycyclic aromatic hydrocarbons (PAHs), and aromatic hydrocarbons. These enzymes have been coupled to screen-printed electrode and have been demonstrated to be able to detect triazine and phenylurea herbicides [79]. Other enzyme inhibitions have been used to detect biotoxins from plant, animals, bacterial, algae, and fungal species (e.g., ricin, botulinum toxins, mycotoxins, cyanobacterial toxins). However, since the identity and specificity of the above toxic compound can be very important during the analysis, other sensor systems such as immunosensors may be preferred to give a better indication to toxin type and identity than the use of enzyme inhibition tests.

The toxicity of heavy metals, biochemically, arises from the strong affinity of these cations for sulphur. Thus, sulphydryl groups, –SH, commonly present in proteins, readily attach themselves to ingested heavy metal cations or other molecules containing metals. The resultant metal sulphur bonding affects the whole protein, whose activity is affected or inhibited, and therefore heavy metals are toxic to human health. The toxicity of heavy metals in nature also depends on their chemical form. Insoluble forms pass through the human body without any effect, while soluble forms can easily be absorbed through biological membranes. Speciation analysis is therefore fundamental for the understanding of metals toxicity to humans. For heavy metals determination, chemical sensors based of striping techniques are usually applied; at Cranfield University we have developed a range of chemical sensors for metal ions detection [80–82] and these are used to quantify accurately the metal ions concentration in the sample. However, for toxicity of the sample as indication of heavy metal ions contamination, biosensors based on enzyme inhibitions are used. Various enzymes such as urease, invertase, xanthine oxidase, peroxidase, glucose oxidase, or alkaline phosphatase have been used in an inhibition sensor format since all of these enzymes are sensitive to heavy metal ions inhibition. Metal ions normally combine with thiol groups present in the enzyme structure, thus resulting in a conformational change, which irreversibly affects the catalytic center. Due to its high sensitivity to heavy metal ions inhibition the enzyme urease have been used extensively to develop sensors to detect these compounds. Assay and sensors based on the inhibition of urease show a high selectivity for the sensitive and effect-based screening of heavy metals [83,84]. Most urease inhibition sensors are based on the measurement of either pH changes [85,86] or ammonia production [87]. Recently, an amperometric screen-printed electrode based sensor was developed for metal ions toxicity detection using the urease enzyme [88,89], in order to produce simple and sensitive sensor system for heavy metal ions detection. In this sensor system the amperometric determination of urease activity has been achieved by coupling the urea breakdown to the reductive amination of α-ketoglutarate to L-glutamate catalyzed by glutamic dehydrogenase. Both NADH and NH_4^+ are required in equimolar amounts for the enzymatic synthesis of glutamate from α-ketoglutarate. NADH consumption is monitored amperometrically using screen-printed three-electrode configuration and its oxidation current is then correlated to urease activity. The presence of heavy metals in the samples inhibits the urease activity, resulting in a lower NH_4^+ production and therefore a decrease in NADH oxidation, which was carried out at +300 mV versus Ag/AgCl on metalized carbon electrode. The linear range obtained for Hg^{2+} and Cu^{2+} was 10–100 µg l^{-1} with a detection limit of 7.2 µg l^{-1} and 8.5 µg l^{-1}, respectively. Cd^{2+} and Zn^{2+} produced enzyme inhibition in the range 1–30 mg l^{-1}, with limits of detection of 0.3 mg l^{-1} for Cd^{2+} and 0.2 mg l^{-1} for Zn^{2+}. Pb^{2+} did not inactivate the urease enzyme significantly at the studied range (up to 50 mg l^{-1}). Coefficients of variation (CV) values were 6%–9% in all cases. Application of the sensor system to leachate samples gave reliable and accurate toxicity assessments when compared to AAS and ICP-MS analysis. This approach proved to be a simple and rapid (15 minutes, including enzyme inhibition time) method for metal ions toxicity assessment. Figure 5.6 shows the portable metal ion sensor system. A multienzyme sensor system has also been used in the development of enzyme inhibition sensor for the detection of heavy metal ions [90].

Limitations for biosensors based on enzyme inhibition is that the assay may need multistep incubation, and the assay formats require the use of substrates/cofactors and mediators depending on the enzyme applied in the sensor format. The irreversible

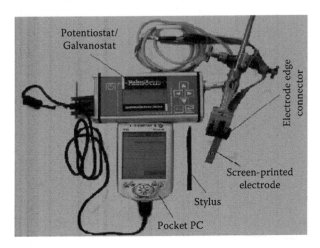

FIGURE 5.6 The portable sensor system for metal ions analysis. Screen-printed sensor (fabricated at Cranfield University) coupled to a portable electrochemical instrument (PalmSense, Palm Instruments BV).

nature of many analyte-enzyme interactions that result in increased sensitivity also renders the biosensor inactive after a single measurement. This may not be a problem if the sensor to be used is a one-shot disposable device, which is usually applied for soil, food, or water grab samples, but for online/in situ systems this sensor will not be suitable if it cannot be reactivated, which the case is in many enzyme inhibition systems. The sensors may not be specific if diverse classes of pollutants inhibit the enzyme and not just one toxic compound. However, the system is good for toxicity assessment, and the type of sensor can be selected depending on its sensitivity for the target analyte or a class of analytes and the analysis requirement. New advances in molecular biology have resulted in engineered enzymes (such as ChE), which can be more sensitive and selective [91]. The use of nanotechnology in the sensor fabrication process such as microarray/nanoarray sensors coupled with artificial intelligence could produce sensors with high sensitivity and selectivity [92]. Examples of enzyme-based biosensors are listed in Table 5.2.

5.6　DNA-BASED SENSORS

DNA damage caused by toxic compounds has been used as a method of detecting these analytes. Nucleic acid-based biosensors, which have potential application for environmental control and toxic compound analysis, have recently been reported [112]. Application areas for DNA biosensors include the detection of chemically induced DNA damage by toxic compounds and the detection of pathogenic microorganisms through the hybridization of species-specific DNA sequences. However, this section will only concentrate on the development of DNA-based sensors for toxic compound detection as microorganism detection is not covered in this chapter.

One of the potential applications of DNA electrochemical sensor is its capability of detecting the presence of toxic analytes (e.g., carcinogens, drugs, mutagenic

TABLE 5.2
Enzyme-Based Toxicity Sensors

Analyte	Enzyme	Transducer	Detection Limit	Reference
Hg^{2+}, Cu^{2+}, Cd^{2+}	Urease	Optical sensor	10 nM, 50 μM, 500 μM	[93]
Hg^{2+}, Cu^{2+}	Glucose oxidase	Amperometric sensor	2.5 μmol L^{-1}–0.2 mmol L^{-1}	[94]
Hg^{2+}	Horse Radish peroxidase	Amperometric sensor	0.1–1.7 ng mL^{-1}	[95]
Heavy metals	L-lactate dehydrogenase/L-lactate oxidase	Oxygen electrode/polysulfone membrane	<100 μM	[96]
Hg^{2+}	Invertase	Platinum electrode	5–10 ng mL^{-1}	[97]
Hg^{2+}	Peroxidase	Glassy carbon electrode/ methylene blue mediator	0.1–0.2 ng L^{-1}	[95]
Pesticides and heavy metals	Alkaline phosphatase acid phosphatase acetylcholinesterase	Three separate enzyme electrodes of pH electrode/cellulose acetate membrane	1 mg L^{-1}	[98]
Hg^{2+}	Glycerol 3-P oxidase	Platinum electrode	0.05–0.4 mg L^{-1}	[99]
Cu^{2+}	Alcohol oxidase		2 mg L^{-1}	
Ni^{2+}	Sarcosine oxidase		1–6 mg L^{-1}	[86]
Hg^{2+}	Urease	pH-sensitive iridium oxide electrode	40.12 μg L^{-1}	[87]
Heavy metals	Urease	Field-effect transducer	3.8 μg L^{-1} Ag	
			41.1 μg L^{-1} Ni	
			127 μg L^{-1} Cu	
Hg^{2+}, Cu^{2+}	Urease/glutamic dehydrogenase	Amperometric screen-printed sensor	7.2 μg L^{-1} Hg^{2+}	[88,89]
			8.5 μg L^{-1} Cu^{2+}	[100]
Aldricarb	Acetyl cholinesterase/choline oxidase	Potentiomentric	10–500 ppb	[74]
		SPE with carbon nanotubes	0.05 μM	
Methyl parathion	Acetyl cholinesterase/choline oxidase	Oxygen electrode	1–50 μg L^{-1}	[74]
Anticholinesterase activity	Acetyl cholinesterase/choline oxidase			

(Continued)

TABLE 5.2
Enzyme-Based Toxicity Sensors (Continued)

Analyte	Enzyme	Transducer	Detection Limit	Reference
Organophosphates	Acetyl cholinesterase	Fiber-optic pH-sensitive electrode (pH indicateur: Thymole blue)	108 µg l^{-1} carbaryl	[76]
Paraoxon	Acetyl cholinesterase	Amperometric sensor	0.5–0.7 nmol l^{-1}	[78]
Paraoxon	Acetyl cholinesterase	Fiber-optic pH-sensitive electrode	152 µg l^{-1}	[100]
Paraoxon	Organophosphorus-hydrolase	Optical/gold nanoparticals	20–240 µ	[101]
Paraoxon carbaryl	Acetyl cholinesterase	Amperometric sensor	1 µg l^{-1} paraoxon 20 µg l^{-1} carbaryl	[102]
Organophosphates pesticides and oxon forms	Butyrylcholinesterase	pH-sensitive membrane on glassy carbon or Au	2 × 10^{-10} M Diazinon	[103]
Paraoxon, diazinon, malathion, coumaphos	Organophosphorus-hydrolase	Optical sensor	7.8 µg l^{-1} 1.25 µg l^{-1}	[104]
Dichlorvos	Acetyl cholinesterase	Amperometric sensor	10^{-12} mol l^{-1}	[105]
Atrazine	Polyphenol oxidase (PPO)	Amperometric sensor	0.05–0.5 ppm	[106]
Parathion	Parathion hydrolase	Amperometric sensor	10–100 ng ml^{-1}	[107]
Cyanide	Tyrosinase	Amperometric sensor	0.1 nmol l^{-1}	[108]
Methyl isothiocyanate	Aldehyde dehydrogenase	screen-printed carbon electrode	100–1000 ppb	[109]
Anatoxin-a(s)	Acetyl cholinesterase	Amperometric sensor		[110]
Anatoxin-a(s)	Acetyl cholinesterase	Amperometric sensor	1–10 µg l^{-1}	[111]

pollutants, etc.), which have binding affinity to the DNA, sensing layer immobilized on the sensor surface. The principle of this sensor is that the binding of small molecules to the DNA sequence can damage the DNA, and this can be detected through the change in the electrochemical signal achieved from the guanine base. Interaction can also be observed through intercalation with organic molecules or electrostatic interaction with the phosphate backbone by positively charged compounds. Several review articles have been reported in the literature describing these types of sensors and their application [113–116]. Mascini and his coworkers have developed a range of DNA biosensors for toxic compounds monitoring for environmental testing such as for water and soils analysis [115,117,118]. The DNA biosensor is assembled by immobilizing double stranded DNA on the surface of a disposable carbon screen-printed electrode. The oxidation signal of the guanine base obtained using square voltammetric scan is used as the analytical signal. In the presence of compounds with affinity for DNA, the guanine oxidation peak is affected, and this is taken as indication of the presence of toxic compounds in the sample. The author applied the sensor for analyzing the presence of toxic compounds in water and wastewater samples. DNA biosensors for the detection of hydrazines [119], aromatic amines [120], heavy metals [121], genotoxicity, and toxicants in water and wastewater [118,122] have been described in the literature. Several studies have also been carried out to compare the DND sensor results with those achieved using other toxicity tests such as the bacterial bioluminescent tests [122,123]. Results indicate the potential of using the sensors to detect toxicants in water samples. Other work described the use of DNA sensors for specific compound detection. Doong et al. [124] developed a sol-gel derived array DNA biosensor coupled with a fluorescence detection system and a robotic pin-printing platform detect polycyclic aromatic hydrocarbons (PAHs) in water samples. The sensor was able to detect naphthalene and phenanthrene in the concentration range of 0–10 mg l^{-1}.

The use of artificial nucleic acid ligands (aptamers), which can be generated against drugs and other toxic compounds in biosensors development, are expanding rapidly [125,126]. These molecules can be isolated from complex libraries of synthetic nucleic acids. Combinatorial searches for in vitro selection of DNA/RNA from a library of 10^{14}–10^{15} random DNA/RNA sequences have been used to develop catalytic DNA sensors [127]. The advantage of in vitro selection of DNA/RNA is that it is able to sample a large pool of sequences, amplify the desired sequence and introduce mutations to improve the performance. Catalytic DNA/RNA molecules that are highly specific for Pb (II) [128], Cu (II) [129] and Zn (II) [130] have been obtained and used for application such as metal ions analysis [131,132]. Lu et al. [127] used in vitro selection and engineering of catalytic DNA that binds to Pb (II) with strong affinity and high selectivity and applied them in the development of a fluorescent sensor for detecting and quantification of lead. The sensor was also demonstrated for sample analysis collected from lake water.

The electrochemical single or double stranded DNA sensors are usually proposed as screening devices for monitoring samples that may contain toxic compounds but not endorsed as toxicity tests. If the samples are positive then better care should be taken to handle and analyze these samples using other toxicity tests. DNA-based sensors have seen great expansion in their use as tools for detecting toxic compounds, and this can be due to their being simple, cheep, and easy to use. The emerging of catalytic DNA molecules and their application in the development of fluorescent and colorimetric sensor systems for toxic compounds such as heavy metal ions analysis have been demonstrated. Catalytic DNA molecules usually form compact globular shape, and when used in sensors they have favorable features such as high sensitivity, specificity, and stability, which make them very promising systems for future sensor applications.

5.7   ELECTRONIC NOSE

Generic electronic noses (E-nose) or odor mapping systems have been defined as "an odor, which is presented to an active material of a sensor which converts a chemical input into an electrical signal." Based on this concept a great deal of research in the development and commercialization of the electronic nose has taken place in recent years, with applications ranging from microbial detection to toxic volatile compound analysis. The sensor instrument mimics the olfactory system in the nose and consists of a sensor array of gas sensors with different selectivity patterns; the circuitry represents the conversion of the chemical reactions on the sensor to electrical signals and the data-analysis software that analyses the signal by pattern recognition methods [44]. These include principal component analysis (PCA), discriminate function analysis (DFA), cluster analysis, and artificial neural networks (ANN). The results are comparative rather than quantitative "fingerprint." Although largely qualitative or semiquantitative, such approach can be ideal for rapid risk assessment of toxic volatile compounds. A variety of sensor array are used, including metal oxides sensors and conducting polymer sensors. The polymers can be highly sensitive but not specific and can respond to volatile compounds with molecular weight ranging from 30 to 300 [132]. Molecules such as alcohols, ketones, fatty acids, and esters give strong response, while fully oxidized species such as CO_2, NO_2, and H_2O have a lower response. The sensor array can also recognize molecules containing sulphur and amine groups. Different types of devices have been developed such as metal oxide silicon field-effect sensors (sensitive to organic compounds), piezoelectric crystals, optical sensors, and electrochemical sensors. The electronic noses have been used in the medical, environmental, and food diagnosis [134–137].

With increasing concern regarding detecting and identifying chemical warfare agents in the air a range of portable E-nose-based instruments have been developed and trained for the detection of toxic compounds in the field. Instrument such as the HAZMATCAD (Microsensor Systems, Inc., Bowling Green, KY), contain three 250 MHz SAW sensors in a handheld portable Chemical Agent Detector instrument. Each sensor is coated with different polymers that provides a multipattern sensor response (fingerprint) to indicate the presence of contaminants in vapor samples. The HAZMATCAD detects and identifies trace amounts of chemical warfare agents, including nerve and blister agents, and can be configured to detect phosgene and/ or hydrogen cyanide. HAZMATCAD Plus supplements the SAW technology™ with electrochemical sensors for additional detection of up to four classes of toxic industrial chemicals, specifically hydride, halogen, choke, and blood agent vapors [9].

Other devices such as the Cyrano Sciences, (Smiths Detection, Edgewood, MD), a handheld device using conductive polymer films deposited in an array on a ceramic substrate and the Micromachined Acoustic Chemical Sensor (Sandia National Laboratories, SNL), which is able to detect VOCs, explosives, illicit drugs, and chemical warfare agents, are also commercially available [32].

E-nose technology is still in development and so far has only had a limited commercial application. This is due to the instrument need to be trained and optimized for the sensing application it is required for and to be able to distinguish between samples before it can be used. The technology is still for qualitative rather than quantitative application. However, the E-nose systems has a recognize advantage in use in areas as safety and quality monitoring.

5.8   ANALYTE-SPECIFIC SENSORS

Analyte specific sensors include sensors that are able to detect and monitor specific toxic compounds or class of compounds. These sensors are not classified as toxicity tests but are able to detect compounds that are known to be toxic. Therefore, these types of sensors are only briefly covered in this chapter as further details are provided in other chapters in the book regarding the detection of toxic analytes.

5.8.1   AFFINITY-BASED SENSORS

Affinity-based sensor systems are usually developed for analyte-specific toxic compounds assessment where the analyte of interest or class of compounds are specifically detected by the sensing layer. Affinity-based sensors are the most applicable technologies that are successful in quantifying and identifying known compounds which are toxic. In recent years phenomenal growth in affinity sensor research and development has been undertaken in toxic compounds detection. By utilizing the immunorecognition properties of antibodies (polyclonal and monoclonal) and the relative affinity and selectivity of these recognition proteins for a specific compound or closely related group of compounds in immunosensors and affinity sensors, the range of analytes that can be diagnosed has been broadened considerably [138,139]. Since most small molecular weight organic pollutants and toxins in the environment have few distinguishing optical or electrochemical characteristics, the detection of stoichiometric binding of these compounds to antibodies is usually carried out using competitive binding assay formats. Immunochemical assays are dominating the market today in contaminant and toxins analysis. The impact of immunoassays on contaminants analysis for environmental and food application is evident in the extensive diversity of kits that are commercially available on the market. Assay kits for the sensing of trace contaminants, including pesticides and herbicides, industrial residues and their degradation products, PCBs, PAHs, microbial toxins, and pathogens are available. The use of antibodies in various immunosensor configurations has also been applied in a wide range of applications and has been reviewed by many authors [5,139–142]. For an immunosensor or immunoprobe, the antibody (Ab) or the antigen (Ag) constitutes the biospecific component in the sensor structure. Many immunosensors have been developed, and with minor adaptation these are replacing immunoassays. The availability of the required antibody can limit the potential for diverse analyte detection by immunosensors. Increased research in the development of specifically tailored antibodies such as plant bodies, recombinant antibodies, catalytic antibodies or enzymes, artificial receptors, and molecularly imprinted polymers should overcome some of these problems. Miniaturization and improved processing power of modern microelectronics have increased the analytical capability of bio- and immunosensors. Research on lab-on-a-chip is exploding and interest from multinational companies in this area is also increasing. The use of antibodies as receptors in sensor configuration combined with a suitable transducer can result in a sensitive and specific affinity sensor.

Affinity sensors can be divided into classes depending on the transducer technology employed, the assay format, or type of application (reusable/regenerable or disposable format configurations). Immunosensors can be divided into direct sensing devices detecting the recognition event and the complex produced between the antibody and the analyte and indirect devices relying on the use of label compounds (enzyme, florescence marker, gold nanoparticles etc.) to create the signal. Detection by electrochemical immunosensors is generally relayed on the use of electroactive labels, usually based on enzyme labeling and amplification techniques. The sensors can be inexpensive, and may achieve very low detection limits (1 ppb). Different types of electrochemical immunosensors have been developed for toxic compounds analysis using different transducers. Disposable amperometric immunosensors for the detection of polycyclic aromatic hydrocarbons (PAHs) using screen-printed electrode have been reported [143]. Others are listed in Table 5.3.

In the field of affinity sensors, optical devices have a clear advantage over electrochemical methods due to their ability to monitor binding reactions directly. The major drawback to the application of optics to chemical sensor applications remains the high cost of many optical systems. Optical techniques such as surface plasmon resonance (SPR) and evanescent wave (EW) have shown promise in providing direct measurement of antigen (Ag)–antibody (Ab) interactions occurring at the surface-solution interface and major advances in this area are taking place. The BIAcore biosensor system based on SPR technology is commercially available from Biacore AB (Uppsala, Sweden) and it has been used in the development of a range of affinity sensors for toxic compounds analysis [104–106].

Piezoelectric transduction approaches and, in particular, surface acoustic wave (SAW) devices have also been widely used to detect antibody binding to an immobilized antigen. The most widely used acoustic transducer is the quartz crystal microbalance (QCM), which measures small changes in surface properties, such as bound surface mass resulting from the binding of molecules to the sensor surface. Again the attraction of piezoelectric sensors is their ability to monitor directly the binding of Ab-Ag reaction encountered in affinity sensing, but they suffer from the disadvantage of having low sensitivity for very small molecules. New commercial instruments are claimed to be more sensitive to low molecular weight compounds. Coupling the sensor with a flow system to concentrate the samples using solid-phase extraction column has increased the sensitivity of these devices. The use of piezoelectric sensor combined with a flow system using a flow cell has been developed and used to detect the toxin microcystin—LR [107,108].

TABLE 5.3
Immunosensors for Contaminant Detection

Analyte	Type of Sensor	Detection Limit	Matrix	Detection Time	Reference
2,4-D dichlorophenoxy acetic acids	Disposable amperometric multi-channel sensor based on screen-printing electrode with acetylcholinesterase as the label	0.01 µg l⁻¹	water	30 minutes	[144]
Dioxin	Nonseparation electrochemical enzyme binding/immunoassay using microporous gold electrodes	0.01 µg l⁻¹	water	–	[145]
Carbaryl insecticide	Flow through column with glass as the solid support to immobilize the antibody	0.029 µg l⁻¹	water	20 minutes	[146]
Okadaic acid toxin	Indirect competitive immunosensor using fiber-optic transduction with flow injection analysis	0.1 µg l⁻¹	Mussel homogenates	20 minutes	[147]
Isoproturon	Label-free immunosensor using reflectometric interference spectroscopy	0.7 µg l⁻¹	water	–	[148]
Diuron	FIA floroimmunosensor for online monitoring system	0.02 µg l⁻¹	water	–	[149]
Staphylococcal enterotoxin B	SPR-detection coupled with identification using mass spectrometry	1 ng l⁻¹	milk and mushroom	–	[150]
Sulfamethazine residues	SPR-based immunosensor	1 ppb	milk	20 minutes	[151]
Insulin-like growth factor-1 (IGF-1)	Automated SPR-based immunosensor	1 µg l⁻¹	milk	–	[152]

Source: From Tothill I. E., On-line immunochemical assays for contaminants analysis, in: *Rapid and On-line Instrumentation for Food Quality Assurance*, ed., Tothill I. E. (Woodhead Publishing Limited and CRC Press LLC., 2003, pp. 14–35).

One of the online immunosensor systems developed in recent years and that has attracted much attention is the prototype FIA River ANALyser (RIANA system), which was developed under the European Commission funding. The RIANA system incorporates a multiple analytes immunoanalysis based on total internal reflection fluorescence with 15 minutes for each analysis [109,110]. The transducer consists of a quartz slide with spatially resolved surface modification for antigen immobilization, along which a coupled laser beam propagates by total internal reflection. The antibodies are labeled with Cy5.5 fluorescent dye, which competes with the free analyte. The system has been applied for the detection of chlorotriazines, atrazine, simazine, and isoproturon. Detection limit for isoproturon in river water was 0.14 $\mu g \ l^{-1}$.

Application areas for affinity-based sensors and immunosensors are specific toxic compounds or class of toxins detection. The most developed and applied recognition layer is antibody based. As the commercial success of immunoassays becomes more evident in health care, food, and environmental monitoring the demand for faster techniques will be sufficient for continued affinity sensors development.

5.9 ADVANCES IN SENSOR TECHNOLOGY

Diver's range of sensing layers is used today for the construction of biosensors for toxins analysis. Biological components (i.e., enzymes, antibodies, cell receptors, DNA, microorganism) offer high sensitivity coupled in most cases with high selectivity but suffer from poor stability. Therefore, the research into the use of biomimics is increasing for analyte specific analysis. The artificial receptor approach is to overcome some of the difficulties encountered using biological components. The search for possible solutions to the low stability of biological molecules has taken several directions. Advances in receptor discovery such as the development of artificial receptors using the combined approach of computer (molecular) modeling and molecularly imprinted polymers (MIP) [111,112] or combinatorial synthesis [113] has increased the range of sensing layers, which can be used for the construction of suitable affinity sensors for toxic compounds analysis. Molecular modeling is a powerful tool to study molecular recognition, the specific interaction between substrate and receptor. The affinity interactions between synthetic receptors and target analytes comprise hydrogen bond interactions, π-stacking interactions, Van der Waals interactions, and electrostatic interactions, which constitute molecular forces, involved in molecular recognition processes. Molecular modeling allows the prediction of ligands that are expected to bind strongly to key regions of biologically important molecules of known structure, so as to inhibit or alter their activity. Artificial receptor design and fabrication is currently the focus of intense research interest and is being used in a wide range of application area, for example, the preparation of selective separation materials, artificial antibodies, and synthetic enzymes. Research activity in the area of artificial receptor discovery has increased in the past few years, in particular, the production of synthetic receptors for medical and environmental diagnostics. The development of artificial receptors for various purposes remains an important challenge. Combinatorial synthesis and molecularly imprinted polymers are two of the most exciting and rapidly growing areas in ligand discovery, which can overcome the stability problems inherent in natural ligands. However, the challenge is to produce receptors that compete with the natural molecules with respect to sensitivity and stability [51]. The use of nanomaterials such as carbon nanotubes and gold nanoparticles in the design and constructions of new sensor platforms has also increased in recent years. Application of these new materials in the development of sensors for general toxicity assessment may be limited, but they will play a major part in the development of sensors for the analysis of specific toxic compounds (analyte specific).

New advances in sensor design and development is based on microchip technology and multiarray sensors. The formulation of arrays of sensors to analyze one sample for several toxic compounds is a very attractive approach. Miniaturization is a key development for making small portable sensors, since the demand for detection of a range of analytes on site and in the field is increasing. Miniaturization of biosensors is also being driven to micro/nanoarray systems and is often referred to as "lab-on-a-chip." The use of photolithography, self-assembly, and microcontact printing offers routes to high-density arrays. The driving force behind this comes from the high throughput-screening programmes. The application of silicon fabrication technology for biosensor array production is receiving great attention from diagnostic companies. The attractive issue here is the on-chip microfabrication and electronic signal amplification and processing. Microfluidic technology, which allows minute quantities of liquids (reagents and sample) to be manipulated and delivered to microchip components, is an essential part of any chip design that utilizes aqueous solutions. Biochips could be designed to detect biological molecules (DNA, RNA, proteins, biotoxins), or nonbiological chemicals. Commercialized microchips are common in research and diagnostic laboratories; however, the support equipment to use the microchips, such as microfluidics stations and chip readers, requires a laboratory setting. Although microchips are a highly developed technology for genomics, they are an emerging technology for other applications, such as proteomics and chemical compounds sensing. The intense interest and promise in the field of micro- and nano-based technologies will significantly contribute to product development and commercialization in diverse areas of application including toxicity testing [163,164]. A wide variety of computational intelligence techniques are being used in conjunction with sensors and biosensors. These include neural networks (ANNs), mathematical modeling, chemometrics, and data mining. Data analysis and bioinformatics are very important to make the most of the data collected from sensor responses. Sensing possibilities based on fingerprinting type approaches are being developed rapidly [165,166].

5.10 CONCLUSION

Portable field sensor devices and online systems have been developed for conducting analyses of grab samples for on-site detection or online monitoring of many possible chemical contaminants and toxicity evaluation. However, online detection technologies for specific chemical contaminants or toxicity require further development to increase their reliability, robustness, sensitivity, and cost-effectiveness. Today there is an increasing need for more sophisticated sensor technologies using artificial intelligence and web-based knowledge systems. New technologies based on micro-chip sensors with smart data management system could revolutionize the chemical/toxicity detection field for a range of application specifically for air monitoring and drinking water safety and security. The scope is very large for sensor application within toxicity monitoring and risk management. A few technologies are mature enough today for real-time application but others require further developments.

REFERENCES

1. ILSI, Early Warning Monitoring to Detect Hazardous Events in Water Supplies. International Life Sciences Institute-Risk Science Institute, 1999. http://rsi.ilsi.org/file/EWM.pdf.
2. Tothill I. E. and Turner A. P. F., Developments in bioassay methods for toxicity testing in water treatment, *Trends Anal. Chem.*, 15(5), 408–418, 1996.
3. Rodriguez-Mozaz S., López de Alda M. J., Marco M.-P., and Barceló D., Biosensors for environmental monitoring, A global perspective, *Talanta*, 65, 291–297, 2005.
4. Logrieco A., Arrigan D. W. M., Brengel-Pesce K., Siciliano P., and Tothill I., DNA arrays, electronic noses and tongues, biosensors and receptors for rapid detection of toxigenic fungi and mycotoxins, A review, *Food Addit. Contam.*, 22(4), 335–344, 2005.
5. Tothill I. E. (ed.), *Rapid and On-line Instrumentation for Food Quality Assurance* (Cambridge: Woodhead, 2003).
6. Grayman W., Deininger R., Males R., and Gullick R., Source water early warning systems, in *Water Supply Systems Security*, ed. Larry Mays (New York: McGraw-Hill and Companies, 2004).
7. Hasan J., States S., and Deininger R., Safeguarding the security of public water supplies using early warning systems: A brief review, *J. Contemp. Water Research and Education*, 129, 27–33, 2004.
8. Farré M. and Barceló D., Toxicity testing of wastewater and sewage sludge by biosensor, bioassays and chemical analysis, *Trends Anal. Chem.*, 22, 299–310, 2003.
9. EPA, Technologies and techniques for early warning systems to monitor and evaluate drinking water quality: State of the art review, *Office of Research and Development, National Homeland Security Research Centre*, 2005.
10. D'Souza S. F., Microbial biosensors, *Biosens. Bioelectron.*, 16, 337–353, 2001.
11. Bentley A., Atkinson A., Jezek J., and Rawson D. M., Whole cell biosensors—electrochemical and optical approaches to ecotoxicity testing, *Toxicology in Vitro.*, 15, 469–475, 2001.
12. Sørensen S. J., Burmølle M., and Hansen L. H., Making bio-sense of toxicity: new developments in whole-cell biosensors, *Curr. Opin. Biotechnol.*, 17, 1–16, 2006.
13. Lei Y., Chen W., and Mulchandani A., A review: Microbial biosensors, *Anal. Chim. Acta.*, 568(1–2), 217–221, 2006.
14. Mikkelsen S. R. and Cortón E., *Bioanalytical Chemistry* (New Jersey: John Wiley and Sons, 2004).
15. Mulchandani A. and Rogers K. R. (eds.), *Enzyme and Micriobial Biosensors: Techniques and Protocols* (Totowa, NJ: Humana Press, 1998).
16. Nikolelis D., Krull U., Wang J., and Mascini M. (eds.), *Biosensors for Direct Monitoring of Environmental Pollutants in Field* (London: Kluwer Academic, 1998).
17. Ramsay G. (ed.), *Commercial Biosensors: Applications to Clinical, Bioprocess and Environmental Samples* (Chichester: Wiley, 1998).

18. Tothill I. E. and Stephens S., Methods for environmental monitoring: Biological methods, in *Analytical Methods for Environmental Monitoring*, ed. R. Ahmad, M. Cartwright and F. Taylor (Harlow: Pearson Education, Chapter 9, pp. 224–259, 2001).

19. Mason J. R., The induction and repression of benzene and catechol oxidizing capacity of *Pseudomonas putida* ML2 studied in perturbed chemostat culture, *Arch. Microbiol.*, 162, 57–62, 1994.

20. Lanyon Y. H., Marrazza G., Tothill I. E., and Mascini M., Flow injection analysis of benzene using an amperometric bacterial biosensor, *Anal. Lett.*, 37(8), 1515–1528, 2004.

21. Lanyon Y. H., Marrazza G., Tothill I. E., and Mascini M., Benzene analysis in workplace air using an FIA-based bacterial biosensor, *Biosens. Bioelectron.*, 20, 2089–2096, 2005.

22. Lanyon Y. H., Tothill I. E., and Mascini M., An amperometric bacterial biosensor based on gold screen-printed electrodes for the detection of benzene, *Anal. Lett.*, 39, 1–13, 2006.

23. Riley M. R., Jordan K. A., and Cox M. L., Development of a cell-based sensing device to evaluate toxicity of inhaled materials, *Biochem. Eng. J.*, 19, 95–99, 2004.

24. Natarajan A., Molnar P., Sieverdes K., Jamshidi A., and Hickman J. J., Microelectrode array recording of cardiac action potentials as a high throughput method to evaluate pesticide toxicity, *Toxicology in Vitro*, 20, 375–381, 2006.

25. Lei Y., Mulchandani P., Chen W., Wang J., and Mulchandani A., A microbial biosensor for *p*-nitrophenol using *arthrobacter* sp., *Electroanalysis*, 15(14), 1160–1164, 2003.

26. Mulchandani P., Carlos M. H., Lei Y., Chen W., and Mulchandani A., Amperometric microbial biosensor for *p*-nitrophenol using *Moraxella* sp.-modified carbon paste electrode, *Biosens. Bioelectron.*, 21, 523–527, 2005.

27. Timur, S., Pazarlio lu N., Pilloton R., and Telefoncu A., Detection of phenolic compounds by thick film sensors based on Pseudomonas putida, *Talanta*, 61, 87–93, 2003.

28. Timur S., Seta L. D., Pazarlio lu N., Pilloton R., and Telefoncu A., Screen-printed graphite biosensors based on bacterial cells, *Process Biochem.*, 39, 1325–1329, 2004.

29. Mazzei F., Botre F., Lorenti G., Simonetti G., Porcelli F., Scibona G., and Botre C., Plant tissue electrode for the determination of atrazine, *Anal. Chim. Acta.*, 316, 79–82, 1995.

30. Wang J., Kane S. A., Liu J., Smyth M. R., and Rogers K. R., Mushroom tissue-based biosensor for inhibitor monitoring, *Food Technology and Biotechnology*, 34, 51–55, 1996.

31. Bhattacharya J., Read D., Amos S., Dooley, S., Killham K., and Paton G. I., Biosensor-based diagnostics of contaminated groundwater assessment and remediation strategy, *Environ. Pollut.*, 134, 485–492, 2005.

32. EPA, A review of Emerging Sensor Technology for facilitating Long-term Ground Water Monitoring of Volatile Organic Compounds (EPA542–R–03–007), 2003, http://www.epa.gov/tio/download/char/542r03007.pdf.

33. EPA Response protocol toolbox: Planning for and responding to drinking water contamination threats and incidents, *Modules* 1–6, 2003/2004.

34. Nunes-Halldorson V. D. and Duran N. L., Bioluminescent bacteria: *Lux* genes as environmental biosensors, *Braz. J. Microbiol.*, 34, 91–96, 2003.

35. Lee J. H., Mitchell R. J., Kim B. C., Cullen D. C., and Gu M. B., A cell array biosensor for environmental toxicity analysis, *Biosens. Bioelectron.*, 21, 500–507, 2005.

36. Tibazarwa C., Corbisier P., Mench M., Bossus A., Solda P., Mergeay M., Wyns L., and van der Lelie D., A microbial biosensor to predict bioavailable nickel in soil and its transfer to plants, *Environ. Pollut.*, 113, 19–26, 2001.

37. Rasmussen L. D., Sorensen S. J., Turner R. R., and Barkay T., Application of a *mer-lux* biosensor for estimating bioavailable mercury in soil, *Soil. Biol. Biochem.*, 32, 639–646, 2000.

38. Lehmann M., Riedel K., Adler K., and Kunze G., Amperometric measurement of copper ions with a deputy substrate using a novel *Saccharomyces cerevisiae* sensor, *Biosens. Bioelectron.*, 15, 211–219, 2000.

39. Lei Y., Mulchandani P., Chen W., and Mulchandani A., Direct determination of p-nitrophenyl substituent organophosphorus nerve agents using a recombinant *Pseudomonas putida* JS444-modified clark oxygen electrode, *J. Agric. Food Chem.*, 53, 524–527, 2005.

40. Wells M., Gosch M., Rigler R., Harms H., Lasser T., and van de Meer J. R., Ultrasensitive reporter protein detection in genetically engineered bacteria, *Anal. Chem.*, 77, 2683–2689, 2005.

41. Bjerketorp J., Hakansson S., Belkin S., and Jansson J. K., Advances in preservation methods: Keeping biosensor microorganisms alive and active, *Curr. Opin. Biotechnol.*, 17, 43–49, 2006.

42. EPA-ETV. U.S. Environmental protection agency, Environmental verification program (2004). http://www.epa.gov/etv/.

43. EPA, Office of Research and Development, Shaw Environmental, Draft Report Evaluation of Water Quality Sensors in Distribution Systems; May 2004 and EPA, Office of Research and Development, Shaw Environmental, Draft Report: *Water Quality Sensor Responses to Chemical and Biological Warfare Agent Simulants in Water Distribution Systems*, July 2004.

44. Tag K., Lehmann M., Chan C., Renneberg R., Riedel K., and Kunze G., Measurement of biodegradable substances with a mycelia-sensor based on the salt tolerant yeast *Arxula adeninivorans* LS3, *Sens. Actuators B*, 67, 142–148, 2000.

45. Liu J., Bjornsson L., and Mattiasson B., Immobilised activated sludge based biosensor for biochemical oxygen demand measurement, *Biosens. Bioelectron.*, 14, 883–893, 2000.

46. Jia J., Tang M., Chen X., Qi L., and Dong S., Co-immobilized microbial biosensor for BOD estimation based on sol-gel derived composite material, *Biosens. Bioelectron.*, 18, 1023–1029, 2003.

47. Liu J., Olsson G., and Mattiasson B., Short-term BOD (BODst) as a parameter for on-line monitoring of biological treatment process Part I. A novel design of BOD biosensor for easy renewal of bio-receptor, *Biosens. Bioelectron.*, 20, 562–570, 2004.

48. Liu J., Olsson G., and Mattiasson B., Short-term BOD (BODst) as a parameter for on-line monitoring of biological treatment process Part II: Instrumentation of integrated flow injection analysis (FIA) system for BODst estimation, *Biosens. Bioelectron.*, 20, 571–578, 2004.

49. Chee G. J., Nomura Y., Ikebukuro K., and Karube I., Development of photocatalytic biosensor for the evaluation of biochemical oxygen demand, *Biosens. Bioelectron.*, 21, 67–73, 2005.

50. Ignatov V., Rogatcheva S. M., Kozulin S. V., and Khorkina N. A., Acrylamide and acrylic acid determination using respiratory activity of microbial cells, *Biosens. Bioelectron.*, 12, 105–111, 1997.

51. Ikebukuro K., Honda M., Nakanishi K., Nomura Y., Masuda Y., Yokoyama K., Yamauchi I., and Karube I., Flow-type cyanide sensor using an immobilized microorganism, *Electroanalysis*, 8, 876–879, 1996.

52. Okochi M., Mima K., Miyata M., Shinozaki Y., Haraguchi S., Fujisawa M., Kaneko M., Masukata T., and Matsunaga T., Development of an automated water toxicity biosensor

using *thiobacillus ferrooxidans* for monitoring cyanides in natural water for a water filtering plant, *Biosens. Bioeng*, 87, 905–911, 2004.

53. Rouillon R., Sole M., Mazzei F., Botre F., Lorenti G., Simonetti G., Porcelli F., Scibona G., and Botre C., Plant tissue electrode for the determination of atrazine, *Anal. Chim. Acta.*, 316, 79–82, 1995.

54. Nandakumar R. and Mattiasson B., A microbial biosensor using *Pseudomonas putida* cells immobilized in expanded bed reactors for the on-line monitoring of phenolic compounds, *Anal. Lett.*, 32, 2379–2393, 1999.

55. Riedel K., Beyersdorf R., Neumann B., and Scheller F., Microbial sensors for determination of aromatics and their chloroderivatives. Part-III: Determination of chlorinated phenols a biosensor containing *Trichosporon beigelli*, *Appl. Microbiol. Biotechnol.*, 43, 7–9, 1995.

56. Riedel K., Hensel J., Rothe S., Neumann B., and Scheller F., Microbial sensor for determination of aromatics and their chloroderivatives. Part-II: Determination of chlorinated phenols using a *Rhodococcus* containing biosensor, *Appl. Microbiol. Biotechnol.*, 38, 556–559, 1993.

57. Peter J., Hutter W., Stollnberger W., and Hampel W., Detection of chlorinated and brominated hydrocarbons by an ion sensitive whole cell biosensor, *Biosens. Bioelectron.*, 11, 1215–1219, 1996.

58. Koenig A., Zaborosch C., Muscat A., Vorlop K. D., and Spener F., Microbial sensors for naphthalene using *Sphingomonas* sp. B1 or *Pseudomonas fluorescens* WW4, *Appl. Microbiol. Biotechnol.*, 45, 844–850, 1996.

59. Koenig A., Zaborosch C., and Spener F., Microbial sensors for PAH in aqueous solution using solubilizers, in *Field Screening Europe*, ed. J. Gottlieb, H. Hotzl, K. Huck, and R. Niessner (Dordrecht, The Netherlands: Kluwer Academic Publishers, 1997), 203–206.

60. Mulchandani A., Mulchandani P., Kaneva I., and Chen W., Biosensor for direct determination of organophosphate nerve agents using recombinant *Escherichia coli* with surface-expressed organophosphorus hydrolase. 1. Potentiometric microbial electrode, *Anal. Chem.*, 70, 4140–4145, 1998.

61. Mulchandani A., Kaneva I., and Chen W., Biosensor for direct determination of organophosphate nerve agents using recombinant *Escherichia coli* with surface-expressed organophosphorus hydrolase. 2. Fiber-optic microbial biosensor, *Anal. Chem.*, 70, 5042–5046, 1998.

62. Polyak B., Bassls E., Novodvorets A., Belkin S., and Marks R. S., Optical fiber bioluminescent whole-cell microbial biosensors to genotoxicants, *Water Sci. Technol.*, 42, 305–311, 2000.

63. Rouillon R., Tocabens M., and Carpentier R., A photochemical cell for detecting pollutant-induced effects on the activity of immobilized cyanobacterium *Synechococcus* sp. PCC 7942, *Enzyme Microb. Technol.*, 25, 230–235, 1999.

64. Biran I., Babai R., Levcov K., Rishpon J., and Ron E. Z., Online and in situ monitoring of environmental pollutants: electrochemical biosensing of cadmium, *Environ. Microbiol.*, 2(3), 285–290, 2000.

65. Tothill I. E. and Turner A. P. F., Biosensors, in *Encyclopaedia of Food Sciences and Nutrition*, ed. Benjamin Caballero, Luiz Trugo, and Paul Finglas, 2nd ed. (London, Oxford, Boston, New York, and San Diego: Academic Press, 2003), pp. 489–499.

66. Gardea-Torresdet J., Darnall D., and Wang J., Bioaccumulation and measurement of copper at an alga-modified carbon paste electrode, *Anal. Chem.*, 60, 72–76, 1988.

67. Shitanda I., Takada K., Sakai Y., and Tatsuma T., Compact amperometric algal biosensors for the evaluation of water toxicity, *Anal. Chim. Acta.*, 530, 191–197, 2005.

68. Tu Y.-F., Fu Z.-Q., and Chen H.-Y., The fabrication and optimization of the disposable amperometric biosensor, *Sens. Actuators B*, 80, 101–105, 2001.

69. Bianco P., Protein modified-and membrane electrodes: strategies for the development of biomolecular sensors, *Rev. Mol. Biotechnol.*, 82, 393–409, 2002.

70. Amine A., Mohammadi H., Bourais I., and Palleschi G., Enzyme inhibition-based biosensors for food safety and environmental monitoring, *Biosens. Bioelectron.*, 21, 1405–1423, 2006.

71. Sanz V. C., Mena M. L., Gonzalez-Cortes A., Yanez-Sedeno P., and Pingarron J. M., Development of a tyrosinase biosensor based on gold nanoparticles-modified glassy carbon electrodes—Application to the measurement of a bioelectrochemical polyphenols index in wines, *Anal. Chim. Acta*, 528, 1–8, 2005.

72. Tothill I. E., Biosensors Developments and potential applications in the agricultural diagnosis sector, *Computers and Electronics in Agriculture*, 30, 205–218, 2001.

73. Andreescu S. and Marty J.-L., Twenty years research in cholinesterase biosensors: From basic research to practical application, *Biomol. Eng.*, 23, 1–15, 2006.

74. Lin Y. H., Lu F., and Wang J., Disposable carbon nanotube modified screen-printed biosensor for amperometric detection of organophosphorus pesticides and nerve agents, *Electroanalysis*, 16, 145–149, 2004.

75. Kok F. N. and Hasirci V., Determination of binary pesticide mixtures by an acetylcholinesterase-choline oxidase biosensors, *Biosens. Bioelectron.*, 19, 661–665, 2004.

76. Andreou V. G. and Clonis Y. D., A portable fiber-optic pesticide biosensor based on immobilized cholinesterase and sol–gel entrapped bromocresol purple for in-field use, *Biosens. Bioelectron.*, 17, 61–69, 2002.

77. Andreescu D., Andreescu S., and Sadik O. A., New materials for biosensors, biochips and molecular bioelectronics, in *Biosensors and Modern Biospecific Analytical Techniques*, ed. L. Gorton (Amsterdam: Elsevier, pp. 285–329, 2005).

78. Joshi K. A., Tang J., Haddon R., Wang J., Chen W., and Mulchandani A., A disposable biosensors for organophosphorus nerve agents based on carbon nanotubes modified thick film strip electrodes, *Electroanalysis*, 17, 54–58, 2005.

79. Koblizek M., Maly J., Masoji dek J., Komenda J., Kucera T., Giardi M. T., Mattoo A. K., and Pilloton R., A biosensor for the detection of triazine and phenylurea herbicides designed using photosystem ii coupled to a screen-printed electrode, *Biotechnol. Bioeng.*, 78(1), 110–116, 2002.

80. Kadara R. O. and Tothill I. E., Stripping chronopotentiometric measurements of lead (II) and cadmium (II) in soils extracts and wastewaters using bismuth film screen-printed electrode assembly, *Anal. Bioanal. Chem.*, 378(3), 770–775, 2004.

81. Kadara R. O. and Tothill I. E., Resolving the copper interference effect on the stripping chronopotentiometric response of lead (II) obtained at bismuth film screen-printed electrode, *Talanta*, 66, 1089–1093, 2005.

82. Md Noh M. F., Kadara R. O., and Tothill I. E., Development of cysteine modified screen-printed electrode for the chronopotentiometric stripping analysis of Cadmium (II) in wastewater and soils extract, *Anal. Bioanal. Chem.*, 382, 1175–1186, 2005.

83. Wittekindt E., Werner M., Reinicke A., Herbert A., and Hansen P., A microtiter-plate urease inhibition assay sensitive, rapid and cost-effective screening for heavy metals in water, *Environ. Technol.*, 17, 597–603, 1996.

84. Brack W., Paschke A., Segner H., Wennrich R., and Schürmann G., Urease inhibition: A tool for toxicity identification in sediment elutriates, *Chemosphere*, 40, 829–834, 2000.

85. Kormos F. and Lengauer A. N., Studies concerning the biomonitoring of the degree of environmental pollution using a urea-sensitive enzymatic sensor, *Lab. Rob. Autom.*, 12, 27–30, 2000.

86. Krawczyk T. K., Moszczy ska M., and Trojanowicz M., Inhibitive determination of mercury and other metal ions by potentiometric urea biosensor, *Biosens. Bioelectron.*, 15, 681–691, 2000.

87. Soldatkin A. P., Volotovsky V., El' skaya A. V., Jaffrezic-Renault N., and Martelet C., Improvement of urease based biosensor characteristics using additional layers of charged polymers, *Anal. Chim. Acta*, 403, 25–29, 2000.

88. Bello-Rodriguez B., Bolbot J. A., and Tothill I. E., Application of urease-glutamic dehydrogenase biosensor to heavy metals screening in water and soil samples from Aznalcollar mine, *Anal. Bioanal. Chem.*, 380, 284–292, 2004.

89. Bello-Rodriguez B., Bolbot J. A., and Tothill I. E., Development of urease and glutamic dehydrogenase amperometric assay for heavy metals screening in polluted samples, *Biosens. Bioelectron.*, 19, 1157–1167, 2004.

90. Mohammadi H., Amine A., Cosnier S., and Mousty C., Mercury–enzyme inhibition assays with an amperometric sucrose biosensor based on a trienzymatic–clay matrix, *Anal. Chim. Acta*, 543, 143–149, 2005.

91. Marques P. R. B. O., Nunes G. S., Andreescu S., and Marty J. L., Comparative investigation between acetylcholinesterase obtained from commercial sources and genetically modified *Drosophila melanogaster*. Application in amperometric biosensors for methamidophos pesticide detection, *Biosens. Bioelectron.*, 20, 824–831, 2004.

92. Mena M. L., Yanez-Sedeno P., and Pingarron J. M., A comparison of different strategies for the construction of amperometric enzyme biosensors using gold nanoparticles-modified electrodes, *Anal. Biochem.*, 336, 20–27, 2005.

93. Tsai H.-C. and Doong R.-A., Simultaneous determination of pH, urea, acetylcholine and heavy metals using array-based enzymatic optical biosensor, *Biosens. Bioelectron.*, 20, 1796–1804, 2005.

94. Malitesta C. and Guascito M. R., Heavy metal determination by biosensors based on enzyme immobilised by electropolymerization, *Biosens. Bioelectron.*, 20, 1643–1647, 2005.

95. Han S., Zhu M., Yuan Z., and Li X., A methylene blue-mediated enzyme electrode for the determination of trace mercury(II), mercury(I), methylmercury, and mercury–glutathione complex, *Biosens. Bioelectron.*, 16, 9–16, 2001.

96. Fennouh S., Casimiri V., Geloso-Meyer A., and Burstein C., Kinetic study of heavy metal salt effects on the activity of L-lactate dehydrogenase in solution or immobilized on an oxygen electrode, *Biosens. Bioelectron.*, 13, 903–909, 1998.

97. Bertocchi P., Ciranni E., Compagnone D., Mageauru V., Palleschi G., Pirvutoiu S., and Valvo L., Flow injection analysis of mercury (II) in pharmaceuticals based on enzyme inhibition and biosensor detection, *J. Pharm. Biom. Anal.*, 20, 263–269, 1999.

98. Danzer T. and Schwedt G., Chemometric methods for the development of a biosensor system and the evaluation of inhibition studies with solutions and mixtures of pesticides and heavy metals. Part 1. Development of an enzyme electrodes system for pesticides and heavy metals screening using selected chemometrics methods, *Anal. Chim. Acta*, 318, 275–286, 1996.

99. Compagnone D., Lupu A. S., Ciucu A., Magearu V., Cremisini C., and Palleschi G., Fast amperometric FIA procedure for heavy metal detection using enzyme inhibition, *Anal. Lett.*, 34, 17–27, 2001.

100. Dong R. A. and Tsai H. C., Immobilization and characterization of sol–gel encapsulated acetylcholinesterase fiber-optic biosensor, *Anal. Chim. Acta*, 434, 239–246, 2001.

101. Simonian L., Good T. A., Wang S.-S., and Wild J. R., Nanoparticles-based optical biosensors for the direct detection of organophosphate chemical warfare agent and pesticides, *Anal. Chim. Acta*, 534, 69–77, 2005.

102. Zhang Y., Muench S. B., Schulze H., Perz R., Yang B., Schmid R. D., and Bachmann T. T., Disposable biosensor test for organophosphate and carbamate insecticides in milk. *Agric. Food Chem.*, 53(13), 5110–5115, 2005.

103. Reybier K., Zairi S., Jaffrezic-Renault N., and Fahys B., The use of polyethylene imine for fabrication of potentiometric cholinesterase biosensors, *Talanta*, 56, 1015–1020, 2002.

104. White B. J. and Harmon H. J., Optical solid-state detection of organophosphates using organophosphorus hydrolase, *Biosens. Bioelectron.*, 20, 1977–1983, 2005.

105. Sotiropoulou S. and Chaniotakis N. A., Lowering the detection limit of the acethyl-cholinesterase biosensor using a nanoporous carbon matrix, *Anal. Chim. Acta*, 530, 199–204, 2005.

106. El Kaoutit M., Bouchta B., Zejli H., Izaoumen N., and Temsamani K. R., A simple conducting polymer-based biosensor for the determination of atrazine, *Anal. Lett.*, 37, 1671–1681, 2004.

107. Sacks V., Eshkenazi I., Neufeld T., Dosoretz C., and Rishpon J., Immobilized parathion hydrolase: An amperometric sensor for parathion, *Anal. Chem.*, 72, 2055–2058, 2000.

108. Shan D., Mousty C., and Cosnier S., Subnanomolar cyanide detection at polyphenol oxidase/clay biosensors, *Anal. Chem.*, 76, 178–183, 2004.

109. Noguer T., Balasoiu A.-M., Avramescu A., and Marty J.-L., Development of disposable biosensor for the detection of metam-sodium and its metabolite MITC, *Anal. Lett.*, 34, 513–528, 2001.

110. Devic E., Li D., Dauta A., Henriksen P., Codd G. A., Marty J.-L., and Fournier D., Detection of anatoxin-a(s) in environmental samples of cyanobacteria by using a biosensor with engineered acetylcholinesterases, *Appl. Environ. Microbiol.*, 68, 4102–4106, 2002.

111. Villatte F., Schulze H., Schmid R. D., and Bachmann T. T., A disposable acetylcho-linesterase-based electrode biosensor to detect anatoxin-a(s) in water, *Anal. Bioanal. Chem.*, 372, 322–326, 2002.

112. Potyrailo R. A., Conrad R. C., Ellington A. D., and Hieftje G. M., Adapting selected nucleic acid ligands (aptamers) to biosensors, *Anal. Chem.*, 70, 3419–3425, 1998.

113. Wang J., Rivas G., Cai X., Palecek E., Nielsen P., Shirashi H., Dontha N., Luo D., Parrado C., Chicarro M., Farias P. A. M., Valera F. S., Grant M., Ozsoz M., and Flair M. N., DNA electrochemical biosensors for environmental monitoring, A review, *Anal. Chim. Acta*, 347, 1–8, 1997.

114. Marrazza G., Chianella I., and Mascini M., Disposable DNA electrochemical biosen-sors for environmental monitoring, *Anal. Chim. Acta*, 387, 297–307, 1999.

115. Chiti G., Marrazza G., and Mascini M., Electrochemical DNA biosensor for environ-mental monitoring, *Anal. Chim. Acta*, 427, 155–164, 2001.

116. Mascini M., Palchetti I., and Marrazza G., DNA electrochemical biosensor, *Fresenius J. Anal. Chem.*, 369, 15–22, 2001.

117. Palanti S., Marrazza G., and Mascini M., Electrochemical DNA probes, *Anal. Lett.*, 29, 2309–2331, 1996.

118. Lucarelli F., Palchetti I., Marrazza G., and Mascini M., Electrochemical DNA biosen-sor as a screening tool for the detection of toxicants in water and wastewater samples, *Talanta*, 56, 949–957, 2002.

119. Wang J., Chicarro M., Rivas G., Cai X., Dontha N., Farias P. A. M., and Shirashi H., DNA biosensor for the detection of hydrazines, *Anal. Chem.*, 68, 2251–2254, 1996.

120. Wang J., Rivas G., Luo D., Cai X., Valera F. S., and Dontha N., DNA modified electrode for the detection of aromatic amines, *Anal. Chem.*, 68, 4365–4369, 1996.

121. Carter M. T., Rodriguez M., and Bard A. J., Voltammetric studies of interaction of metals chelates with DNA: 2. Tris-chelated complexes of Co(III) and Fe (III) with 1,10-phenantroline and 2,2'-bipyridine, *J. Am. Chem. Soc.*, 111, 8901–8911, 1998.

122. Lucarelli F., Kicela A., Palchetti I., Marrazza G., and Mascini M., Electrochemical DNA biosensor for analysis of wastewater samples, *Bioelectrochemistry*, 58, 113–118, 2002.

123. Tencaliec A. M., Laschi S., Magearu V., and Mascini M., A comparison study between a disposable electrochemical DNA biosensor and a *Vibrio fischeri* based luminescent sensor for the detection of toxicants in water samples, *Talanta*, 69, 365–369, 2006.

124. Doong R. A., Shih H., and Lee S., Sol-gel-derived array DNA biosensor for the detection of polycyclic aromatic hydrocarbons in water and biological samples, *Sens. Actuators B*, 111–112, 323–330, 2005.

125. Luzi E., Minunni M., Tombelli S., and Mascini M., New trends in affinity sensing: aptamers for ligand binding, *Trends Anal. Chem.*, 22, 810–818, 2003.

126. Tombelli S., Minunni M., and Mascini M., Analytical applications of aptamers, *Biosens. Bioelectron.*, 20, 2424–2434, 2005.

127. Lu Y., Liu J., Li, Jing, Bruesehoff P. J., Pavot C. M.-B. and Brown A. K., New highly sensitive and selective catalytic DNA biosensors for metal ions, *Biosens. Bioelectron.*, 18, 529–540, 2003.

128. Pan T. and Uhlenbeck O. C., A small metalloribozymic with two-step mechanism, *Nature*, 358, 560–563, 1993.

129. Carmi N., Shultz L. A., and Breaker R. R., In vitro selection of self-cleaving DNAs, *Chem. Biol.*, 3, 1039–1046, 1996.

130. Li J., Zheng W., Kwon A. H., and Lu Y., In vitro selection and characterization of a highly efficient Zn (II) dependent RNA cleaving deoxyribozyme, *Nucleic. Acids. Res.*, 28, 481–488, 2000.

131. Li J. and Lu Y., A highly sensitive and selective catalytic DNA biosensor for lead ions, *J. Am. Chem. Soc.*, 122, 10466–10467, 2000.

132. Bruesehoff P. J., Li J., Augustine A. J., and Lu Y., Improving metal ion specificity during in vitro selection of catalytic DNA, *Comb. Chem. High. Throughput. Screening.*, 5, 327–335, 2002.

133. Reineccius G., Instrumental means of monitoring the flavor quality of foods. *Acs. Symp. Series.*, 631, 241–252, 1996.

134. Gardner J. W. and Bartlett P. N., *Electronic Noses: Principle and Applications* (Oxford: Oxford University Press, 1999).

135. Tothill I. E., Piletsky S., Magan N., and Turner A. P. F., New biosensors, in *Instrumentation and Sensors for the Food Industry*, eds. Erika Kress-Rogers and Cristopher J. B. Brimelow, 2nd ed. (Cambridge: Woodhead Publishing Limited, pp. 760–775, 2001).

136. Pearce T. C., Schiffman S. S., Nagle H. T., and Gardner J. W. (eds), *Handbook of Machine Olfaction: Electronic Nose Technology* (Chichester: Wiley, 2002).

137. Turner A. P. F. and Magan N., Electronic noses and disease diagnostics, *Nature Reviews*, 2, 1–6, 2004.

138. Malan P. G., Immunological biosensors, in *The Immunoassay Handbook*, ed. David Wild, 2nd ed. (London, U.K.: Nature Publishing Group, pp. 229–239, 2001).

139. Tothill I. E., On-line immunochemical assays for contaminants analysis, in *Rapid and On-Line Instrumentation for Food Quality Assurance*, ed. I. E. Tothill (Cambridge: Woodhead Publishing Limited and CRC Press LLC. 2003), pp. 14–35.

140. Gizeli E. and Lowe C. R., Immunosensors, *Curr. Opin. Biotechnol.*, 7, 66–71, 1996.

141. Wittmann C. and Schmid R. D., Bioaffinity sensors for environmental monitoring, in *Handbook of Biosensors and Electronic Noses—Medicine, Food, and the Environment*, ed. E. Kress-Rogers, pp. 333–349. (Boca Raton, FL: CRC Press, 1997).

142. Mascini M., Affinity electrochemical biosensors for pollution control, *Pure. Appl. Chem.*, 73, 23–30, 2001.

143. Fähnrich K. A., Pravda M., and Guilbault G. G., Disposable amperometric immunosensor for the detection of polycyclic aromatic hydrocarbons (PAHs) using screen-printed electrode, *Biosens. Bioelectron.*, 18, 73–82, 2003.

144. Kalab T. and Skládal P., Disposable multichannel immunosensors for 2,4-D dichlorophenoxyacetic acids using acetylcholinesterase as an enzyme label, *Electroanalysis*, 9(4), 293–297, 1997.

145. Ducey M. W., Smith A. M., Guo X. A., and Meyerhoff M. E., Competitive nonseparation electrochemical enzyme binding immunoassay (NEEIA) for small molecule detection, *Anal. Chim. Acta.*, 357(1–2), 5–12, 1997.

146. Gonzalez Martinez M. A., Puchades R., and Maquieira A., Reversibility in heterogeneous flow immunosensing and related techniques, *Food Technol. Biotechnol.* 35(3), 193–204, 1997.

147. Marquette C. A., Coulet P. R., and Blum L. J., Semi-automated membrane based chemiluminescent immunosensor for flow injection analysis of okadaic acid in mussels. *Anal. Chim. Acta.*, 398, 173–182, 1999.

148. Haake H.-M., de best L., Irth H., Abuknesha R., and Brecht A., Label-free biochemical detection coupled on-line to liquid chromatography, *Anal. Chem.*, 72, 3635–3641, 2000.

149. Krämer P. M., Baumann B. A., and Stoks P. G., Prototype of a newly developed immunochemical detection system for the determination of pesticide residues in water, *Anal. Chim. Acta.*, 347(1–2), 187–198, 1997.

150. Nedelkov D., Rasooly A., and Nelson R. W., Multitoxin biosensor-mass spectrometry analysis: A new approach for rapid, real-time, sensitive analysis of staphylococcal toxins in food, *Int. J. Food Microbiol.*, 60, 1–13, 2000.

151. Sternesjo A., Mellgren C., and Bjorck L. Determination of sulfamethazine residues in milk by a surface plasmon resonance-based biosensoassay, *Anal. Biochem.*, 226, 175–181, 1995.

152. Guidi A., Laricchia-Robbio L., Gianfaldoni D., Revoltella R., and Del Bono G., Comparison of a conventional immunoassay (ELISA) with a surface plasmon resonance-based biosensor for IGF-1 detection in cow's milk, *Biosens. Bioelectron.*, 16, 971–977, 2001.

153. Cornell B. A., Optical biosensors: Present and future, in *Membrane Based Biosensors*, eds. F. Lighler and C. Rowe Taitt (Amsterdam: Elsevier, Chapter 12, 457–495, 2002).

154. Homola J., Dostalek J., Chen S., Rasooly A., Jiang S., and Yee S. S., Spectral surface plasmon resonance biosensor for detection of staphylococcal enterotoxin B in milk, *Int. J. Food Microbiol.*, 75, 61–69, 2002.

155. Lotierzo M., Henry O. Y. F., Piletsky S., Tothill I. E., Cullen D., Kania M., Hock B., and Turner A. P. F., Surface plasmon resonance sensor for domoic acid based on grafted imprinted polymer, *Biosens. Bioelectron.*, 20, 145–152, 2004.

156. Chianella I., Lotierzo M., Piletsky S., Tothill I. E., Chen B., and Turner A. P. F., Rational design of a polymer specific for microcystin-LR using a computational approach, *Anal. Chem.*, 74, 1288–1293, 2002.

157. Chianella I., Piletsky S., Tothill I. E., Chen B., and Turner A. P. F., Combination of solid phase extraction cartridges and MIPbased sensor for detection of microcystin-LR, *Biosens. Bioelectron.*, 18, 119–127, 2003.

158. Mallat E., Barceló D., Barzen C., Gauglitz G., and Abuknesha R., Immunosensor for pesticide determination in natural waters, *Trends in Anal. Chem.*, 20(3), 124–132, 2001.

159. Mallat E., Barzen C., Abuknesha R., Gauglitz G., and Barceló D., Part per trillion level determination of isoproturon in certified and estuarine water samples with a direct optical immunosensors, *Anal. Chim. Acta.*, 426, 209–216, 2001.

160. Mosbach K. and Ramstrom O., The emerging technique of molecular imprinting and its future impact on biotechnology, *Biotechnology*, 14, 163–170, 1996.

161. Masque N., Marce R. M., Borrull F., Cormack P. A. G., and Sherrington D. C., Synthesis and evaluation of a molecularly imprinted polymer for selective on-line solid phase extraction of 4-nitophenol from environmental water, *Anal. Chem.*, 72, 4122–4126, 2000.

162. Chen B., Bestetti G., and Turner A. P. F., The synthesis and screening of a combinatorial library for affinity ligands for glycosylated haemoglobin, *Biosens. Bioelectron.*, 13, 779–785, 1998.

163. Feeney R. and Kounaves S. P., Microfabricated ultramicroelectrode arrays: Developments, advances and applications in environmental analysis, *Electroanalysis*, 12, 677–684, 2000.

164. Kojima K., Hiratsuka A., Susuki H., Yano K., Ikebukuro A., and Karube I., Electrochemical protein chip with arrayed immunosensors with antibodies immobilised in a plasma polymerized film, *Anal. Chem.*, 75, 1116–1122, 2003.

165. Braggins D., Fingerprint sensing and analysis, *Sensor Review*, 21(4), 272–277, 2001.

166. Chang S., Cheng Y., Larin K. V., Mao Y., Sherif S., and Flueraru C., Optical coherence tomography used for security and fingerprint-sensing applications, *IEEE Trans. Image Process.*, 2(1), 48–58, 2008.

6 Application of Electronic Noses and Tongues

Anil K. Deisingh*
Caribbean Industrial Research Institute
University of the West Indies
St. Augustine, Trinidad and Tobago, West Indies

CONTENTS

* In memory of Dr. Anil Deisingh.

6.1 INTRODUCTION

Electronic noses and tongues are arrays of sensors used to characterize complex samples, with the former being arrays of gas sensors while the latter are composed of liquid sensors [1]. These devices are composed of a chemical sensing system and a pattern recognition (PR) system [usually an artificial neural network (ANN)]. The array sensing system allows different properties to be measured simultaneously [2]. Each chemical, which reaches the sensor array, will produce a characteristic pattern and therefore a database of patterns will be built up for a series of chemicals [2].

The widest application of electronic noses and tongues is in the food industry in areas as varied as quality control, process operations, taste studies, and identification of flavor and aroma [3]. All of these areas are discussed in this chapter. In addition, electronic noses are finding use in environmental applications such as the analysis of fuel mixtures, identification of toxic wastes, and the detection of oil leaks [2]. Electronic tongues are also now being applied in environmental areas such as analysis of natural waters and detection of heavy metals. Both devices are also being more widely used in clinical and pharmaceutical applications, and several of these will be highlighted.

6.2 SENSOR DEVELOPMENT

Cattrall [4] has defined a chemical sensor as "a device, which responds to a particular analyte in a selective way by means of a reversible chemical interaction and can be used for the quantitative or qualitative determination of the analyte." This involves the transduction of chemical and/or physical properties at an interface into usable information [5]. The two main features are the transducer, which converts chemical information into an electrical signal, and the interface, which allows selectivity and/ or sensitivity to the chemical detection process (Figure 6.1).

The main categories of transducers, which have been employed to date, include [3] the following:

1. Electrochemical, such as ion-selective electrodes (ISE), ion-selective field affects transistors (ISFET), solid electrolyte gas sensors, semiconductor-based gas sensors, and conducting polymer sensors. Most electrochemical sensors are based on potentiometry, voltammetry, or amperometry although coulometry and conductimetry have also been utilized.

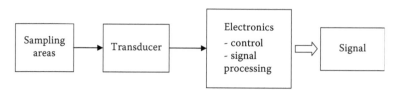

FIGURE 6.1 Schematic of a typical sensor device.

2. Piezoelectric, such as surface acoustic wave (SAW) sensors. In general, piezoelectricity is defined as "electric polarization" produced by mechanical strain in crystals belonging to certain classes, the polarization being proportional to the strain and changing sign with it [5]. A mechanical deformation gives rise to polarization of a dielectric material, which, in turn, generates an electric potential difference across the material. Piezoelectric materials tend to be sensitive to changes in mass, density, or viscosity, and it is because of these features that frequency can be used as a sensitive transduction parameter [6]. The most widely used piezoelectric material is quartz.

3. Optical—examples being fiber optics and devices making use of the principles of absorbance, reflectance, luminescence, and surface plasmon resonance (SPR). With fiber optics, evanescent wave technology is of prime importance. If the refractive index of one material is less than the other, there is a change of phase in the reflected ray with a corresponding energy loss. The lost energy is released as an evanescent wave, which decays rapidly outside the first medium and penetrates 1 Å wavelength. In addition, any modulation of the evanescent wave via absorption, polarization, or refraction leads to changes in the properties of light transmitted through the fiber [5]. This allows measurements to be made through the wall of a fiber optic.

4. Thermal systems—where the heat of a chemical reaction involves the analyte and, is monitored with a transducer. The example can be a thermistor.

A subclass of chemical sensors is the biosensors. These biological sensors incorporate a biological sensing element positioned close to the transducer to give a reagentless sensing system specific for the target analyte [6]. Table 6.1 gives a summary of some of the main developments in the sensor field, starting with the glass pH electrode, which appeared in 1930.

6.3 BASIC SENSOR THEORY

Since sensors measure the effect of changes in analyte concentration at the sample-transducer interface, it is important to consider calibration of the sensor response. Measurements can be either static or dynamic, with the latter being based on either transient or steady state conditions.

Static measurements do not involve a flow of sample since a fixed amount of analyte is introduced into the sample chamber. After a suitable period to allow equilibration, the steady-state signal is read. In some cases, such as slow processes involving enzymes, the kinetics of the signal rise to the steady-state value is monitored [5].

Dynamic responses involve exposure of the sensor to a continuous stream of sample. The concentration of the sample may either be constant (dynamic steady-state) or varied during the measurement cycle (transient).

Although static measurements allow for simpler handling, they can take longer to reach equilibrium and may be prone to sample depletion errors. Dynamic steady-state measurements are not subject to sample depletion since the sample is continuously renewed. However, larger volumes of sample are required, and there is also the possibility of dynamic delay errors arising as a result of the sample concentration varying faster than the response time of the sensor [5].

TABLE 6.1

Major Developments in the Sensor Area [3,7]

Year	Development
1930	Glass pH electrode
1956	Clark oxygen electrode
1959	Piezoelectric mass deposition sensor (quartz crystal microbalance)
1961	Solid electrolyte sensor
1962	Enzyme electrode (the first biosensor)
	Metal oxide semiconductor gas sensor (Taguchi sensor)
1964	Piezoelectric bulk acoustic wave (BAW) chemical vapor sensor
1966	Glucose sensor
	Fluoride ion-selective electrode
1970	Ion-selective field effect transistor (ISFET)
	Fiber optic gas sensor
1975	Palladium gate FET hydrogen sensor (MOSFET)
1977	Enzyme FET biosensor (ENFET)
1979	Surface acoustic wave (SAW) vapor/thin film sensor
1980	Liquid-phase BAW operation
1982	Surface plasmon resonance (SPR) sensor
1984	Evanescent wave fiber optic sensor
	First working artificial nose
1986	BAW liquid-phase immunosensor
1996	Electronic tongue first reported
1999	Electronic nose was used to identify the presence of pathogens in wine
2000	Launch of first commercially available electronic tongue
1990s onward	Developments in miniaturization (lab-on-a-chip) and sensor arrays

6.4   SENSOR MATERIAL AND TECHNOLOGIES

Three major types of sensor materials have been used in the development of electronic noses and tongues, and each are described in the following sections.

6.4.1   METAL OXIDE SENSORS

Oxides of semiconductors are of low cost and are widely used for the detection of gases such as carbon monoxide. These show high gas sensitivity, fast response time, and long-term stability [8]. The process of oxidizing and/or reducing gases by the oxides involves a change in conductivity of the oxide due to catalytic reduction and oxidation reactions taking place at the surface. These reactions are controlled by the electronic structure of the oxide system, the chemical composition, crystal structure, and orientation of the surfaces of the oxide phases exposed to the gas. Oxides that are used include tin (VI) oxide and titanium (VI) oxide with an increasing trend to use nanostructured oxides, for example, SnO_2 nanobelts [8].

6.4.2 Conducting Polymer Sensors

Conducting polymers are popular in the development of gas- and liquid-phase sensors, with polypyrrole and polyaniline being the most widely used [3]. Common features of materials used to fabricate conducting polymers include the ability to make them through either chemical or electrochemical polymerization or the ability to change their conductivity through oxidation or reduction.

Persaud and Travers [9] have summarized the main reasons for the use of conducting polymers as odor-sensing devices as follows:

1. The sensors display rapid adsorption and desorption phenomena at room temperature.
2. Power consumption is low.
3. They are sensitive to humidity.
4. They are not easily inactivated by contaminants.
5. Specificity is achieved by modifying the structure of the polymer.

6.4.3 Acoustic Wave Sensors

Piezoelectric sensors are generally based on the use of AT-cut quartz crystals (+ 35°15' orientation of the plate relative to the crystal plane) because of their excellent temperature coefficients [3]. The generated acoustic wave depends on the crystal cut, thickness of the material used, and the geometry and configuration of the metal electrodes used to produce the electric field [5].

These devices are based on the variation of the fundamental oscillating frequency (Δf) of a very thin quartz crystal due to the adsorption of gas analyte molecules on its surface. This has the effect of changing the oscillating mass (Δm) and is based on the Sauerbrey equation [8]. However, there are deficiencies with this model especially if a liquid phase system is being investigated. This is because of numerous other interfacial factors that determine the behavior of the oscillating thickness shear mode (TSM) sensor, including hydrophobicity/hydrophilicity, slip, thickness of crystal, coupling, and roughness of the surface [10].

6.4.4 Newer Sensing Technologies

The use of polymer-coated cantilevers such as microfabricated beams of silicon is becoming more popular as the basis of nanomechanical sensors [11]. These devices detect physical and chemical interactions between the reactive layer on the surface and the environment [8]. When the polymer interacts with a gaseous species, it swells and causes the cantilever to bend as a result of surface stresses when used in the static mode. In the dynamic mode, the cantilever acts as a microbalance, which responds to changes in resonance frequency. Savran's group at Purdue University has been researching the micromechanical detection of proteins by use of aptamer-based receptor molecules [12].

Ion-mobility spectrometry (IMS) is also being applied to vapor-sensing devices. This has the ability to separate ionic species at atmospheric pressure although research

TABLE 6.2

Main Sensing Methods in Electronic Noses and Tongues [14]

Electronic Nose	Electronic Tongue
Conducting polymer	Potentiometric sensors
Metal oxide silicon field effect transistor (MOSFET)	Optical sensors
Surface acoustic wave	Biosensors
Optical sensors	Conductivity measurements
	Voltammetric measurements

is going on to develop low-pressure IMS devices to detect organic vapors in air. Analysis is based on analyte separations from ionic motilities rather than ionic masses. Advantages of detection at atmospheric pressure include lower power requirements, lighter weight, the possibility of smaller analytical units, and easier use [13]. Table 6.2 summarizes the main sensing methods applied to electronic noses and tongues.

6.5 SENSATIONS OF TASTE AND SMELL

6.5.1 TASTE

Taste is mainly a function of the taste buds, of which there are, approximately, 10,000 in the tongue with a few on the soft palate, inner surface of the cheek, pharynx, and epiglottis of the larynx [15]. There are five primary sensations of taste: sweet, sour, salty, bitter, and umami. Table 6.3 provides various examples of these.

Each of the basic taste sensations has a different threshold level, with the bitter taste having the lowest while sweet and salty are about the same. Table 6.4 gives further details of the threshold levels. The low level for bitterness is to be expected since this sensation provides a protective function [16].

Finally, it is important to note that there are hundreds of different types of tastes that are the result of varying degrees of stimulation of two, three, four, or five of the primary sensations at the same time.

TABLE 6.3

Taste Sensations [15,16]

Sensation	May Be Elicited By
Sweet	Sugars, glycols, alcohols, aldehydes, esters, amino acids
Sour	Acids, e.g., citric, acetic
Salty	Ionized salts, e.g., table salt
Bitter	Quinine, caffeine, aspirin, nicotine, strychnine
Umami	Monosodium glutamate (MSG), disodium inositate in meat and fish, disodium guanylate in mushrooms

TABLE 6.4

Threshold Levels for Basic Sensations [16]

Taste	Threshold Level
Sour (hydrochloric acid)	0.0009 M
Salty (sodium chloride)	0.01 M
Sweet (sucrose)	0.01 M
Bitter (quinine)	0.000008 M
Umami	Not applicable

6.5.2 SMELL

The receptor cells for the smell sensation are the olfactory cells, which are bipolar nerve cells, of which there are about 100 million in the olfactory epithelium [16]. Humans can recognize and discriminate volatile compounds with high sensitivity and accuracy. Some odors are detected at parts per trillion levels, with even stereo-isomers being differentiated [17].

Unlike taste, it has not been easy to classify smells into primary sensations. On the basis of psychological tests and action potential studies, seven different primary classes of olfactory stimulants have been identified [16]:

1. Camphoraceous
2. Musky
3. Floral
4. Pepperminty
5. Ethereal
6. Pungent
7. Putrid

However, researchers have suggested that the number can be as high as 50 different primary sensations.

Odorant molecules are generally light (molecular masses up to 300 Da), small, polar, and often hydrophobic [18].

6.6 DATA PROCESSING

Figure 6.2 shows the structure of a typical electronic nose/tongue.

The PR system allows the devices to be capable of recognizing simple or complex odors or tastes. PR techniques are used for data processing, and the data generated by each sensor are processed by a PR algorithm before the results are analyzed. Advantages of this approach include the following:

1. A reduction in complexity of the sensor coating.
2. The ability to characterize complex mixtures without the need to identify and quantify individual components.
3. It can be exploited for structure-activity studies [3].

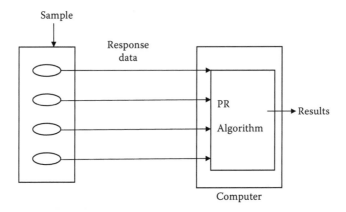

FIGURE 6.2   A typical electronic nose/tongue.

Several techniques are available, either separately or together, to perform PR after measurements are obtained. These include principal components analysis (PCA), cluster analysis (CA), and ANN.

6.6.1   PRINCIPAL COMPONENTS ANALYSIS

PCA is a method that is used to reduce the dimensionality of the parameter space, and it can also reveal the parameters that determine some inherent structure in the data, which may be interpreted in chemical terms [19]. Thus, PCA algorithms are used to project the data sets into two dimensions (termed the principal components), which allow minimum loss of information [8].

6.6.2   CLUSTER ANALYSIS

CA is similar to PCA and is based on the assumption of a close position of similar samples in multidimensional pattern space. Any similarity between two close samples is calculated as a function of the distance between them and displayed on a dendrogram.

6.6.3   ARTIFICIAL NEURAL NETWORK

ANNs are computer programs based on a simplified model of the brain, and they reproduce its logical operation using a collection of neuron-like entities to carry out processing [20]. These programs are multipurpose and, with suitable training, a single program can solve several problems.

Advantages of ANN include the following: it can handle noisy or missing data, no equations are involved, a network can deal with previously unseen data as soon as training has been completed, a large number of variables can be manipulated, and there is good accuracy [3].

There are three main stages involved in the development and use of ANNs [14]:

1. The learning stage: the number of neurons, layers, and type of architecture, transfer function, and algorithm are established, after which the network is allowed to achieve the desired outputs linked to an input.
2. The validation stage: this is achieved by the verification of the capability of the network by using different data from that utilized during the learning stage.
3. The production stage: the network has the ability to provide outputs corresponding to any input.

6.7 COMMERCIALLY AVAILABLE INSTRUMENTS

Many commercial instruments are now available, although the majorities are electronic noses. A growing number of companies are involved in the development of instruments and these include, but are not limited to [21], Agilent Technologies, Alpha M.O.S., Bloodhound Sensors, Cyrano Sciences Inc., Illumina Inc., Marconi Applied Technologies, Osmetech PLC, and Sensobi Sensoren GmbH.

A few typical instruments are briefly described, but the naming of a particular company does not imply any preference by the author for this instrument over any other (named or unnamed). These have been chosen merely to show the different transducer mechanisms employed.

Sensobi Sensoren GmbH has several new products available, each of which can be customized, if necessary. There are systems for continuous breath analysis (8–16 sensors), detection of volatile components in air (1–16 sensors), detection of organic traces in liquids (16 sensors), and for pressure measurements within narrow cavities for medical applications having <4 sensors [22].

The Bloodhound BH 114 instrument is based on conducting and semiconducting polymer sensors, and it has a sampling time of less than a minute [23]. The sensor arrays produce signals from volatile organic compounds and are stable and reversible, thus allowing the instrument to be used over long periods of time. In addition, a recalibration system is available and the instrument weighs less than 3 kg. Similarly, the Osmetech eNose is based on conducting polymer technology, whereby the total headspace of a sample is dynamically measured [24]. There is no surface contamination since the sensors do not come into direct contact with the sample.

Presearch is a British-based company, which offers a variety of instruments from different manufacturers. These include the Prometheus SAS-FMS system, which is billed as the first odor and volatile organic compounds (VOC) analyzer, which combines a sensitive fingerprint mass spectrometer (FMS) with a sensor array system (SAS). This system allows direct analysis of samples without pretreatment [25]. In addition, the company offers the Astree II e-tongue sensor array, which consists of seven sensors on a probe, which can be automatically lowered into a vessel containing the sample to be analyzed. The process is electrochemically based on the response of the sensors being dependent on the type and quantity of molecules present in the sample [26].

6.8 APPLICATIONS OF ELECTRONIC NOSES

6.8.1 FOOD AND BEVERAGE ANALYSES

Electronic noses are becoming more widely used in food and beverage analyses, especially in the areas of flavor, identification, and classification, and in various quality aspects. Some of these applications are discussed later.

6.8.1.1 Flavor and Aroma

Changes in aroma characteristics during the preparation of red ginseng were investigated by using an electronic nose containing metal oxide (MO) sensor arrays. Sensory evaluation and GC-MS were also utilized during the study [27]. It was reported that the MO sensor responses for fresh ginseng were higher than those for steamed ginseng and most responses for red ginseng were also higher than those for steamed ginseng using an electronic nose. The sensory panels found that fresh ginseng had strong, fresh, earthy, floral flavor notes while steamed ginseng had moderate, earthy flavor notes and red ginseng had strong fragrant and sweet notes. The sweet aroma compound from red ginseng was identified as 3-hydroxy-2-methylpyran-4-one.

In an interesting study, Italian researchers have studied the aging of white truffle with solid-phase microextraction (SPME) GC-MS and an electronic nose [28]. The changes in the aromatic compounds in the headspace of white truffles after storage at 4°C for a few days were monitored. The electronic nose showed a high sensitivity for the aroma compounds and is also able to detect gases from the truffles when up to 10 mg are used. Moreover, the results from both methods show strong correlation for truffle aging monitoring.

Characterization of alcoholic beverages by electronic noses is difficult due to the masking effect of ethanol. However, Ragazzo-Sanchez and coworkers [29] have proposed the use of back-flush GC as a tool for the pretreatment of vapor samples before analysis. The dehydration and dealcoholization step is conducted in parallel with electronic nose detection so that analysis time is reduced. This approach allowed the detection of five molecules responsible for off-flavors in wine. PCA allowed for easy discrimination between controls and off-flavor samples.

Electronic noses have also been used to analyze meat flavor. Warmed-over flavor (WOF) in beef was studied by 32 conducting polymer sensors [30]. The meat was processed by vacuum cook-in-bag/tray technology (VCT) and stored in a refrigerator. The VCT process involves treatment at 50°C for 390 minutes. The authors indicated that electronic noses can be readily utilized for WOF identification in beef and can be a useful addition to sensory analysis.

6.8.1.2 Identification and Classification

Walte and Munchmeyer [31] have described the analysis of alcoholic beverages by electronic nose technology. A selective enrichment procedure was performed prior to detection as this allows better correlation to human taste. A major problem with electronic noses is the drift in the signals. To compensate for this, a zero gas and a differential measuring technique were used. Another problem is that the composition of the headspace is monitored rather than the sample itself. The concentration of the headspace is related to the vapor pressure and to the liquid-phase concentration of the substance. Thus, more volatile compounds are present in greater quantities in the headspace and may not truly represent the sample itself [3].

Many methods have been used for the determination of the floral and geographical origin of honey, and these include analysis of pollen content, sensory analysis, volatile compounds, phenols, and markers [32]. Some of these, however, are time and labor intensive and/or require technical personnel. With this in mind, Benedetti and colleagues have used an electronic nose equipped with 10 MOSFET's and 12 MOS sensors to generate a pattern of the volatile compounds present in 70 honey samples [32]. The responses were evaluated by PCA and ANN and it was claimed that good results were obtained and that the electronic nose was useful for the characterization of honey. In addition, the technique does not require isolation of the volatile components.

An electronic nose has been used successfully to discriminate among different *Allium* (onion) species [33]. The commonly used tests (pyruvic acid content, thiosulphinate determination) have limitations, especially when large numbers of samples are to be analyzed. The electronic nose was employed to discriminate flavor characteristics of garlic, leek, shallot, bulb onion, and spring onion. PCA and other statistical evaluation suggested similarities in headspace volatiles for shallot, spring, and bulb onions and differences for leek and garlic.

6.8.1.3 Other Quality Aspects

A Bloodhound BH 114 electronic nose was used to differentiate between different types of bread spoilage [34]. Microbial and enzymatic spoilage caused by lipoxygenase can be differentiated from one another and from unspoilt bread after 48 hours by using cluster analysis. This is achieved before visible spoilage signs. Analysis of the bread samples with GC-MS identified the spoilage volatiles and the volatiles from the unspoilt bread.

Assessment of rancidity in potato crisps is usually done by the Rancimat or acid degree value tests, but these are labor-intensive. Electronic noses based on mass spectrometry or gas sensors were used to assess rancidity without oil extraction steps [35]. This removes the need for sample preparation, avoids the use of solvents, and reduces the analysis time from several hours to about 25 minutes. The nose based on MS had a 100% success rate in determining rancidity while that based on GC was 68%.

Labreche and coworkers have determined the shelf life of milk by storing products at ambient temperature and at a constant 5°C [36]. Samples were withdrawn at different times and analyzed by the electronic nose. It was reported that significant parameters were determined that corresponded to milk aging processes.

The ability of an electronic nose to assess the microbiological quality of raw poultry meat as a function of storage time and temperature was investigated [37]. Chicken pieces were stored for up to 2 days at 13°C or up to 5 days at 4°C. Saline rinses of meat samples and their dilutions were analyzed by an electronic nose containing 12 MO sensors. PCA confirmed that the electronic nose could differentiate volatile compounds associated with the meat samples stored at the two temperatures. This technology allows the determination of the length of the sample storage as well as deviation from refrigeration temperature.

6.8.2 Environmental Applications

British and Italian researchers have reported on the use of an electronic nose for the detection of moulds in libraries and archives [38]. The aim was to ascertain whether the device could be suitable for detecting mould activity on paper. It was found that it was possible to discriminate "in vitro" between affected and unaffected (by mould) paper samples at both 100% and 75% relative humidity by measuring the odor fingerprint. Three different species of actively growing fungi were detected and cluster analysis allowed differentiation between specific species. However, PCA indicated that only samples analyzed at 100% RH could be separated, suggesting that further research is required before electronic nose technology could be applied.

Canhato and Magan [39] have also described the detection of microbial and chemical contamination of potable water. Two electronic noses consisting of conducting polymer sensor arrays were compared for the early detection and discrimination of bacterial species, fungal spores, and trace concentrations of pesticides. Using PCA and CA allowed differentiation of the bacterial and fungal species after 24 hours of incubation at 25°C. However, this was not possible with the pesticides.

Portable electronic noses were used for outdoor air monitoring of sewage odors. One conducting polymer system and one MOS nose were used to compare results [40]. To improve classification ability, the effect of environmental parameters in different data preprocessing algorithms was evaluated. The algorithm that allowed the least humidity, temperature, and day correlation was selected. PCA was used to determine discriminating ability. The results indicate that MOS sensors were much better at discriminating the different odors when compared with the conducting polymer sensors. It was also shown that the location of samples for MOS sensors depended on the quality of water.

Zampolli and coworkers have developed a selective hybrid microsystem based on GC to enhance indoor air quality monitoring applications such as CO and NO_2 [41]. In the miniaturized GC, a solid-state sensor acts as the detector along with a silicon micromachined packed GC column, a zero grade air unit, a minipump, and a mini-valve. It was possible to detect benzene, toluene, and m-xylene in both synthetic and real indoor air down to 5 ppb.

Polypyrrole-based electronic noses have been used for the detection of both toxic and nontoxic substances in environmental analyses. Such substances include ammonia, oxides of nitrogen, carbon monoxide, phenol, and benzene. Applications have been in water, wastewater, and sewage. The interested readers can refer to the paper by Ameer and Adeloju, which gives a detailed overview of this area [42].

Finally, Figueiredo and Stentiford have evaluated nose technology for detecting the onset of anaerobic conditions during composting [43]. Closed reactors were set up in water baths in a laboratory to enable their operating temperature to be changed. In the report, the temperature was 40°C and the waste used was the vegetable fraction of food waste. In addition, each reactor was fitted with an aeration system for use when required. Headspace samples were analyzed and the odor was assessed during the biodegradation process. Oxygen was measured in the headspace gas and the redox potential was measured in the solid material. The reactors were operated aerobically and then the air was turned off, after which the changes in headspace gases and redox were monitored as the conditions turned anaerobic. It was reported that the electronic nose could detect an early transition from aerobic to anaerobic conditions but that further trials are required.

6.8.3 CLINICAL AND PHARMACEUTICAL APPLICATIONS

Tanaka and colleagues have used an electronic nose to assess oral malodor and to examine the association between oral malodor strength and oral health status [44]. Twenty-nine healthy adults and 49 patients were assessed by organoleptic tests, electronic nose, and volatile sulfur compound (VSC) concentrations. The authors concluded that the electronic nose provided objective halitosis-related measurements, but those obtained by a multiple linear regression method showed only relative but not absolute values. However, it is important to consider variables such as probing depth of gums, tongue coating, and levels of plaque for an informed judgment to be made. In a follow-up paper, the researchers set out to assess malodor expressed as an absolute value whereby 66 patients were evaluated. This time there was an association between percentage of teeth with pocket depth ≥4 mm, tongue coating score, and plaque levels with the electronic nose malodor intensity scores in multiple logistic regression analyses [45].

Breath analysis was also used to detect the presence of *Helicobacter pylori*, which has been implicated as a major cause of gastritis, peptic ulcer, and gastric cancer [46]. It was possible to predict the presence of the bacteria without the need for performing the standard ^{13}C-urea breath tests. Breath samples from 11 patients and 22 healthy volunteers were collected and immediately analyzed by the electronic nose consisting of eight thickness shear-mode sensors. Data analysis was by linear discriminate analysis (LDA), and it was found that 87.5% of the patients were correctly classified.

Headspace analysis was used to investigate urine samples from patients suffering from kidney diseases with some of the samples containing traces of blood [47]. The results indicated that it may be possible to distinguish the samples containing blood from the rest, and a linear correlation between the first three principal components and the blood content was found. In addition, when the electronic nose was used with a neural network, there was good performance in measuring pH and specific weight of the samples.

Flavors are widely used in pharmaceutical solutions to mask drug bitterness. Zhu's group [48] has used an MOS electronic nose to perform headspace analysis of these formulations. The method was able to qualitatively distinguish six common flavors (raspberry, red berry, strawberry, pineapple, orange, and cherry) in placebo mixtures. The instrument was also able to identify unknown flavors. It was also indicated that the instrument could be used to identify different flavor raw materials. Moreover, the electronic nose was used for quantitative analysis of flavors in an oral solution. Data processing and identification were done by PCA, discriminant factorial analysis (DFA), and partial least squares.

An electronic nose consisting of 32 polymer carbon black composite sensors was used to identify two species of *Staphylococcus aureus* bacteria [methicillin-resistant *S. aureus* (MRSA) and methicillin-susceptible *S. aureus* (MSSA)]. These bacteria are responsible for ear, nose, and throat infections. Swab samples were collected from patients and by PCA, Fuzzy C Means, and self-organizing map network analyses; it was possible to identify the bacteria subclasses with up to 99.69% accuracy. The authors, however, point out that this type of bacteria analysis is complex [49].

6.9   APPLICATIONS OF ELECTRONIC TONGUES

Electronic tongues are not as widely used as electronic noses, but their popularity is slowly increasing for a variety of applications.

6.9.1   FOOD AND BEVERAGE ANALYSIS

Voltammetric sensors based on chemically modified electrodes (conducting polymers, phthalocyanine complexes) with improved cross-selectivity were developed for the discrimination of bitter solutions [50]. The performance and capability were tested by using model solutions of bitterness such as magnesium chloride, quinine, and four phenolic compounds responsible for bitterness in olive oils. The sensors gave electrochemical responses when exposed to the solutions. A multichannel taste sensor was constructed using the sensors with the best stabilities and cross-selectivities and PCA of the signals allowed distinct discrimination of the solutions.

In recent work, this same research group has reported on the use of modified carbon paste electrodes for the discrimination of vegetable oils [51]. The oils were used as an electroactive binder material of carbon paste electrodes and the responses of these electrodes immersed in various solutions were used to discriminate the oils. The polyphenol content of olive oil allows it to be differentiated from sunflower or corn oils. In addition, the voltammograms are influenced by the pH and the nature of the ions present, resulting in characteristic signals for PCA; even olive oils of different quality (extra virgin, virgin, lampante, and refined) could be discriminated.

An electronic tongue based on polymeric membrane ion-selective electrodes was used for the qualitative analysis of different brands of orange juice, tonic, and milk [52]. The tests were performed by using products of the same brand but with different manufacture dates. PCA and back propagation neural network methods allowed 90%–100% discrimination of the various beverages. Similarly, beverages such as mineral water and apple juice were classified by the electronic tongue [53]. In this case, however, a reduced sensor array was used, which was efficient at discriminating the products being studied.

Electronic tongues have also been used to detect off-flavor and bitterness in beer [54]. In the former, low concentrations of dimethyl sulfide (0.01–0.25 mg/L), formed during worth production or by bacteria during fermentation, were easily detected. In the latter, bitterness is measured in terms of BU (calculated from UV analysis of isooctane at 275 nm). Usually, American beers range from 6–8 BU with European beers being much higher. A range of beers with 12–42 BU were prepared and analyzed. The r^2 value was better than 0.99, and no sample preparation was required.

Teas have been successfully discriminated by the use of electronic tongues based on voltammetric and potentiometric [55,56] principles. The voltammetric system used three different metallic working electrodes in tandem with a waveform to separate nine different teas. The waveforms utilized for this purpose were large amplitude pulse voltammetry (LAPV), small amplitude pulse voltammetry (SAPV), and staircase voltammetry. The best discrimination was obtained with a combination of LAPV and staircase [55]. The potentiometric tongue was composed of polymeric sensors based on PVC and aromatic polyurethane (ArPU) matrices doped with various membrane active components, such as polypyrrole and polyaniline. The system was able to discriminate between black and green teas and natural coffee. Individual components such as caffeine, catechines, and sugar were determined.

An electronic tongue based on dual shear horizontal surface acoustic wave (SH-SAW) devices was developed to discriminate between the basic tastes of sour, salt, bitter, and sweet [57]. Sixty MHz SH-SAW delay line sensors were fabricated and placed below a miniature PTFE housing containing the test liquid. All the tastes were correctly classified without the need for a selective biological or chemical coating.

The freshness of milk has been monitored by both voltammetric-based sensors [58] and by a disposable system [59]. The former consisted of a reference electrode, an auxiliary electrode, and five wires of different metals (gold, iridium, palladium, platinum, and rhodium) as working electrodes. Pulsed voltammetry was the method of detection before PCA was applied. The models were able to predict the course of bacterial growth (spoilage) in the milk. The latter involved the development of a disposable screen-printed multichannel electronic tongue composed of lipids as transducers. This device was used to measure the electrical potential resulting from the interaction of lipid membranes and samples. Both UHT (ultra-high temperature) and pasteurized milk were tested, and the taste sensor was able to discriminate between fresh and spoilt milk. Moreover, the deterioration of stored milk can be followed by PCA.

Finally, an electronic tongue based on 30 nonspecific potentiometric sensors was developed to qualitatively and quantitatively monitor a batch fermentation process of starting culture for light cheese production [60]. Process control charts were developed and these allowed discrimination of samples run under "abnormal" conditions from those of "normal" operating parameters. Average prediction errors were in the range 5%–13% based on test set validation. In addition, correlation between peptide profiles determined by high performance liquid chromatography (HPLC) and the electronic tongue was established. The authors indicate that this technology is promising for fermentation process monitoring and for the quantitative analysis of growth media.

6.9.2 ENVIRONMENTAL USES

An electronic tongue consisting of potentiometric sensors was developed for the qualitative analysis of natural waters [61]. An array containing a series of electrodes (RuO_2, C, Ag, Ni, Cu, Au, Pt, Al, Sn, Pb, and graphite) was used to investigate seven samples of mineral water, tap water, and "osmotized" water. Qualitative analysis of the samples of water was done by fuzzy neural networks with a differentiation accuracy of 93%.

A very interesting application is the supervision of rinses in a washing machine [62]. A voltammetric electronic tongue was used to apply a potential pulse train over two electrodes to measure the current. The rinses from 20 machine wash runs were investigated, and PCA and soft-independent modeling of class analogy (SIMCA) were used to classify the rinses. With PCA, only one rinse was wrongly classified while SIMCA produced no wrong classifications, although 38% of the rinses were assigned to more than one class.

A flow-injection electronic tongue based on potentiometric sensors was used to determine nitrate in the presence of chloride [63]. Four sensors were employed, and with data treatment it was possible to quantify the concentration of nitrate between 0.1 and 100 mg/L. This was achieved without the need to eliminate chloride ions. The authors reported that this approach is more sensitive than the direct determination of nitrate using ion-selective electrodes.

An integrated electronic tongue consisting of a multiple light-addressable potentiometric sensor (MLAPS) and two sets of electrochemical electrodes was developed to simultaneously detect Fe (III) and Cr (VI) ions and other heavy metals, respectively [64]. The MLAPS was based on chalcogenide thin film while the electrochemical electrodes used stripping voltammetry. It has been suggested by the authors that these methods are suitable for the detection of heavy metals in wastewater or seawater.

6.9.3 CLINICAL AND PHARMACEUTICAL APPLICATIONS

An electronic tongue based on pulsed voltammetry was used to detect six microbial species (1 yeast, 2 bacteria, and 3 molds). The electrode array was dipped into malt extract growth medium and a voltage applied over the electrodes in pulses of different amplitude [65]. The resultant current data were evaluated with PCA and SIMCA. PCA was performed on all growth stages (lag, logarithmic, and stationary) but no recognition of species was obtained in the lag phase. However, species were recognized in the logarithmic and stationary phases. SIMCA was able to predict species to the correct classes.

A microcontroller-based electronic tongue capable of discriminating herbal remedies containing *Eurycoma longifolia* has been described [66]. The liquid sample is sensed by disposable screen-printed lipid-membrane sensors and classified by ANN. It was claimed that excellent recognition results were obtained and that the system was able to discriminate between samples containing the active herbal ingredient and those which did not.

A quantitative approach to optimize taste masking in a lyophilized orally disintegrating tablet has been described [67]. Lyophilized tablets disintegrate rapidly when placed on the tongue and organoleptic properties—taste, mouth feel, and appearance—are critical to the success of a product. The tongue was made of seven sensors to detect dissolved organic and inorganic compounds, and measurement was potentiometric with readings taken against an Ag/AgCl reference electrode. PCA and SIMCA allowed comparisons of bitterness and discrimination among differences in the formulation to be made. Similarly, measurement of bitterness masking efficiency was performed with an Astree electronic tongue, and the results indicate that this system can be used to evaluate the range in which masking agents need to be added [68].

6.10 INTEGRATION OF ELECTRONIC NOSES AND TONGUES

In recent years, there have been several attempts to integrate electronic noses and tongues to improve discriminating abilities. Winquist et al. described such a combination where the nose consisted of an array of gas sensors with different selectivity patterns while the tongue was based on pulsed voltammetry [69]. PCA allowed good discrimination between samples such that classification properties were improved with the combined system.

DiNatale and coworkers have used a combined nose/tongue system to extract chemical information from liquid samples through the analysis of the solution and its headspace [70]. Both devices were based on the use of metalloporphyrins, which allowed uniformity of results. This system was tested for both food and clinical analyses, and there was an increase in the amount of information obtained compared with the individual devices.

Finally, Taguchi's group developed an integrated system for the detection of gases and volatile liquids [71]. The detection was based on changes in electrical resistance, which occurs when polymer-coated microelectrodes were exposed to the different samples. In this investigation, pH and sodium chloride were detected, and it is claimed that this approach may even be used to detect color for display device applications. This may lead to further development of an electronic eye.

6.11 CONCLUDING REMARKS

This chapter discussed the applications of electronic noses and tongues in such areas as food and beverage analysis, environmental applications, and clinical and pharmaceutical uses. Each technique still has many developments to undergo, but the advantages have already been established. For the nose, these include high sensitivity and correlation to human sensory panels [72]. Similarly, the tongue can also reproduce the taste sense of humans to a high degree [73].

However, several problems still exist. These include sensor drift, which leads to the inability to provide proper calibration. This is of special concern to quality control laboratories and is one of the reasons for the general absence of these instruments in these laboratories [3]. Limitations to the use of the electronic nose include loss of sensitivity in the presence of water vapor and high concentrations of individual components such as alcohol, relatively short life of some sensors, and the inability to obtain quantitative data for aroma differences [72]. Each device also still needs considerable method development, but progress is being made at a rapid rate. Finally, sensor arrays and PR tend to predict the quality of a sample without providing hard data with respect to composition and concentration [74].

ACKNOWLEDGMENTS

I would like to thank Ms. Satie Siewah for her assistance with providing the figures.

REFERENCES

1. Stetter J. R. and Penrose W. R., Understanding chemical sensors and chemical sensor arrays (electronic noses): past, present and future, *Sensors Update*, 10, 189, 2002.
2. Giese J., Electronic noses, *Food Technol.* 54(3), 96–98, 2000.
3. Deisingh A. K., Stone D. C., and Thompson M., Applications of electronic noses and tongues in food analysis, *Int. J. Food Sci. Technol.*, 39, 587–604, 2004.
4. Cattrall R. W., *Chemical Sensors* (Oxford: Oxford University Press, 1997, 1–2).
5. Thompson M. and Stone D. C., *Surface-launched Acoustic Wave Sensors* (New York: Wiley-Interscience, 1997, 15–25).
6. Hall E. A. H., *Biosensors* (Milton Keynes: Open University Press, 1990).
7. Rivera C. T., The nose knows, *Med Hunters*, www.medhunters.com/articles/theNose Knows.html (accessed April 29, 2005).
8. Gouma P. and Sberveglieri G., Novel materials and applications of electronic noses and tongues, *MRS Bull.*, 29(10), 697–702, 2004.
9. Persaud K. C. and Travers P. J., Arrays of broad specificity films for sensing volatile chemicals, in *Handbook of Biosensors and Electronic Noses: Medicine, Food and Environment,* ed. E. Kress-Rogers (Boca Raton: CRC Press, 1997, 563–589).
10. Thompson M., Kipling A. L., Duncan-Hewitt W. C., Rajakovic L. V., and Cavic B. A., Thickness shear-mode acoustic wave sensors in the liquid phase, *Analyst*, 116, 881–890, 1991.
11. Lang H. P., Hegner M., Meyer E., and Gerber C., Nanomechanics from atomic resolution to molecular recognition based on atomic force microscopy technology, *Nanotechnol.*, 13, R29–R36, 2002.
12. Savran C. A., Knudsen S. M., Ellington A. D., and Manalis S. R., Micromechanical detection of proteins using aptamer-based receptor molecules, *Anal. Chem.*, 76, 3194–3198, 2004.
13. Graseby Ionics http://www.graseby.com (accessed April 20, 2003).
14. Ciosek P., Electronic tongue/electronic nose, http://csrg.ch.pw.edu.pl/prepares/pciosek/etong.html (accessed April 29, 2005).
15. Marieb E., The special senses, in *Human Anatomy and Physiology*, 4th ed. (Menlo Park, CA: Benjamin Cummings, 1998, 537–540).
16. Guyton A. C., The chemical senses—taste and smell, in *Textbook of Medical Physiology*, 7th ed. (Philadelphia, PA: WB Saunders, 1986).
17. Breer H., Sense of smell: Signal recognition and transduction in olfactory receptor neurons, in *Handbook of Biosensors and Electronic Noses: Medicine, Food and Environment,* ed. E. Kress-Rogers (Boca Raton, FL: CRC Press, 1997, 521–532).
18. Craven M. A., Gardner J. W., and Bartlett P. N., Electronic noses-development and future prospects, *Trends Anal. Chem.*, 15, 486, 1996.
19. Adams M. J., *Chemometrics in Analytical Spectroscopy* (Cambridge, MA: Royal Society of Chemistry, 1995, 70–71).
20. Cartwright H. M., *Applications of Artificial Intelligence in Chemistry* (Oxford: Oxford Science Publications, 1993).
21. http://www.nose-network.org/review/commercial (accessed June 14, 2005).
22. http://www.nose-network.org/review/commercial/sensobi.asp (accessed June 14, 2005).
23. Gibson T. D., Hubert J. N., Prosser O. C., and Pavlou A. K., Not to be sniffed at, *Microbiology Today*, 27, 14, 2000.
24. http://www.osmetech.com/enose.htm (accessed June 14, 2005).
25. http://www.presearch.co.uk/pages/products/show-product.asp?id=1118 (accessed June 14, 2005).

26. http://www.presearch.co.uk/pages/products/show-product.asp?id=1119 (accessed June 14, 2005).

27. Lee S. K., Kim J. H., Sohn H. J., and Yang J. W., Changes in aroma characteristics during the preparation of red ginseng estimated by electronic nose, sensory evaluation and gas chromatography/mass spectrometry, *Sens. Actuators B*, 106(1), 7, 2005.

28. Falasconi M., Pardo M., Sberveglieri G., Battistutta F., Piloni M., and Zironi R., Study of white truffle aging with SPME-GC-MS and the Pico 2-electronic nose, *Sens. Actuators B*, 106(1), 88, 2005.

29. Ragazzo-Sanchez J. A., Chalier P., and Ghommidh C., Coupling gas chromatography and electronic nose for dehydration and desalcoholization of alcoholized beverages: Application to off-flavor detection in wine, *Sens. Actuators B*, 106(1), 253, 2005.

30. Grigioni G. M., Margaria C. A., Pensel N. A., Sanchex G., and Vaudagma S. R., Warmed-over flavour analysis in low-temperature-long time processed meat by an electronic nose, *Meat Science*, 56, 221, 2000.

31. Walte A. and Munchmeyer W., Novel electronic nose for the analysis of alcoholic beverages, in *Fontiers of Flavor Science*, ed. P. Schieberle and E. K.-H. Garching (Germany: Deutsche Forschungsanstalt fur Lebensmittelchemie, 2000, 144–147).

32. Benedetti S., Mannino S., Sabatini A. G., and Marcazzan G. L., Electronic nose and neural network use for the classification of honey, *Apidologie*, 35, 1, 2004.

33. Abbey L., Aked J., and Joyce D. C., Discrimination amongst *Alliums* using an electronic nose, *Ann. Appl. Biol.*, 139(3), 337, 2001.

34. Needham R., Williams J., Beales N., Voysey P., and Magan N., Early detection and differentiation of spoilage of bakery products, *Sens. Actuators B*, 106(1), 20, 2005.

35. Vinaixa M., Vergara A., Duran C., Llobet E., Badia C., Brezmes J., Vilanova X., and Correig X., Fast detection of rancidity in potato chips using e-noses based on mass spectrometry or gas sensors, *Sens. Actuators B*, 106(1), 67, 2005.

36. Labreche S., Bazzo S., Cade S., and Chanie E., Shelf life determination by electronic nose: Application to milk, *Sens. Actuators B*, 106(1), 199, 2005.

37. Boothe D. D. H. and Arnold J. W., Electronic nose analysis of volatile compounds from poultry meat samples, fresh and after refrigerated storage, *J. Sci. Food Agric.*, 82(3), 315, 2002.

38. Pinzari F., Fanelli C., Canhato O., and Magan N., Electronic nose for the early detection of moulds in libraries and archives, *Indoor Built Environ.*, 13(5), 387, 2004.

39. Canhato O. and Magan N., Electronic nose technology for the detection of microbial and chemical contamination of potable water, *Sens. Actuators B*, 106(1), 3, 2005.

40. Nake A., Dubreuil B., Raynaud C., and Talou T., Outdoor in situ monitoring of volatile emissions from wastewater treatment plants with two portable technologies of electronic noses, *Sens. Actuators B*, 106(1), 36, 2005.

41. Zampolli S., Elmi I., Sturmann J., Nicoletti S., Dori L., and Cardinali G. C., Selectivity enhancement of metal oxide gas sensors using a micromachined gas chromatographic column, *Sens. Actuators B*, 105(2), 400, 2005.

42. Ameer Q. and Adeloju S. B., Polypyrrole-based electronic noses for environmental and industrial analysis, *Sens. Actuators B*, 106(2), 541, 2005.

43. Figueiredo S. A. B. and Stentiford E. I., Evaluating the potential of an electronic nose for detecting the onset of anaerobic conditions during composting, *Bioproc. Solid Waste Sludge*, 2(1), 1, 2002.

44. Tanaka M., Anguri H., Nonaka A., Kataoka K., Nagata H., Kita J., and Shizukuishi S., Clinical assessment of oral malodor by the electronic nose system, *J. Dent. Res.*, 83(4), 317, 2004.

45. Nonaka A., Tanaka M., Anguri H., Nagata H., Kita J., and Shizukuishi S., Clinical assessment of oral malodor intensity expressed as absolute value using an electronic nose, *Oral Dise.*, 11, 35, 2005.

46. Romano M., Scarpa A., Sinopoli S., Amarri S., and Macagnano A., *Helicobacter pylori* identification by the analysis of breath with an electronic nose, http://www.technobiochip.com/www_en/a3pubb.htm. *ISOEN 2002*, Ninth International Symposium on Olfaction and Electronic Noses, Rome (Italy), September 29–October 2.

47. Di Natale C., Mantini A., Macagnano A., Antuzzi D., Paolesse R., and D'Amico A., Electronic nose analysis of urine samples containing blood, *Physiol. Meas.*, 20, 377, 1999.

48. Zhu L., Seburg R. A., Tsai E., Puech S., and Mifsud J.-C., Flavor analysis in a pharmaceutical oral solution formulation using an electronic nose, *J. Pharm. Biomed. Analysis*, 34, 453, 2004.

49. Dutta R., Morgan D., Baker N., Gardner J. W., and Hines E., Identification of *Staphyloccous aureus* infections in hospital environment: electronic nose-based approach, *Sens. Actuators B*, 109(2), 355, 2005.

50. Apetrei C., Rodriguex-Mendez M. L., Parra V., Gutierrez F., and de Saja J. A., Arrays of voltammetric sensors for the discrimination of bitter solutions, *Sens. Actuators B*, 103(1–2), 145, 2004.

51. Apetrei C., Rodriguez-Mendez M. L., and de Saja J. A., Modified carbon paste electrodes for discrimination of vegetable oils, *Sens. Actuators B*, doi: 10.1016/j.snb.2005.03.041

52. Ciosek P., Augustyniak E., and Wroblewski W., Polymeric membrane ion-selective and cross-sensitive electrode-based electronic tongue for qualitative analysis of beverages, *Analyst*, 129, 639, 2004.

53. Ciosek P., Brzozka Z., and Wroblewski W., Classification of beverages using a reduced sensor array, *Sens. Actuators B*, 103(1–2), 76, 2004.

54. Tan T., Schmitt V., and Isz S., Electronic tongue: A new dimension in sensory analysis, *Food Technol.*, 55(10), 44, 2001.

55. Ivarsson P., Holmin S., Hojer N. E., Krantz-Rulcker C., and Winquist F., Discrimination of tea by means of a voltammetric electronic tongue and different applied waveforms, *Sens. Actuators B*, 76(1–3), 449, 2001.

56. Lvova L., Legin A., Vlasov Y., Cha G. S., and Nam H., Multicomponent analysis of Korean green tea by means of disposable all-solid-state potentiometric electronic tongue microsystem, *Sens. Actuators B*, 95(1–3), 391, 2003.

57. Sehra G., Cole M., and Gardner J. W., Miniature taste sensing system based on dual SH-SAW sensor device: An electronic tongue, *Sens. Actuators B*, 103(1–2), 233, 2004.

58. Winquist F., Krantz-Rulcker C., Wide P., and Lundstrom I., Monitoring of freshness of milk by an electronic tongue on the basis of voltammetry, *Meas. Sci. Technol.*, 9, 1937, 1998.

59. Sim M. Y. M., Shya T. J., Ahmad M. N., Shakaff A. Y. M., Othman A. R., and Hitam M. S., Monitoring of milk quality with disposable taste sensor, *Sensors*, 3, 340, 2003.

60. Esbensen K., Kirsanov D., Legin A., Rudnitskaya A., Mortensen J., Pedersen J., Vognsen L., Makarychev-Mikhailov S., and Vlasov Y., Fermentation monitoring using multisensor systems: feasibility study of the electronic tongue, *Anal. Bioanal. Chem.*, 378, 391, 2004.

61. Martinez-Manez R., Soto J., Garcia-Breijo E., Gil L., Ibanez J., and Llobet E., An "electronic tongue" design for the qualitative analysis of natural waters, *Sens. Actuators B*, 104(2), 302, 2005.

62. Ivarsson P., Johansson M., Hojer N.-E., Krantz-Rulcker C., Winquist F., and Lundstrom I., Supervision of rinses in a washing machine by a voltammetric electronic tongue, *Sens. Actuators B*, 108(1–2), 851, 2005.

63. Gallardo J., Alegret S., and del Valle M. A., Flow-injection electronic tongue based on potentiometric sensors for the determination of nitrate in the presence of chloride, *Sens. Act. B: Chemical*, 101(1–2), 72, 2004.

64. Men H., Zou S. F., Li Y., Wang Y., Ye X., and Wang P., A novel electronic tongue combined MLAPS with stripping voltammetry for environmental detection, *Sens. Actuators B*, 110(2), 350, 2005.

65. Soderstrom C., Winquist F., and Krantz-Rulcker C., Recognition of six microbial species with an electronic tongue, *Sens. Actuators B*, 89(3), 248, 2003.

66. Abdul Rahman A. S., Sim Yap M. M., Shakaff A. Y., Ahmad M. N., Dahari Z., Ismail Z., and Hitam M. S., A microcontroller-based taste sensing system for the verification of *Eurycoma longifolia*, *Sens. Actuators B*, 101(1–2), 191, 2004.

67. Murray O. J., Dang W., and Bergstrom D., Using an electronic tongue to optimize taste-masking in a lyophilized orally disintegrating tablet formulation, *Pharmaceutical Technology*, http://www.cardinal.com/pts/content/aboutus/whoweare/broch/CH-PJS-E-Tongue.pdf (accessed June 28, 2005).

68. Pharmaceutical application note # 02, Measurement of the bitterness masking efficiency, http://www.presearch.co.uk (accessed June 28, 2005).

69. Winquist F., Lundstrom I., and Wide P., The combination of an electronic tongue and an electronic nose, *Sens. Actuators B*, 58(1–3), 512, 1999.

70. Di Natale C., Paolesse R., Macagnano A., Mantini A., D'Amico A., Legin A., Lvova A., Rudnitskaya A., and Vlasov Y., Electronic nose and electronic tongue integration for improved classification of clinical and food samples, *Sens. Actuators B*, 64(1–3), 15, 2000.

71. Talaie A., Lee J. Y., Eisazadeh H., Adachi K., Romagnoli J. Y., and Taguchi T., Towards a conducting polymer-based electronic nose and tongue, *Iranian Polymer Journal*, 9, 3, 2000.

72. Harper W. J., The strengths and weaknesses of the electronic nose, *Adv. Exp. Med. Biol.*, 488, 59, 2001.

73. Toko K., Electronic tongue, *Biosens. Bioelectron.*, 13, 701, 1998.

74. Krantz-Rulcker C., Stenberg M., Winquist F., and Lundstrom I., Electronic tongues for environmental monitoring based on sensor arrays and pattern recognition: A review, *Anal. Chim. Acta.*, 426, 217, 2001.

7 Applications of Sensors in Food and Environmental Analysis

Richard O'Kennedy, W. J. J. Finlay, Paul Leonard
Applied Biochemistry Group, School of Biotechnology
Biomedical Diagnostics Institute
National Centre for Sensor Research
Dublin City University, Dublin, Ireland

Stephen Hearty
Applied Biochemistry Group, School of Biotechnology
National Centre for Sensor Research
Dublin City University, Dublin, Ireland

Joanne Brennan, Sharon Stapleton,
Susan Townsend, Alfredo Darmaninsheehan
Applied Biochemistry Group, School of Biotechnology
Cambridge Antibody Technology
Cambridge, U.K.

Andrew Baxter, Claire Jones
Xenosense
Belfast, N. Ireland

CONTENTS

7.1   INTRODUCTION

The basic principal of any sensor system is illustrated in Figure 7.1. It consists of a sensing element that detects the target analyte. This detection event must then be transduced into a signal that can be recorded in a readout device. The sensing element can consist of a variety of different binding ligands, as summarized in Table 7.1. The term "biosensor" covers a very wide range of technologies, but biosensors may be divided into several groups, depending on the method of signal transduction [1–4]. Technologies that are currently utilized are listed in Table 7.2. Sensors, and biosensors in particular, have a number of advantages and disadvantages when applied to the analysis of food and environmental samples (Table 7.3).

Current studies focusing on biosensing methods aim to achieve more sensitive, robust, reproducible, user-friendly, and portable analytical systems. To achieve these goals almost every biosensing format has utilized immunoglobulins (antibodies) as the molecular recognition element, leading to the adoption of the term "immunobiosensor."

Antibodies are the key recognition elements of the immune system and various antibodies and antibody-derived fragments have been produced, with the capability to detect a remarkable variety of analytes [5]. Antibody structures are

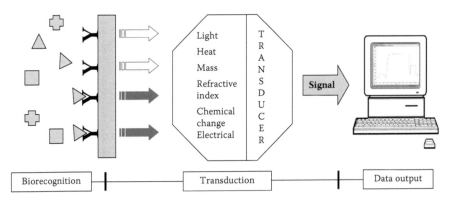

FIGURE 7.1   Basic concept of a biosensor illustrating analyte recognition followed by transduction and signal generation.

TABLE 7.1
Sensing/Biorecognition Elements Commonly Used in Sensor Systems

	Sensing/Biorecognition Element
1	Antibodies and antibody derivatives, e.g., complete, biospecific, fragmented, or genetically engineered antibodies (Fab fragment, scFv, or diabody)
2	Enzymes, e.g., an enzyme specific for particular analyte (glucose oxidase, specific for glucose)
3	Receptor, e.g., a component from the antennae of an insect or a cell membrane receptor
4	A living cell that reacts to a particular stimulus or that registers the presence of a toxic material
5	Nucleic acid-based probe, e.g., DNA or RNA, or derivatives such as a peptide nucleic acids and aptamers
6	Plastibody or chemically generated recognition surface, e.g., "antibody-like" molecularly imprinted polymer

TABLE 7.2
Examples of Transducer Systems

Type	Example
Electrochemical	Conductimetric
	Potentiometric
	Voltametric
FET-based	Field effect transistor
Optical	Surface plasmon resonance (SPR)
Thermal	Calorimetry
Surface acoustic wave	Rayleigh surface waves
Piezoelectric	Electrochemical quartz crystal microbalance

illustrated in Figure 7.2, which shows the structure of a typical antibody molecule and the various fragments (Fab, scFv, diabody, etc.) that are commonly genetically engineered. The use of immunoglobulins for analyte detection began with simple immunochemical techniques such as radial immunodiffusion (RID) and was followed by faster, more sensitive techniques such as radio immunoassay (RIA), enzyme-linked immunosorbent assay (ELISA), and fluorescence-based ELISA. More recent technologies include both "real-time" methods and higher density multiplexed "arrays." However, despite decades of intensive research, the development of optimal antibodies for each highly specialized function is still laborious, difficult, and has many potential pitfalls. Several aspects of antibody generation have been greatly accelerated and expanded in scope by the advent of recombinant antibody technology.

TABLE 7.3

Advantages and Disadvantages of Sensor-Based Systems for Use in Food and Environmental Analysis

Advantages	Disadvantages
Capacity for "real-time" measurement. This may be of major importance in "large-scale" production or water monitoring	High costs of instrumentation, maintenance, and consumables may be problematic
Specificity can be tailored to requirements by correct choice of biorecognition entity	Tests need to be developed for many relevant analytes
High sensitivity even in complex matrices, e.g., food	Sample pretreatment and analyte extraction often necessary
Multiple recognition components available (Table 7.1)	New methodologies for sample handling and processing need to be developed and optimized. This may involve use of solvents, antibody, or other extraction columns
Biosensor can perform single independent measurements with "throw-away" element or may be regenerable for multiple analyses, thus greatly reducing cost	
Capacity for "online" and "at-site" measurement	
Capacity for continuous measurement with wireless linked reporting of results	Microbial detection often requires use of growth-enrichment media before sufficient numbers are reached to allow accurate detection
Multiple different analyte determinations on a single sensor chip feasible and major advantage for future	Some sensor systems are large, complex, and expensive
Ease of use based on simplicity of design or due to user-friendly computer-based control	
Relatively robust	
Capability of incorporation into robotic highly controlled formats with high throughput	
Ideal systems for use as a first screening for rapid detection of contaminants or conditions which can subsequently be confirmed using more in-depth detection with conventional established methodology, for example, LC/MS	
Have major potential to revolutionize many aspects of food/water quality control	

7.2 MEASURING MULTIPLE ANALYTES IN COMPLEX MATRICES

Initially, one of the main areas of activity in biosensor research was the development of sensors designed for use in diagnostics, for example, enzyme-based systems used in glucose measurement for diabetics. A key goal in the development of diagnostic immunobiosensors is the sensitive quantification/recognition of proteins and small molecules (or haptens), with low molecular weights of less than 1,000 Da, across very broad dynamic ranges of concentration (potentially differing by as much as 10^4). The difficulty in reaching this goal is exacerbated by the need for highly specific identification of single biomarkers in extremely complex clinical samples. Exactly the same problems occur in complex food matrices. Indeed, given the complexity of food matrices and also those associated with environmental and water analysis; we can expect even greater problems with interferants and nonspecific effects. Clinical analyses are generally confined to relatively well-defined volumes, for example, a urine,

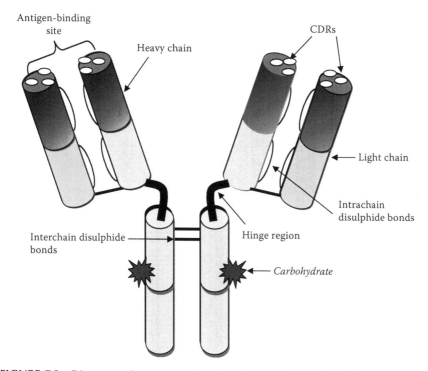

FIGURE 7.2 Diagrammatic representation of an immunoglobulin G (IgG) molecule. The IgG molecule is composed of two identical light chains and two identical heavy chains. The light chains are composed of a variable (V_L) and constant (C_L) domain. The heavy chain consists of one variable (V_H) and three constant (C_{H1}, C_{H2}, and C_{H3}) domains with a hinge region connecting the C_{H1} and C_{H2} regions. The heavy and light chains are connected via disulphide bonds; disulphide bonds are also present in the constant and variable regions. The complementarity-determining regions (CDRs) at the amino terminal of the variable domains confer antigenic specificity and contain considerable amino acid sequence variation.

blood, or salivary sample from a specific individual or animal, this is certainly not the case with large food volumes/environmental or oceanic samples, for example, a harvested crop, a food processing run, or a large body of water. In many cases, some sample cleanup procedures may be necessary to extract (and often also to concentrate) the proposed analyte so that it can be accurately determined. In these diverse application areas, the antibody–antigen interaction is potentially the most generally applicable method for highly specific molecular recognition. However, the optimal antibody formats for immunobiosensor applications are yet to be fully defined. What is clear, however, is that antibodies which are "monoclonal" in nature, that is, mono-specific in nature (hybridoma monoclonals, single-clone recombinant antibodies), may be preferable to traditionally produced polyclonal antibodies which are inherently polyspecific and heterogeneous.

Food related, environmental, and clinical immunobiosensors constitute the most challenging areas in biosensor research to date. First, to achieve the ideal of a multiplexed system in which one can profile several biomarkers simultaneously in a highly complex matrix, such as serum or food, requires highly defined recognition agents (typically antibodies) which have been extensively characterized and have no cross-reactivity with any of the closely related (possibly numerous), but irrelevant molecules which may also be present. Second, for the immunobiosensor to be automated for more rapid throughput, microfluidic systems must be specially designed to take into account the viscosity and so forth, of the analyte matrix. Third, the detection method must be of sufficient sensitivity, reproducibility, and linearity to produce accurate quantification of all the analytes examined. Currently, there are a huge number of actual and potential areas for the application of sensors in food and environmental analysis and these are listed in Tables 7.4 and 7.5.

TABLE 7.4
Application Areas for Sensors in Food and Environmental Analysis

Target	Examples	Area
Pathogens	*Clostridium botulinum*	Food, water, etc.
	Staphylococcus aureus	
	Bacillus cereus	
	E. coli	
	Salmonella spp.	
	Listeria monocytogenes	
	Shigella spp.	
Spoilage indicators	Lactic acid bacteria	Food, milk, silage
Viruses	Norovirus	Food (causes gastroenteritis)
Marine toxins	Azaspiracid	Water
	Domoic acid	Shellfish
	Brevetoxin	
	Tetrodoxin	
Phenols	Nonyphenols	Shellfish, fruit, wine, cereals, milk
Mycotoxins	Fumonisin A	Foodstuffs
	Ochratoxin	Infant foods
	Deoxynivalenol	Infant formula
	Aflatoxins, especially B1	
Biogenic Amines	Histamine	Foods, e.g., cheese, fish, beer,
	Tyramine	some fermented foods, and
	Putrescine	red wine
	Cadaverine	Spoiled foods
Prions	BSE	Meat, meat products, and gelatin
Antibiotics	Beta lactams, e.g., ampicillin	Milk, chickens, honey
Dioxins	Polychlorinated biphenyls (PCB's)	Foods
Allergens	Very varied, e.g., lactose, gluten	Cereals, milk, nuts, fish, seeds, gums
Citrate	Citrate	Fruits, vegetables, processed foods

Target	Examples	Area
Sugars	Glucose	Raw materials for foods, drinks
	Sucrose	Detection of silage effluents
	Lactose	
	Pyruvate	
Estrogenic compounds	Estrone	Effluents
Sweeteners	Aspartamate	Beverages, dairy products
Organic Residues	Benzene	Soft drinks
		Effluents
Veterinary Drug	Group A	
Residues	Substances causing anabolic effects,	Meat
	e.g., stilbenes, steroids, beta agonists	Meat products
	Group B	
	Veterinary Drugs and contaminants,	Meat
	e.g., sulphonamides, antihelminthics,	Meat products
	coccidiostats, nonsteroid	
	anti-inflammatory drugs, levamisole	
Pesticides	Organophosphates	
	Malathion	Fruit, vegetables
	Fenitrothion	Cereals
	Pirimiphos-methyl, Chloropyrifos	Meats
	Pyrethroids	Fruit
	Carbamates	
	Carbofurans	Fruit
	Triazines	Fruit, water
	Atrazine	
	Simazine	
Coumarins	Coumarin	Vanilla extracts
	Warfarin	Foods
Vitamins	B12	Foods
	Folic acid	Infant formula and
	Riboflavin	supplements
	Vitamin D	
GMO Status	Genetically modified material	Foods, e.g., meats, fruits,
		cereals, environment
Gas levels	CO_2	Foods
	O_2	Food storage
	Ethylene	Packaging
		Water
		Environment
Temperature	Inappropriate temperatures	Fruit storage
		Meat storage
		Transportation
		Production of
	Required processing temperatures	Beverages
		Food fermentation
		Waste treatments

TABLE 7.5

Relative Advantages and Disadvantages of Different Antibody Preparations

Characteristics	Polyclonal	Monoclonal	Recombinant
Relative ease of production	++++	+++	+++
Economy of production	++	+++	++++
Stability for use	++++	+++	+++
Ease of immobilization	++++	++++	+++++
Potential sensitivity	++++	++++	++++++
Ease of labeling	++++	+++	+++++
Ease of incorporation into biosensors, biochips, and arrays	++	+++	++++
Capacity for improvement/refinement of specificity and affinity	–	–	++++

7.3   PATHOGENIC MICROBES AND TOXINS

Pathogenic microbes and toxins are of considerable interest in biosensor develop-ment. Heightened awareness of food poisoning risk and also the possibility of bioter-rorist attacks have led to a requirement for high-speed portable sensors for microbial detection. While traditional antibody-based and nucleic acid-based detection of microbes have greatly evolved to reduce assay processing time, they (with the excep-tion of real-time Polymerase Chain Reaction PCR) still lack the ability to detect microorganisms in "real-time," taking up to several hours to yield results, often after lengthy sample preenrichment. The need for more rapid, reliable, specific, and sen-sitive methods of detecting a target analyte, at low cost, has focused considerable research on microbial detection by SPR or antibody array-based immunobiosensors, especially for applications outside the laboratory environment. Since its inception, in the 1970s, Hazard Analysis at Critical Control Point (HACCP) methodology has evolved as the leading food safety strategy used by the food industry. HACCP identi-fies where potential contamination, time, and temperature problems can occur (the critical control points). However, key technologies needed to successfully implement any HACCP program are "real-time" microbial detection, traceability, and source identification. Biosensors offer the potential of detecting pathogens with the microbe-antibody binding event being observed as it happens, notwithstanding the fact that to date most sensor-based methods still require a time-consuming preenrichment step in order to detect low numbers of pathogens in food, water, and the environment.

Advances in antibody production, and specifically the recent emergence of phage-displayed peptide biosensors [6,7], now offer increased possibilities for the rapid detection of pathogens with immunosensor systems. Surface plasmon resonance-based biosensor studies involve the detection of proteins [8], hormones [9], toxins [10], and drugs [5]. One of the earliest attempts to use SPR for direct detection of microbes [11] achieved a detection limit for $E.\ coli$ O157:H7 of 5–7 x 10^7 CFU/mL. Fratamico and coworkers used a sandwich assay format in which they immobilized an anti-*E. coli* O157:H7 monoclonal antibody on the surface of the sensor chip. They then passed various concentrations of cells over the surface and amplified the signal with the injection of polyclonal anti-*E. Coli* antibodies. Koubova et al. [12] later described the detection of *Salmonella enteritidis* and *Listeria monocytogenes* at concentrations down to 10^6 cells/mL. Koubova and coworkers used a surface plasmon resonance (SPR) biosensor based on prism excitation of surface plasmons and spectral interrogation to directly detect cells on antibody-immobilized surfaces. Various cell concentrations were injected over the surface and the shift in resonant wavelength recorded. Bokken et al. [13] described the detection of *Salmonella* with a Biacore surface plasmon resonance-based biosensor. Anti-*Salmonella* antibodies immobilized to the biosensor surface were allowed to bind injected bacteria, followed by a pulse with soluble anti-*Salmonella* immunoglobulins to intensify the signal. The target organism was successfully detected at 1.7×10^5 CFU/mL. Recently, Leonard et al. [14] have simplified the assay format and have shown that by coupling a purified recombinant version of a major *L. monocytogenes* surface antigen to the SPR surface in Biacore analysis, as few as 2×10^5 cells/ml could be reliably detected with coefficients of variation between 2.5% and 7.7%. These recently reported publications highlight the potential of SPR-based biosensors for the detection of pathogens. Other sensor formats such as electrochemical sensors [3] have been reported in the literature with limits of detection as low as approximately 10^3 cells/mL [15,16]. However, further characterization of these formats has not been performed and, therefore, the reproducibility of these assays and regeneration of the sensor surfaces would need to be determined before an accurate limit of detection can be assessed.

SPR has been used very effectively to quantify enterotoxin B of *S. aureus* down to femtomolar levels [17], but the majority of currently reported studies have concentrated on the production of fluorescently probed antibody microarrays that are based on the combination of immunoassay and DNA array technologies [18–20]. These arrays have proven to be highly effective for the quantification of *S. aureus* enterotoxin B, as well as anthrax, diphtheria, and tetanus toxins, with sensitivities in the low nanomolar range [18]. Given the success of these simple array sensors, novel rapid variants are now also being examined for their ability to function in toxin detection. One such example is planar waveguide arrays [21], in which the influence of spot size on reaction kinetics was effectively characterized for sensor optimization [22]. A number of separate assay formats for planar waveguide technology (fluorescence enhanced by wave excitation) have also been investigated including direct, competitive, and displacement assays, in which measurements were obtained in "real-time" [23]. These combined assays have now been integrated into a prototype portable biosensor for on-site biohazard identification [24].

7.4 DRUGS, DRUG METABOLITE, AND TOXIN DETERMINATION IN FOOD AND ENVIRONMENTAL SAMPLES

Food contamination is an increasingly important public health concern often championed by intensive media coverage. The presence of unwanted chemicals in food, for example, antibiotic residues, hormones, drugs, natural toxins, and so forth, can also effect international trade and damage a company's image. Food ingestion is considered to be a major route of exposure to many contaminants known to have harmful effects (e.g., carcinogens).

A large range of methods of detection are currently available for drug analysis including radioimmunoassay, thin-layer chromatography, chemical ionization, and gas chromatography coupled with mass spectrometry (MS), high-performance liquid chromatography (LC), capillary electrophoresis, and solid-phase microextraction. Many, if not all, of these techniques are highly sensitive for many different small molecular weight analytes, but they are methods that are highly specialized, time-consuming, low throughput, and often expensive. Current detection techniques employed for the detection of contaminants often involve LC-MS-based techniques. This usually requires extensive sample pretreatment and extraction processes (e.g., liquid extraction). Due to the bulky nature of the equipment, relatively low throughput and the expertise required to carry out such analyses, it is not suitable for "field" measurements. Such techniques are therefore usually carried out as a confirmatory test. Biosensor-based methods for drug analysis are, by contrast, an attractive approach for the development of higher-throughput "first-screen" systems that are very sensitive and highly specific [25]. This methodology is well suited for single measurements of analytes, but it is also very desirable to simultaneously and rapidly quantify multiple analytes from a similarly small sample size. In order to improve on existing techniques and for the development of rapid "on-the-spot" testing for drugs or other contaminants, the integration of immunobiosensing techniques into a miniaturized portable device is required, for example, in a biochip format.

Veterinary medicine residues such as antibiotic residues in milk [26] have also been measured. The application of SPR-based biosensing has included the determination of sulphadiazine and sulphametazine residues in pork [27], streptomycin residues in whole milk [28], clenbuterol residues in bovine urine [29], the β-adrenergic compound, zilpaterol [30], 4-nonyphenols in shellfish [31], nicarbazin in liver and eggs from poultry [32], levamisole in liver and milk [33], *veterinary drugs and growth-promoting agents* [34], and multi-β-agonist residues in liver [35].

The observation that isoform-specific antidrug residue scFv antibodies can be isolated from large phage libraries of naive human scFv repertoires [36] further underlines the ability of antibodies to create accurate and rapid analyses for drug residues, and also the ability of recombinant antibody methods to generate those antibodies.

Several studies have described the production of antibodies to toxins, that are capable of detecting very small quantities of contaminants, such as aflatoxins in grain and nut products [37–39] and shellfish poisoning toxins [40,41].

There are many chemical contaminants that may be present in a vast array of different foodstuffs. This highlights the need for an extremely flexible, sensitive, and rapid analytical methodology capable of detection of multianalyte in various food samples. Such systems should be capable of detecting the residues of the parent compound and its metabolites in various sample matrices. This may be facilitated in the future by multiplex antibody microarrays, possibly integrated with label-free methods. However, within the European Union (EU) there is no obligation to use standardized methods in residue control studies on food-producing animals. Techniques are required to satisfy new performance characteristics, limits, and other criteria, opening the door for modern analytical approaches such as antibody microarrays [42].

Monoclonal antibodies to many pesticides, for example, atrazine, have been generated. The presence of these molecules in the environment poses a huge risk for public and environmental health since they are known to have mutagenic, teratogenic, and carcinogenic properties. Some of these pesticides have, therefore, recently been evaluated in a "reverse phase" antibody microarray format [43]. This format employs protein-conjugated haptens, which are arrayed by covalent linkage onto glass slides. Fluorescently labeled monoclonal antibodies are passed over the immobilized pesticide-conjugate chip in the presence of control standards or sample and the concentration of antibody binding is determined by scanning with a typical microarray scanner. The assay is, therefore, simply a miniaturized and integrated version of the competition ELISA microplate analyses in common use for the detection of many different haptens. The detection limit using the microarray assay format is slightly improved in comparison to plate immunoassay: the limits of detection were found to be 5 ng/mL and 8 ng/mL for 2,6-dichlorobenzamide (BAM) and atrazine, respectively, while with the ELISA technique, detection limits are 20 ng/mL for BAM and 24 ng/mL for atrazine. Using gas chromatography/MS, the detection limit was 10 ng/mL for both analytes [43].

A novel quartz crystal microbalance method has been described, which measures the concentration of the antibiotic chloramphenicol, via antichloramphenicol antibodies, which were covalently coupled to the monolayer on a gold surface [44]. While this approach shows some promise, the system described was of low sensitivity, detecting only in the μM range. The development of highly sensitive recombinant antibody fragments to atrazine [45] and their potential for expression in multiple different structural formats [46] will greatly aid the development of rapid biosensors for environmental contaminants in the future.

7.5 "REAL-TIME" BIOSENSORS

The current literature suggests the emergence of two relatively specific types of immunobiosensor: those which are made for the detection of limited numbers of analytes, but which are performed in "real-time" and those which can identify higher numbers of analytes, but are not measured in "real-time."

7.5.1 Surface Plasmon Resonance-Based Sensors

Although traditional antibody-based detection has been successively improved, leading to greatly decreased assay times, the techniques still lack the ability to detect analytes in "real-time" and typically require several hours of processing to yield confirmation of results. SPR sensors are one of the best characterized formats that offer the potential of analyte recognition in "real-time," using a combination of biological receptor (antibody, enzyme, nucleic acid, etc.) coupled with a physical or physicochemical transducer which allows one to observe a specific biological event (e.g., antibody–antigen interaction) as it happens. SPR allows the detection of a broad spectrum of analytes in complex sample matrices, and have shown great promise in areas such as clinical diagnostics, food analysis, bioprocess, and environmental monitoring. The phenomenon of SPR has shown excellent biosensing potential and many commercial SPR systems are now available such as Biacore® and Spreeta™ [4,47,48].

7.5.1.1 History of Surface Plasmon Resonance Sensing

Optical SPR was first reported in 1968 by Otto [49,50]. The technique found its first practical application as a tool for characterizing solid phase media, such as thin films [51,52] and monitoring processes at electrochemical interfaces [53]. It was first employed in a sensing format for gas detection [54] and subsequently, was adapted for "immunobiosensing" of a specific antibody–antigen interaction [55]. These developments employed the "Kretschman configuration" [56,57], which is more appropriate than the "Otto configuration" for solution-phase sensing, typically required in clinical, food, and environmental diagnostics. This technology was refined and optimized for biosensing applications and led to the commercialization of the Biacore® SPR biosensor in 1990 [58,59] for generic biospecific/biomolecular interaction analysis (BIA) [60].

7.5.1.2 Theory of Surface Plasmon Resonance-Based Biosensing

An appreciation of the physico- and electrochemical basis of SPR sensing [61,62] requires a basic understanding of total internal reflection (TIR), evanescent wave, and surface plasmon theory. Figures 7.3 and 7.4 illustrate the basis of the SPR-based sensor, Biacore®.

Total internal reflection (TIR) occurs at the interface between two nonabsorbing materials when light impinges on the interface at or above an angle specified as the critical angle (θ_c). Although no refraction occurs and the intensity of the reflected light is equal to the incident light intensity, a weak electrical field intensity component (evanescent wave) leaches perpendicularly into the lower refractive index medium and penetrates to a depth of approximately one wavelength, its amplitude decaying exponentially with increasing distance from the interface. If the interface is coated with a nonmagnetic, noble metal film that is much thinner (depth = 50 nm) than the wavelength of the incoming visible light (λ = 400–800 nm, 760 nm in Biacore®), a p-polarized charge density oscillation manifests at the metal/ambient dielectric interface. This is referred to as a surface plasmon wave. Gold thin films are preferred due to their demonstrated chemical resistance and tendency to produce strong plasmons and consequently, sharp reflectance minima [63].

Under conditions of TIR and when the energy and momentum of the incident light exactly matches that of the surface plasmon wave (i.e., when the vector component of the p-polarized incident light parallel to the interface, "K_X," is equal to the surface plasmon wave vector, "K_{SP}"), a resonance condition is achieved. Resonance results in an increase in energy "coupling" from the incident light, culminating in an enhanced evanescent wave field that is concomitantly reported as a decrease in the intensity of the reflected light. This can be "tuned" by varying the wavelength or angle of the incident light. The main disadvantage of the latter method is the requirement for complex, bulky, and slow optical angle scanning components. However, this is overcome by employing a wedge-shaped incident beam [64,65] that comprises a continuum of predefined incident angles and a photodiode detector array that can accurately extrapolate the angle at which the reflected light intensity is minimal (θ_{SPR}). Thus, when the prism, incident wavelength, metal properties, and temperature are kept constant, θ_{SPR} varies only in response to changes in the refractive index of the ambient medium (n_2), which is directly influenced by biospecific binding events

P-polarized incident light

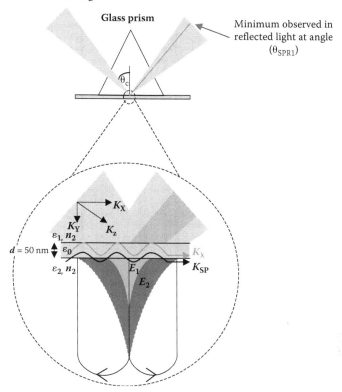

FIGURE 7.3 Evanescent wave theory and surface plasmon resonance at a metal-dielectric interface. ε_1 and ε_2 refer to the dielectric constants of the prism and ambient media, respectively. These constants determine the complex indices of refraction of each media (N; where $N = n - ik$; $i = \sqrt{-1}$; $k =$ the extinction coefficient and $n =$ refractive index), according to $\varepsilon = N^2$. Total internal reflection occurs at $\theta \geq \theta_c = \arcsin(\sqrt{\varepsilon_2}/\sqrt{\varepsilon_1})$, where $\sqrt{\varepsilon_1} > \sqrt{\varepsilon_2}$ and thus, $n_1 > n_2$ and an evanescent wave (E_1) extends across the metal film, decaying exponentially into the ambient medium. When the p-polarized incident light vector (K_X), parallel to the incident plane, matches exactly the surface plasmon wave propagation vector (K_{SP}), energy from the incident light is coupled into the surface plasmon wave and leads to an enhancement of the net evanescent field (E_2) and reduction in the intensity of the reflected light. That is, surface plasmon resonance occurs when $K_X = K_{SP}$, where; $K_X = (\omega/c).\varepsilon_1^{1/2} \sin\theta_{SPR}$ and $K_{SP} = (\omega/c).\{\sqrt{(\varepsilon_0)(\varepsilon_2)/(\varepsilon_0+\varepsilon_2)}\}$. $\omega =$ frequency of the incident light and surface plasmon wave ($2\pi c/\lambda$), $\varepsilon_0 =$ dielectric constant of the metal and $c =$ speed of light in a vacuum.

at the metal surface that occur within the approximate 300 nm range of the enhanced evanescent field.

The refractive index change can be directly related to the progress of the binding event [66] by

$$\Delta C = \Delta n_1 (\delta C / \delta n_1) \qquad (7.1)$$

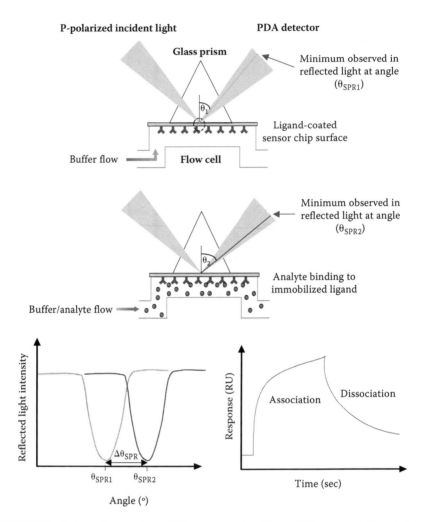

FIGURE 7.4 Schematic overview of BIA process using Biacore™ instrumentation. A light-emitting diode (LED) emits light at 760 nm, which is then p-polarized and focused as a wedge-shaped beam (of multiple, defined angles) onto the sampling area. Under ambient conditions, with continual buffer flow over the sensor chip surface, a reflectance dip (the SPR minima) or "shadow," is detected by a PDA detector, for θ_{SPR1}. When analyte (e.g., antigen) that is complementary to the immobilized ligand (e.g., antibody), is passed over the sensor chip surface it binds to the ligand and causes an increase in the mass concentration at the surface of the sensor chip and thus, an increase in the ambient refractive index close to the surface. This is detected as a shift in the angular position of the "shadow" (θ_{SPR2}). $\Delta\theta_{SPR}$ is measured in response units (RU), and it is monitored in real-time, with an adjustable data collection frequency of 1–10 Hz, on a sensorgram.

where $\delta C/\delta n_1$ is $5-7 g/cm^3$, Δn_1 is refractive index of sample, and ΔC is protein concentration.

The change in position of θ_{SPR} is quantified in RUs and is monitored in "real-time" as a function of time in a sensorgram.

There are a variety of Biacore® sensor chips available (http://www.biacore.com) for general or specific applications. The majority incorporates a carboxymethylated (CM) dextran surface. This functions as an extended 3D-coupling matrix or "hydrogel" [67] and increases both the binding capacity and dynamic range of the biosensor. The most commonly used sensor chip is the "CM5" chip, which possesses a hydrogel ~100 nm thick. For such a surface, a 1000 RU change corresponds to a 0.1° shift in θ_{SPR} and a change in surface concentration of ~1 ng/mm² [66]. Assuming a 100 nm thick hydrogel, this corresponds to an effective surface concentration of 10 mg/mL. Generally the measured change in surface refractive index is independent of the protein/peptide type, although it can vary for more complex molecules, such as glycoproteins, lipids, and nucleic acids [66,68,69]. The hydrogel also serves to create a hydrophilic environment with a low propensity for nonspecific binding. The extensive carboxylation provides convenient sites for covalent attachment of amine containing molecules.

SPR biosensors have a number of advantages over traditional immunoassays and these were highlighted by Canziani et al. [70], as follows:

1. SPR technology can be used to measure complex formation without labeling of the reactants.
2. Complex formation can be monitored in real-time, providing detailed information about the kinetics and equilibrium dissociation constants.
3. Samples from crude preparations may be analyzed.

In addition, the Biacore® system is fully automatic and incorporates a reliable microfluidic system that facilitates accurate and precise sample delivery and flow-rate manipulation. Significantly, it also permits multichannel analysis and thus, reference-subtraction [71], which is useful for comparative analyses and is a prerequisite for kinetic and affinity estimations. A significant advantage of SPR over optical detection techniques is that the incident light energy does not actually penetrate the bulk sample and thus, measurements can be made equally on colored or turbid solutions and on clear samples [72]. Typically Biacore® and SPR/evanescent wave-based technologies have been routinely used for analysis of small molecules. With specific reference to foodstuffs, these include hormones [9], antibiotic residues [73], and small molecules that are indicative of microbial contamination such as microbial toxins [10,46,74].

For a comprehensive treatment of the general principles and applications of Biacore®, the reader is directed to the dedicated text of Nagata and Handa [75]. The technology and instrumentation underpinning SPR-based biosensors itself is well established. The most crucial determinant of assay performance undoubtedly surrounds the biospecificity component or, more specifically, the choice of ligand and the immobilization strategy. This is limited by the availability of suitably specific and stable receptor ligands. The role of ligand can be filled by molecules such as antibodies, natural biological receptors, or nucleic acids [76].

In practice, immunosensing using SPR typically follows one of two formats:

1. The binding ligand (antibody) is either covalently or reversibly linked to the dextran layer and the sample is then passed over this fictionalized layer to observe any binding interaction [77].
2. Purified analyte is either covalently or reversibly linked to the dextran layer and the sample is passed over this fictionalized layer in the presence of the corresponding binding ligand. The quantity of analyte present is quantified as a reciprocal of the inhibition of the standardized binding interaction. This method is commonly applied in the measurement of small molecules whose binding to immobilized antibodies would provide insufficient mass change to be measured by the instrument.

7.5.2   Quartz Crystal Microbalance Immunobiosensors

Quartz Crystal Microbalance (QCM) sensors detect changes in mass adsorption at an interface and may represent an alternative sensor technology for the study of biospecific interactions in "real-time" [78]. The operating principle of these sensors is based on changes of frequency in acoustic shear waves in the substrate of the sensor. When the QCM system is used in piezoelectric detection mode, the resulting frequency will shift in direct proportion to molecular mass adsorbed at the surface of the sensor [79].

Bizet et al. [80] demonstrated the ability of QCM to investigate antibody–antigen interactions, but also pointed out the potential problem of mass "overestimation" in their system. These inaccuracies may be due to several uncharacterized factors such as hydrophilicity, surface charge, meniscus, temperature viscosity, and buffer composition [81]. However, it has subsequently been demonstrated that increased resonance frequency in more refined QCM systems reduced these effects considerably and increased signal-noise ratio up to Σsix-fold [82]. Kaiser et al. [78] have also very effectively demonstrated that a QCM-based sensor could be used to characterize the interaction between steroids and antisteroid antibodies, in "real-time." The sensor was capable of rapidly calculating both binding kinetics and antigen specificity, with the data being closely correlated to that generated by SPR and ELISA. QCM has also been effectively employed in nonimmuno interaction analyses, such as the investigation of the estrogen receptor–DNA binding kinetics in the presence and absence of estrogenic ligands [83].

Gerdon et al. [84] have combined a sensitive QCM system with protein A-mediated antibody presentation and antigen-coated nanoparticles in the fluid phase. Shen et al. [85] have recently developed a QCM-based nonlabeled biosensor ("piezoimmunosensor") with a modified single-chain antibody on its surface. It exhibited high selectivity, sensitivity, and stability and was also regenerable for at least one cycle of reuse.

7.5.3   Recombinant Antibodies (and Recent Advances)

In addition to the two-biosensor examples outlined above, protein arrays are the most heavily studied of the multianalyte immunobiosensing platforms and, therefore, many variations have been described. However, the majority of the protein arrays described to date have been antibody arrays. Most antibody and protein arrays vary fundamentally only in their surface chemistry (discussed below) and to a lesser extent in their detection method. However, the assays invariably adopt one of three configuration formats:

1. Competitive [86]
2. Sandwich [87]
3. Inhibition [42]

All immunobiosensors (including the protein arrays described to date), would benefit considerably from a greater level of sensitivity and broader dynamic range. While increased sensitivity is needed in the signal transduction mechanisms of sensors, the primary limitation for producing optimal conditions in all immunobiosensors is the affinity and, most importantly, the specificity of the antibodies, which comprise the biorecognition element [88]. Poorly selected/developed antibodies with suboptimal characteristics will limit sensor performance. In Table 7.5, the relative merits and problems associated with various antibody preparations are shown. Antibodies with poor affinity will have insufficient sensitivity to detect small quantities of analyte, thereby reducing the dynamic range of detection. In addition, those antibodies which have not been carefully screened or designed or do have high specificity can greatly reduce the reliability of the sensor. Studies have shown that antibody recognition of proteins [89,90] can be highly promiscuous. Interestingly, studies have also shown that hapten recognition is further complicated by antibody conformational change in response to antigen contact [91] and apparently specific hydrogen bonding [92]. These studies illustrate the care and diligence required in designing the molecular recognition portion of immunobiosensors.

The main antibody formats currently available for use in immunobiosensors are polyclonal, monoclonal, and recombinant antibodies (typically, engineered antibody fragments, as shown in Figure 7.5). Immune sera from rabbits, sheep, or other mammals may be produced in relatively large quantities, but they are heterogenous and do not offer the uniformity of monoclonal antibodies. Immune sera are also polyclonal and polyspecific, rendering them suboptimal for the measurement of single components in complex matrices. Monoclonal antibodies are potentially monospecific and are generated by immortalized cell lines (hybridomas), rendering them more homogenous than immune sera. However, the laborious nature of monoclonal antibody generation, coupled with their mainly murine host dependency, has led to the investigation of recombinant antibodies, to produce optimized biorecognition elements for multiple applications [93].

Recombinant antibodies offer a number of potential benefits over monoclonal antibodies, as they can be derived from almost any animal in which the immunoglobulin DNA sequences have been described. This aspect of recombinant antibody technology is of particular benefit, as biomarkers for disease diagnosis are often proteins and peptides, which are highly conserved in animals closely related to humans. This can be a particular problem in mice, which frequently develop tolerance for the immunizing antigen, rendering traditional monoclonal antibody development to

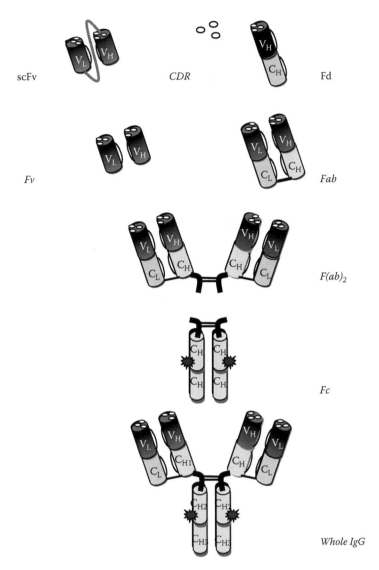

FIGURE 7.5 Diagramatic illustration of antibody and antibody fragments which can be produced by genetic, chemical, or enzymatic means. The antibody may be broken up into either Fab (single antigen-binding fragment) or F(ab')$_2$ (two antigen-binding fragments), and Fc (crystalline fragments) regions. Fab fragments may be further broken up into Fv (variable fragment), scFv (variable fragment, stabilized with a synthetic linker), and CDR regions, which are the smallest fragment capable of antigen binding.

some proteins very difficult. The use of animal models for immunization which are more evolutionarily distant to human than mice, such as chickens [93–96], rabbits [97–101] and llamas [102,103], has proven to be a very effective method of providing specific recombinant antibodies to antigens previously regarded as "problematic" when relying on mouse immunizations.

The availability of recombinant antibody technology also facilitates the development of higher affinity antibodies to molecules of interest in food quality and environmental testing. As rabbits and sheep are renowned for their high-affinity serum immunoglobulin response to peptides and hapten antigens, a number of groups have investigated their use in recombinant antibody development, successfully producing antibodies of excellent sensitivity [44,100,104]. In a recent study, recombinant antibodies were generated to quantify the paralytic toxin domoic acid in shellfish. This study has further simplified antibody development for small toxin molecules by employing the very simple chicken immunoglobulin system [40]. Recombinant antibodies against many antigens of diagnostic and environmental interest have also been developed from very large libraries of naive, semisynthetic, or fully synthetic antibodies. These libraries are typically human in origin, were predominantly developed with the intent to create therapeutic antibodies and have produced many antibodies with the ability to recognize proteins and peptides. They have also yielded some useful antibodies for the detection of smaller molecules like marine toxins [105] and herbicides [106].

Recombinant antibodies are encoded, selected, and expressed in multiple structural formats, which can each have considerably different biophysical attributes. The simplest (commonly used) structural format is the single-chain fragment (scFv) in which the V-regions of antibodies are linked by a flexible peptide linker (Figure 7.5). The most common linkers are based on glycine-serine repeat structures and can be of different lengths, depending on the intended valency of the molecule. When stability and folding of the scFv is unfavorable, methods have been developed to select the highest stability, nonrepetitive linker, by introducing random mutations in the linker followed by aggressive reselection of the scFv by phage display [107]. Long linkers of 18–21 amino acids favor the production of scFv, which is predominantly monomeric [108–110]. However, shorter linkers of, for example, seven amino acids do not allow sufficient effacement of the V_H and V_L regions in any single chain, for the monomer to function. Therefore, scFv selected in this format invariably form bivalent dimers [108,111,112], or "diabodies," which often have increased avidity for antigen over the monomeric forms typically observed in long-linker systems.

Large, high-avidity trimers and tetramers have also been formed by shortening the linker length to three amino acids, or even using no linker at all [111,113]. The stability characteristics of short-linker, multimerizing scFv forms have not yet been fully established, but one study has suggested that dimeric scFv fragments can dissociate into monomeric chains when at low concentration [114]. The stability of scFv dimerization can be further increased by expression with fusion protein tails which naturally interact, such as leucine-zipper regions, Jun and Fos regions and the constant regions of human IgG [115–117].

The second most commonly used recombinant antibody format is the Fab (fragment antigen binding) molecule (Figure 7.5). This structure is comprised of the complete immunoglobulin light chain, expressed in conjunction with the V_H-C_{H1} region [118,119], and obligatory forms monomeric, monovalent fragments. The Fab is the most "natural" of the recombinant antibody fragments and it has recently been shown that the presence of the constant regions help to stabilize antibody variable regions, which might not function as well when expressed in the monomeric scFv format

[120]. This stabilization effect appears to be mediated by the C-region scaffolds on which the V-regions sit and, critically, by the disulphide bonds that covalently join the C-regions. It has recently been shown that the Fab antibody format is likely to be the most reliable and sensitive format for use in small molecule competition assays. The strict monovalency of the Fab format can lead to 2–12 fold improvements in assay sensitivity in both ELISA and competition SPR analyses [121].

The most robust and heavily used technology for the development of recombinant antibody fragments is the phage display method [122], which has become a powerful tool for generating useful antibodies with novel characteristics (Figure 7.6). Phage display of recombinant antibodies typically involves a multistep process beginning with the amplification of V-region cDNA from the lymphoid tissue of humans or animals. The V-regions are then assembled into the desired structural format in a phage or "phagemid" vector (a plasmid containing a cloning region for fusion of the antibody fragment to a phage coat protein), before "brute-force" cloning into E. coli by electroporation. The resulting population of antibody fragments, harbored by the bacteria, is then regarded as an antibody "library." Specific antibody fragments are then selected by a process often labeled "biopanning." Coinfection of the E. coli library with helper phage such as VCS-M13 then leads to the expression of the harbored antibody fragments fused to a phage coat protein. The concurrent packaging of the phagemid into phage displaying the antibody-coat protein fusion, leads to effective genotype-phenotype linkage. By selecting the library of phage-antibodies through binding to, for example, immobilized or labeled antigen, nonspecific antibodies are removed and antigen-specific antibodies are readily obtained. The biopanning method may provide a considerable advantage when trying to develop highly specific antibodies, as not only can multiple species be used to provide novel epitope recognition, but molecules closely related to the antigen of interest can be included in the panning process to actively deplete antibodies with poly-specific binding characteristics [123,124].

Several variations on this theme of recombinant antibody generation have been developed since the advent of phage display, including yeast display [125], bacterial surface display [126,127], retroviral display [128], and ribosomal display [129,130]. All apart from ribosome display are cell-based methods, which apply the same basic fundamental principles as phage display. However, in yeast and bacterial display, selection is usually achieved via flow cytometry, rather than solid-phase selection. Each of these methods have independently been shown effective for the production of high-affinity antibodies, but it can be difficult to produce high-diversity antibody libraries using the yeast and retroviral methods, due to the low-efficiency transformation of the cellular host. Bacterial display does not have this problem, but relies on antigen penetration of the bacterial periplasm and has mostly been shown to efficiently produce antibodies to fluorophore-labeled hapten molecules [131]. Recently, however, the additional permeabilization of the bacterial outer membrane with Tris-EDTA-lysozyme has allowed fluorophore-labeled protein molecules up to 240 kDa in size to be used as antigen in selections [132]. This modification has led to excellent outcomes in the affinity maturation of antibodies [132]. Ribosome display, however, is a completely *in vitro* system that relies on the translation of artificially produced antibody mRNA (usually of extremely high diversity), which encodes for

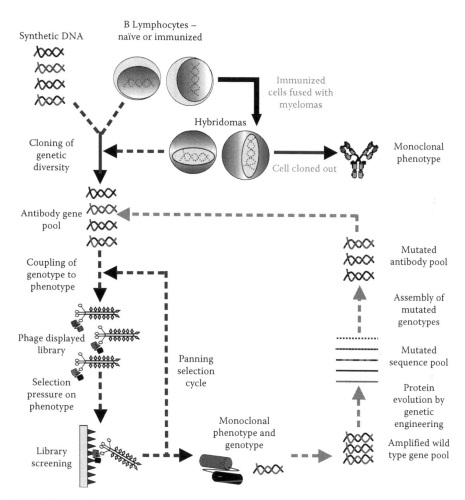

FIGURE 7.6 Schematic overview of major steps involved in the production of phage-display libraries (dotted arrows), from initial cloning of genetic diversity and production of antibody gene pool by extraction of mRNA from populations of hybridomas (black arrows) or B-cells taken from the spleen of a naive/preimmunized animal. Antibody (i.e., Fab or scFv) cDNA gene constructs are then ligated into phagemid vector and transfected into competent bacterial cells (e.g., *E. coli* XL1 blue). Antibody fragment is coexpressed with phage coat protein and preferentially packaged into phage particles using a helper phage (e.g., M13-K007) in effect coupling genotype and phenotype. Phage-displayed antibodies are then subjected to panning selection cycle and subsequently, monoclonal antibody obtained by isolating infected single *E. coli* colonies on selective agar. Also shown (dash arrows) are the basic steps involved in the production of mutant libraries.

a linker peptide after the antibody fragment, but does not contain a stop codon. Therefore, on translation, the protein is produced but remains linked to the ribosome and the encoding RNA as a large complex [130]. Specific antibodies can then be isolated using a typical biopanning method, with retrieval of binding clones by RT-PCR. Despite the potential of ribosome display as a phage alternative, its uptake has been slow due to the inherent limitations such as unknown complex stability and high cost. It is expected, therefore, that phage display will remain the core technique in the production of recombinant antibodies for the foreseeable future.

7.5.3.1 Stability/Affinity/Specificity of Antibody Fragments in Biosensors

Few examinations of antibody fragment stability in biosensors have been made, but their function in some formats was addressed. Conventional (nonrecombinant) methods of generating antibody fragments involve cleavage of the IgG molecule with pepsin producing a $(Fab')_2$ fragment. The disulfide bonds within the $(Fab')_2$ fragment can then be cleaved by a mild reducing agent to form cysteine thiols, which can be used for the subsequent covalent linkage to a surface. Vikholm [133] employed SPR technology to compare the attachment of $(Fab')_2$ and Fab fragments of antihuman IgG onto gold. Fragments were immobilized directly onto the gold surface and remaining binding sites blocked with a nonionic hydrophilic repelling polymer of N-[tris(hydroxymethyl)methyl]acrylamide. This study indicated that Fab fragments bound in a more orientated manner than the larger $(Fab')_2$ fragments, resulting in a fourfold increase in the binding capacity of the Fab fragment layer compared to that of the $(Fab')_2$ layer. In 2006, Vikholm-Lundin et al. [134] further improved the immobilization of Fabs onto gold by addressing the influence of buffer conditions, polymer length, and sensitivity of the layer in buffer and serum. Karyakin et al. [135] have also added to the interest in antibody fragment use by reporting that half-IgG fragments immobilized on gold supports exhibited affinity constants 30 times higher than those measured for similar intact full-length antibodies.

Early reports using recombinant antibodies in immunobiosensor applications were relatively basic. Nielsen et al. [87] used a single scFv as a secondary antibody in antibody array sandwich ELISA analysis, but the main interest in recombinant antibodies is in their potential as highly defined capture and recognition agents. A number of groups have focused on the use of large naive human antibody libraries in their search for antiprotein and hapten antibody fragments for biosensors. Steinhauer et al. [136] isolated a set of scFvs against an antigen of interest (a human cytokine) and then investigated the relationship between their encoding framework region families and their stability and function when coated directly onto nitrocellulose-coated protein chips. The human V_H3-23/V_L1-47 framework was identified as the best scaffold for recombinant antibody microarrays, of the limited number of antibodies, which they isolated [136]. This variable heavy and light chain combination was found to be the most stable for chip development, suggesting that future research should focus on building a library based on that framework region scaffold, to tailor for on-chip stability [136]. However, subsequent reports from the same group [137] have shown that these resulting recombinant human scFv function best only when captured by conventional antibodies, in an oriented fashion, via a peptide tag.

Peluso et al. [138] have compared randomly versus specifically oriented capture of both full-sized antibodies and natural Fab' fragments using two types of streptavidin-coated monolayer surfaces. Up to 10-fold improvements were observed for oriented over randomly associated capture agents and SPR revealed a dense monolayer of Fab' fragments that were on average 90% active when specifically oriented. Randomly attached Fab's could not be packed at such a high density and generally also had a lower specific activity. In a further recent development, Shen et al. [85] have shown that by introducing a cysteine residue into an scFv linker sequence, the scFv antibodies self-assemble into a densely packed, covalently linked monolayer on the QCM gold surface. The scFv molecules therefore exhibit defined orientation and high areal density, with up to 35 times as many binding sites per square centimeter as surfaces labeled with whole antibody. This novel scFv construct was found to function better on the gold surface than the natural Fab fragment and whole antibody. The resulting immunosensor accurately quantified the concentration of rabbit Fab fragment in a serum matrix. These findings strongly suggest that antibody fragments, in the same sense as full-length antibodies, require oriented anchoring to the chip surface such that their binding sites are oriented toward the solution phase, if they are to be optimally active. When coupled in this fashion they can exhibit better surface coverage and functional activity than their full-length counterparts.

7.5.4 IMMOBILIZATION: CURRENT METHODS AND POSSIBLE FUTURE DIRECTIONS

The immobilization of antibodies onto the sensor surface is a major challenge [139]. This is illustrated in Figure 7.7. Correct orientation requires the antibody binding site(s) to be fully available for interaction with the associated antigen. A number of methods previously used during DNA microarray chip fabrication, such as covalent coupling of antibodies directly to silanized glass surfaces, have been adapted for antibody array chip development. These methods have mostly proved to be unfavorable, however, because unlike DNA, antibodies are highly complex in their structure and chemistry. Nonspecific adsorption or covalent coupling of antibodies onto solid supports can cause them to denature and hence, their antigen-binding function is reduced or even lost [87]. Indeed, key studies have shown that the covalent coupling of antibodies to sialinized glass is such a harsh method that as few as 1 in 20 [87] or <60% [86] of antibodies function effectively in the resulting chip formats. Susceptibility to denaturation may be increased in recombinant antibody formats such as scFv, due to the inherent flexibility of their linker region and the lack of costabilizing constant regions. Therefore, it is logical that new immobilization strategies employed in biosensor fabrication should preserve the structure and activity of the antibody to the greatest extent possible, particularly when using recombinant antibody fragments. These strategies should ideally also involve orientation of the antibody to maximize binding site contact with the analyte matrix [137].

In recent years, a number of different types of chip surfaces have been investigated; including glass or plastic slides, porous gel pad slides, filters, microwells, and multiple polymer surfaces [140–142]. However, the basis for antibody immobilization in these strategies has almost universally been nonspecific, that is, not leading to the oriented presentation of the antibody. The most frequently used method utilizes

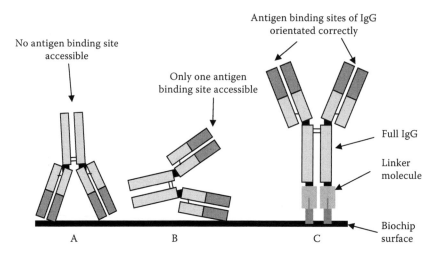

FIGURE 7.7 Antibody orientation strategy for biochip development. Random immobilization by adsorption or covalent attachment leads to nonaccessible (A) or only partially accessible (B) antigen-binding domains. In (C) correct orientation of antibody will give maximum activity and this is achieved using a linker molecule (e.g., Protein A).

poly-L-lysine for covalent antibody immobilization. This method has a number of advantages including high signal intensity, low background, and highly homogenous spot morphology. It has, therefore, often allowed detection of antigens below 1 ng/mL [142]. Angenendt et al. [141] evaluated a number of different surface chemistries for the modification of glass supports. Chemically modified glass slides, including dendrimer and poly(ethylene glycol)–epoxy slides, a number of commercially available substrates including amine, epoxy, sialinated, and FAST (nitrocellulose-coated) slides and the more traditional polystyrene cell culture slides, were all investigated. Dendrimer slides were found to exhibit excellent signal-to-concentration ratio for quantitative measurement of proteins in complex samples, while PEG-epoxy slides were the surface coating of choice for quantitative measurement of antibodies. Results obtained with the use of FAST slides also showed very good detection limits, as corroborated by Wingren et al. [137]. Kim et al. [143] modified glass surfaces using several hydrophilic polymers for the fabrication of protein chips. Glass surface-bound polymer layers have the advantage of minimizing nonspecific adsorption and, in this case, reduced nonspecific adsorption by 10%–60% compared to silanized slides. Rucker et al. [20] have also developed antibody arrays on epoxide fictionalized glass slides, which are capable of detecting toxins in both buffer and serum, at limits comparable with those attained using conventional sandwich immunoassays.

While glass has proven to be the most popular conventional platform for protein microarrays, a number of other physical substrates have also been evaluated. These have included gold, silica, and polyvinylidene fluoride and nitrocellulose membranes. Nitrocellulose has been employed in array development as it can retain large quantities of antibody, is inexpensive, and disposable. However, it is not suitable for kinetic studies due to the slow diffusion of proteins through its pores. It is also not compatible with MS analysis and can cause light scattering [144]. Array surfaces utilizing metal or silica are usually photolithographically etched and the proteins chemically cross-linked to the surface by means of aminosilanes, aldehydes, epoxides, or *N*-succinimidyl-4-maleimidobutyrate [145]. To preserve antibody activity, high-porosity gels, which facilitate protein diffusion and support the 3D structure of proteins, have also been used for immobilization [86,146]. Angenedt et al. [140] compared 11 different array surfaces including plastic, gel-coated and nongel coated chemically modified glass slides. Results indicated that polyacrylamide-coated slides were found to be most suitable for detection of very low concentrations of antigen. However, although gel pads fixed on glass slides or glass slides coated with a hydrogel offer numerous advantages, fabrication is complicated and impractically long washing times are necessary [88,142].

Despite the extensive studies outlined above, none fully examined the concept of antibody functional orientation. Therefore, several groups have focused on developing methods for site-specific immobilization and resulting orientation of antibodies. One of the most common methods used is biotinylation of capture antibodies and their subsequent immobilization onto streptavidin-coated supports. This approach, in particular, appears to have great potential. Wacker et al. [147] employed a novel modification to the conventional biotin-streptavidin strategy by combining it with DNA-directed immobilization (DDI). An anti-rabbit IgG-DNA conjugate was generated from biotinylated anti-rabbit IgG and a DNA-streptavidin conjugate. This was then capable of hybridizing to complementary captured oligonucleotides spotted on dendrimer-activated glass slides. This DNA-directed immobilization method was compared with direct spotting of antibodies onto chemically activated glass slides and with immobilization of biotinylated antibodies on streptavidin-coated slides. All three antibody arrays permitted the detection of antigen concentrations as low as 150 pg/mL. However, results obtained found that DDI and direct spotting of antibody led to higher fluorescence intensities and improved spot homogeneity. Overall, DDI proved to be the most reproducible and economical method, with 100-fold lower antibody concentration being required on the array.

Peluso et al. [138] have employed biotin-avidin coupling to compare random and specifically orientated antibody immobilization strategies. Randomly biotinylated IgG, orientated IgG (biotinylated via carbohydrate on the Fc domain), oriented Fab fragments (biotinylated on the hinge region terminus), and randomly biotinylated Fab fragments were all investigated. The Fab fragments used in this study were natural and generated by enzymatic cleavage, rather than recombinant, but they nonetheless indicated that specifically orientated antibody fragments exhibited better binding capacity (up to 10-fold higher) compared to that of their randomly orientated counterparts. A twofold increase in loading density was also observed when Fabs were specifically immobilized as opposed to randomly attached. The authors further report that this is the highest reported density of active antibodies on a planar surface. This increase in surface density is a major advantage, as previous site-specific immobilization methods have resulted in a lower surface coverage compared to those employing direct, random surface attachment. Site-specific methods, which utilize protein A, G, or other Fc binding proteins are established for controlled specific orientation of antibodies on fictionalized surfaces [148, 149–151]. However, methods utilizing these intermediate proteins require two immobilization steps, often suffer from increased variability and are unlikely to be a useful substrate for orientating scFv or Fab molecules, both of which lack the Fc domain. In addition, antibody fragments from sources other than humans and mice can exhibit poor binding to these molecules, if any. The recent development of commercially available bacterial expression vectors which introduce biotin (AviTag, Roche) or biotin-like binding peptides (StrepTag, Qiagen) into specific sites on the termini of recombinant proteins, suggests that biotin-avidin coupling may become a simple and rapid method for high stability, oriented tethering of recombinant antibody fragments.

Several other possible orientation/tethering methods have also become available in recent years [152]. With the development of new protein expression vectors, it is now possible to engineer tags into the antibody sequence to facilitate site-directed orientation. The incorporation of a histidine (His) tag allows the immobilization onto nickel-coated slides. However, this method is dependent on pH and the tag may not always be accessible. Others include N-terminal hemagglutinin (Ha-tag) and C-terminal (myc-tag) tags, *E. coli* maltose binding protein (MBP), FLAG peptide, and glutathione S-transferase (GST), most of which have been evaluated for protein purification. Wingren et al. [137] systematically compared the use of C-terminal His- or myc-tags for the generic immobilization of multiple different scFvs onto nickel-coated slides, plus silanized glass and FAST slides that had been pre-spotted with anti-Histag antibodies. On antibody coated, silanized slides the limits of detection (LODs) were similar to those observed on nickel slides. However, on FAST slides, the LODs were up to 1,000 times lower. This result was reliant on using the anti-His antibody, which functioned best on FAST slides. This method is of interest as it allows functional orientation of the scFv, but it is still somewhat limited by the requirement to precoat any surface of interest with an intact full-length IgG.

A number of other technologies are of particular interest to circumvent this problem, such as the method of Blank et al. [153], which involves constructing and synthesizing antibody fragments as fusions to two chitin-binding domains (scFv-cbd$_2$). The expression of these fusion proteins in *E. coli* facilitated the simple and effective purification of their cognate antigens by scFv-cbd$_2$ immobilized tightly on chitin-coated beads. Importantly, there was no need for prepurification of the scFv-cbd$_2$ proteins from the *E. coli* lysate before column immobilization. The possibility of introducing chitin layers into a sensor format, however, has not yet been evaluated. A second interesting possibility is the expression of recombinant antibody fragments with DNA binding domains such as protamine [154]. This method had been used to achieve antibody-mediated plasmid delivery into cells, but the affinity of protamine for DNA is unknown at present.

One further method that removes the requirement for a tethering protein on the sensor surface is the use of "calixerenes" [155], which are bifunctional linker molecules that allow extremely high-affinity labeling of biomolecules to a surface without covalent bonding. One company (BodiTech Med) has demonstrated how to immobilize proteins while preserving their folded structural integrity by coating novel calixarene derivatives, designed for the immobilization of proteins, on the surface of a glass slide (designated as a "Proteochip") and also on QCM sensors [156]. In their first study, this group demonstrated the handling simplicity of ProteoChip, with no requirement for covalent linker chemistry to effectively immobilize antibody and probe for its function. They also showed the detection sensitivity of the chip to be as low as 1–10 fg/mL of analyte protein, one of the lowest detection sensitivities ever reported for a protein microarray chip, theoretically equivalent to one antibody molecule per spot. The company has now recently compared the sensitivity and the specificity of the linker molecules with those of five other protein attachment agents on glass slides using prostate-specific antigen (PSA) and associated antibodies as a model system [151]. The Proteochip showed a superior sensitivity and a much lower detection limit than chips prepared by any other method and antibodies were found to be ordered in a more vertical orientation than when using a covalent bond agent.

Other less conventional methods of site-specific immobilization include the use of functional lipid bilayer membrane vesicles (liposomes). Liposomes have a number of structural advantages, including high surface density; good reproducibility, stability, and can also contain functional molecules, such as biotin. A liposome layer containing two functional molecules, biotin and N-(10,12-pentacosadiynoic)-acetylferrocene (Fc-PDA) grown on a gold layer was evaluated by Lee et al. [157]. Atomic force microscopy (AFM) studies showed a rather uniform surface layer, allowing for efficient immobilization of biotinylated anti-HSA antibodies. The protein assay based on the functional liposome-modified fold electrodes demonstrated good sensitivity, specificity and stability, with the antibody layer retaining activity after 1 week. To ascertain whether any of these approaches is truly viable in recombinant antibody-based immunobiosensors will require considerable further study, but the adoption of the best principles outlined in this section might greatly improve their construction in the future.

7.6 STRATEGIES FOR DEVELOPING SENSOR-BASED METHODOLOGIES FOR FOOD OR ENVIRONMENTAL ANALYSIS

The reader will now have a good appreciation for many aspects of sensors as applied to food/environmental analysis. However, the question often arises as to how one would develop a method for a new analyte or an existing target that needs an improved approach. Experience has clearly demonstrated to us that there are a number of key questions that must be answered before attempting to develop such strategies. Scott [2] has also examined many of the relevant aspects in relation to food analysis.

Key initial questions for developing sensor-based methodologies for food or environmental analysis:

1. Is there any existing method that can generate the required results?
2. What are they key advantages/disadvantages of the method?
3. What specific advantages will the sensor provide?
4. Is the time-frame and cost of development really justifiable?

Based on the material already discussed in the chapter and careful research the answers should emerge. If the result is affirmative then the following list of issues should be critically evaluated:

1. What is the sensitivity required?
2. What is the ideal time frame for the generation of results?

3. Where will the device be used? Will it be lab-based or "in the-field?"
4. Who will use it? Will they be highly trained or should it be "user-friendly?"
5. How much will the device cost? While costs need to be minimized, the instrumentation must be capable of monitoring with the required sensitivity stability, reproducibility and robustness.
6. Is it a "once-off" test for each sensor "chip" or is it reuseable?
7. For reuseable systems, how often is it necessary to regenerate them and how stable are they? Can they be easily stored to provide a realistic shelf life?
8. What controls/standards will be necessary and how will they be incorporated? Will the device be self-calibrating?
9. Should the device provide a single result or multiple analyte determinations? Generally this appears to be an area where the market is demanding greater analyte capacity.
10. Choice of biorecognition element. What biorecognition element will be used? Is it tailored for the application? The use of materials (e.g., antibodies) produced for other assay formats often results in poor-quality sensors, as the properties of the antibody may not be customized for the format chosen.
11. How specific does the recognition unit need to be? Should it be class specific, that is, for a group of structurally similar components, or should it be specific for one analyte?
12. Should an assay format be used with both class and analyte-specific biorecognition units?
13. Will mixed biorecognition units be used, for example, antibodies and nucleic acid probes? This complicates the system but very often may offer far greater analytical potential. A key issue that must be addressed is how the biorecognition unit will be immobilized to optimize its performance [139].
14. Food and environmental samples often comprise very complex matrices. Will the analyte be preextracted prior to analysis? Are there many natural interferants present? Are there any nonspecific binding effects? These can often be overcome by pretreatment of the sample, blocking of the chip for prevention of nonspecific binding, or incorporation of blocking agents as part of the sample treatment strategy. Sometimes dilution can further reduce interference/nonspecific binding effects. Varying salt concentration and pH are also worth investigation to optimize assay performance.
15. Should the sample be filtered/treated to remove large particles, cells, or other potential interferants? Many samples may require carefully optimized strategies for pretreatment prior to analysis. This is not just a requirement for sensor-based analysis and much food/environmental samples have well-established protocols for cleanup. Such elements can affect the analysis as they may block microfluidic devices or cause nonspecific binding to a number of commercially available systems. These factors are critical in sample preparation and should be investigated, where relevant. However, where possible for ease of analysis/reduction of costs pretreatments need to be kept to a minimum.

16. The performance characteristics of the biosensor-based assay need to be carefully determined and validated in terms of its precision, sensitivity, specificity, accuracy, robustness, stability, and so forth.

17. If the system can provide a generic platform that can be used with many different analytes, this is a major advantage.

Finally, while the points outlined are key for the development of successful biosensor-based assays and will help in the planning and execution of the research and development, it is also very important that the cost-effectiveness of both the work and the device be carefully considered at all stages. The markets for sensors vary and their requirements are distinctly different. This provides opportunities for a number of manufacturers who focus on different market segments. However, one of the major difficulties in the past was that sensors were hailed, sometimes over-optimistically, as providing the "holy grail" for analysis. While some were very successful commercially (e.g., Medisense glucose monitor, Biacore) others failed to achieve expectations. Recent developments in technology (e.g., antibody arrays, antibody engineering, microfluidics, and nanotechnology) should provide considerable scope for new formats with attendant opportunities. Nevertheless, without very considerable hard work, resources, ingenuity, and a realization of the time required, the potential opportunities and outcomes will not be successfully achieved.

7.7 FUTURE APPLICATIONS OF SENSORS IN FOOD AND ENVIRONMENTAL MONITORING

The number of potential future applications of biosensors in food and environmental monitoring is clearly very large, and few examples are given in Table 7.6. There are great opportunities for the exploitation of the unique features that sensors may provide, to solve complex analytical problems. It is our view that a combination of optimized sample handling/extraction, judicious selection of the sensor format, the

TABLE 7.6
Areas for Research and Development

	Research and Development Areas
1	Use of multiassays with simultaneous measurement of temperature profile, chemical components, microbial contamination, etc.
2	Development of microfluidics-based systems with small sample volumes and reduction of extensive sample cleanup requirements prior to analysis
3	Need for well defined biochemical markers of quality for a variety of foods
4	Low cost sensor-based analytical systems with automatic monitoring and reporting, e.g., via wireless-based systems
5	Regulatory Agency approval for use of sensor technologies
6	"Household"-based sensors for food/water/environmental monitoring, e.g., quality evaluation of drinking water, freshness of foods in fridge, etc.
7	Monitors of waste effluent

capacity for multiple simultaneous analyses with appropriate controls, and, lastly, the incorporation of microfluidic devices will yield a generation of sensors that can overcome the major obstacles presently encountered in food and environmental analysis.

ACKNOWLEDGMENTS

The financial support of Enterprise Ireland, Department of Agriculture and Food, Science Foundation Ireland, and Fusion is gratefully acknowledged. We wish to thank Michelle Meehan for her help in the preparation of this chapter.

REFERENCES

1. McCormack T., Keating G., Kilhard A., Manning B. M., and O'Kennedy R., in *Principles of Chemical and Biological Sensors,* ed. D. Diamond (New York: Wiley, 1998).
2. Scott A. O. (ed.), *Biosensors for Food Analysis* (Cambridge: The Royal Society for Chemistry, 1998).
3. Leonard P., Hearty S., Brennan J., Dunne L., Quinn J., Chakraborty T., and O'Kennedy R., Advances in biosensors for detection of pathogens in food and water, *Enz. Microbial Technol.*, 32, 3–13, 2003.
4. O'Kennedy R., Leonard P., Hearty S., Daly S., Dillon P., Brennan J., Dunne L., et al., Advances in biosensors for detection of pathogens in food and water, in *Rapid Methods for Biological and Chemical Contamination in Food and Feed*, ed. A. van Amerogen, D. Barry, and M. Lauwaars (The Netherlands: Wageningen Academic, 2005) 85–104.
5. Dillon P. P., Daly S. J., Manning B. M., and O'Kennedy R., Immunoassay for the determination of morphine-3-glucuronide using a surface plasmon resonance-based biosensor, *Biosens. Bioelectron.*, 18, 217–227, 2003.
6. Goldman E. R., Pazirandeh M. P., Mauro J. M., King K. D., Frey J. C., and Anderson G. P., Peptide-displayed peptides as biosensor reagents, *J. Mol. Recognition*, 13, 382–387, 2000.
7. Goodridge L. and Griffiths M., Reporter bacteriophage assays as a means to detect foodborne pathogenic bacteria, *Food Res. Int.*, 35, 863–870, 2002.
8. Nieba L., Nieba-Axmann S. E., Persson A., Hamalainen M., Edebratt F., Hansson A., Lidholm J., Magnusson K., Karlsson A. F., and Plückthun A., BIAcore analysis of histidine-tagged proteins using a chelating NTA sensor chip, *Anal. Biochem.*, 252, 217–228, 1997.
9. Gillis E. H., Gosling J. P., Sreenan J. M., and Kane M., Development and validation of a biosensor-based immunoassay for progesterone in bovine milk, *J. Immunol. Meth.* 267, 131–138, 2002.
10. Daly S. J., Keating G. J., Dillon P. P., Manning B. M., and O'Kennedy R., Development of surface plasmon resonance immunoassay for aflatoxin B1, *J. Agric. Food Chem.*, 48, 5097–5104, 2000.
11. Fratamico P. M., Strobaugh T. P., Medina M. B., and Gehring A. G., Detection of Escherichia coli O157:H7 using a surface plasmon resonance biosensor, *Biotechnol. Tech.*, 12, 571–576, 1998.
12. Koubova V., Brynda E., Karasova L., Skvor J., Homola J., Dostalek J., Tobiska P., and Rosicky J., Detection of foodborne pathogens using surface plasmon resonance biosensors, *Sens. Actuators B*, 74, 100–105, 2001.
13. Bokken G. C. A. M., Corbee R. J., van Knapen F., and Bergwerff A. A., Immunochemical detection of Salmonella group B, D and E using an optical surface plasmon resonance biosensor, *FEMS Microbiol. Lett.*, 222, 75–82, 2003.

14. Leonard P., Hearty S., Wyatt G., Quinn J., and O'Kennedy R., Development of a surface plasmon resonance-based immunoassay for *Listeria monocytogenes*, *J. Food. Protect.*, 68(4), 728–735, 2005.

15. Eshkenazi R. I., Neufeld I. T., Opatowsky T., Shaky J. S., and Rishpon J., Recombinant single chain antibodies in bioelectrochemical sensors, *Talanta*, 55, 899–907, 2001.

16. Susmel S., Guilbault G. G., and O'Sullivan C. K., Demonstration of labeless detection of food pathogens using electrochemical redox probe and screen printed gold electrodes, *Biosens. Bioelectron.*, 18(7), 881–889, 2003.

17. Naimushin A. N., Soelberg S. D., Nguyen D. K., Dunlap L., Bartholomew D., Elkind J., Melendez J., and Furlong C. E., Detection of Staphylococcus aureus enterotoxin B at femtomolar levels with a miniature integrated two-channel surface plasmon resonance (SPR) sensor, *Biosens. Bioelectron.*, 17(6–7), 573–584, 2002.

18. Rowe C. A., Tender L. M., Feldstein M. J., Golden J. P., Scruggs S. B., MacCraith B. D., Cras J. J., and Ligler F. S., Array biosensor for simultaneous identification of bacterial, viral, and protein analytes, *Anal. Chem.*, 71(17), 3846–3852, 1999.

19. Ligler F. S., Taitt C. R., Shriver-Lake L. C., Sapsford K. E., Shubin Y., and Golden J. P., Array biosensor for detection of toxins, *Anal. Bioanal. Chem.*, 377(3), 469–477, 2003.

20. Rucker V. C., Havenstrite K. L., and Herr A. E., Antibody microarrays for native toxin detection, *Anal Biochem.*, 339(2), 262–270, 2005.

21. Rowe-Taitt C. A., Hazzard J. W., Hoffman K. E., Cras J. J., Golden J. P., and Ligler F. S., Simultaneous detection of six biohazardous agents using a planar waveguide array biosensor, *Biosens Bioelectron.*, 15(11–12), 579–589, 2000.

22. Sapsford K. E., Charles P. T., Patterson C. H. Jr, and Ligler F. S., Demonstration of four immunoassay formats using the array biosensor, *Anal. Chem.*, 74(5), 1061–1068, 2002.

23. Sapsford K. E., Liron Z., Shubin Y. S., and Ligler F. S., Kinetics of antigen binding to arrays of antibodies in different sized spots, *Anal Chem.*, 73(22), 5518–5524, 2001.

24. Sapsford K. E., Shubin Y. S., Delehanty J. B., Golden J. P., Taitt C. R., Shriver-Lake L. C., and Ligler F. S., Fluorescence-based array biosensors for detection of biohazards, *J. Appl. Microbiol.*, 96(1), 47–58, 2004.

25. Yau K. Y. F., Lee H., and Hall J. C., Emerging trends in the synthesis and improvement of hapten-specific recombinant antibodies, *Biotechnol. Adv.*, 21, 599–637, 2003.

26. Anodón A. and Martínez-Larrañaga M. R., Residues of antimicrobial drugs and feed additives in animal products: regulatory aspects, *Livestock Production Science*, 59, 183–198, 1999.

27. Bjurling P., Baxter G. A., Caselunghe M., Jonson C., O'Connor M., Persson B., and Elliott C. T., Biosensor assay of sulfadiazine and sulfamethazine residues in pork, *Analyst*, 125(10), 1771–1774, 2000.

28. Baxter G. A., Ferguson J. P., O'Connor M. C., and Elliott C. T., Detection of streptomycin residues in whole milk using an optical immunobiosensor, *J. Agric. Food. Chem.*, 49(7), 3204–3207, 2001.

29. Haughey S. A., Baxter G. A., Elliot C. T., Persson B., Jonson C., and Bjurling P., Determination of clenbuterol residues in bovine urine by optical immunobiosensor assay, *J. AOAC. Int.*, 84(4), 1025–1030, 2001.

30. Shelver W. L., Keum Y. S., Li Q. X., Fodley T. L., and Elliott C. T., Development of an immunobiosensor assay for the β-adrenergic compound zilpaterol, *Food and Agricultural Immunology*, 16(3–4), 199–211, 2005.

31. Samsanova J. V., Uskova N. A., Andresyuk A. N., Franek M., and Elliott C. T., Biacore biosensor immunoassay for 4-nonyphenols: assay optimization and applicability for shellfish analyses, *Chemosphere*, 57(8), 975–985, 2004.

32. McCarney B., Traynor I. M., Fodley T. L., Crooks S. R. H., and Elliott C. T., Surface plasmon resonance biosensor screening of poultry, liver and eggs for nicarbazin residues, *Anal. Chim. Acta*, 483(1–2), 165–169, 2003.

33. Crooks S. R. H., McCarney B., Traynor I. M., Thompson C. S., Floyd S., and Elliott C. T., Detection of levamisole residues in bovine liver and milk by immunobiosensor, *Anal. Chim. Acta*, 483(1–2), 181–186, 2003.

34. Stolker A. A. M., Brinkman U. A. Th., Chromatog J., *Analytical strategies for residue analysis of veterinary drugs and growth-promoting agents in food-producing animals—A review*, 1067, 15–53, 2005.

35. Traynor I. M., Crooks S. R. H., Bowers J., and Elliott C. T., Detection of multi-beta-agonist residues in liver matrix by use of a surface plasmon resonance biosensor, *Anal. Chim. Acta*, 483(1–3), 187–191, 2003.

36. Moghaddam A., Borgen T., Stacy J., Kausmally L., Simonsen B., Marvik O. J., Brekke O. H., and Braunagel M., Identification of scFv antibody fragments that specifically recognise the heroin metabolite 6-monoacetylmorphine but not morphine, *J. Immunol. Methods.*, 280(1–2), 139–155, 2003.

37. Ferguson L. R., Natural and human-made mutagens and carcinogens in the human diet, *Toxicology*, 181–182, 79–82, 2002.

38. Chemburu S., Wilkins E., and Abdel-Hamid I., Detection of pathogenic bacteria in food samples using highly-dispersed carbon particles, *Biosens. Bioelectron.*, 21(3), 491–499, 2005.

39. Than K. A., Stevens V., Knill A., Gallagher P. F., Gaul K. L., Edgar J. A., and Colegate S. M., Plant-associated toxins in animal feed: Screening and confirmation assay development, *Animal Feed Sci. Technol.*, 121, 5–21, 2005.

40. Kawatsu K., Hamano Y., and Noguchi T., Production and characterization of a monoclonal antibody against domoic acid and its application to enzyme immunoassay, *Toxicon*, 37, 1579–1589, 1999.

41. Finlay W. J. J., Shaw I., O'Reilly J., and Kane M., Generation of high affinity chicken scFv fragments for measurement of the Pseudonitzschia pungens toxin, domoic acid, *Appl. Environ. Microbiol.*, 72(5), 3343–3349, 2006.

42. Christiansen C. B. V., Arrays in biological and chemical analysis, *Talanta*, 56, 289–299, 2002.

43. Belleville E., Dufva M., Aamand J., Bruun L., Clausen L., and Christensen C. B. V., Quantitative microarray pesticide analysis, *J. Immunol. Meth.*, 286, 219–229, 2004.

44. Park I. S., Kim D. K., Adanyi N., Varadi M., and Kim N., Development of a direct-binding chloramphenicol sensor based on thiol or sulfide mediated self-assembled antibody monolayers, *Biosens Bioelectron.*, 19(7), 667–674, 2004.

45. Charlton K., Harris W. J., and Porter A. J., The isolation of super-sensitive anti-hapten antibodies from combinatorial antibody libraries derived from sheep, *Biosens. Bioelectron.*, 16(9–12), 639–646, 2001.

46. Rau D., Kramer K., and Hock B., Cloning, functional expression and kinetic characterization of pesticide-selective Fab fragment variants derived by molecular evolution of variable antibody genes, *Anal. Bioanal. Chem.*, 372(2), 261–267, 2002.

47. Homola J., Dostálek J., Chen S., Rasooly A., Jiang S., and Yee S. S., Spectral surface plasmon biosensor for detection of staphylococcal enterotoxin B in milk, *Int. J. Food Microbiol.*, 75, 61–69, 1999.

48. Quinn J. and O'Kennedy R., Transduction patterns and biointerfacial design of biosensors for "real-time" biomolecular interaction analysis, *Anal. Letts.*, 32(8), 147, 1999.

49. Otto A., Excitation of non-radioactive surface plasma waves in silver by the method of frustrated total internal reflection, *Z. Phys.*, 216, 398–410, 1968.

50. Otto A., A new method for excitating nonradiactive surface plasma oscillations, *Phys. Stat. Sol.*, 26, K99–K101, 1968.

51. Raether H., Surface plasma oscillations and their applications, *Phys Thin Films*, 9, 145–251, 1977.

52. Pockrand I., Swalen J. D., Gordon J. G., and Philpott M. R., Surface plasmon spectroscopy of organic monolayer assemblies, *Surface Sci.*, 74, 237–244, 1978.

53. Gordon II J. G. and Ernst S., Surface plasmons as a probe of the electrochemical interface, *Surface Sci.*, 101, 499–506, 1980.

54. Nylander C., Leidberg B., and Lind T., Gas detection by means of surface plasmon resonance, *Sens. Actuators*, 3, 79–88, 1982.

55. Liedberg B., Nylander C., and Lundström I., Surface plasmon resonance for gas detection and biosensing. *Sens. Actuators* 4, 299–304, 1983.

56. Kretschmann E. and Raether H., Radiative decay of non-radiative surface plasmons excited by light, *Z. Naturforsch*, 23(A), 2135–2136, 1968.

57. Kretschmann E., The determination of the optical constants of metals by excitation of surface plasmon resonance, *Z. Phys.*, 241, 313–324, 1971.

58. Löfås S., Malmqvist M., Rönnberg I., Stenberg E., Liedberg B., and Lundström I., Bioanalysis with surface plasmon resonance, *Sens. Actuators B*, 5, 79–84, 1991.

59. Liedberg B., Nylander C., and Lundström I., Biosensing with surface plasmon resonance—how it all started, *Biosens. Bioelectron.*, 10, i–ix, 1995.

60. Malmqvist M., BIACORE: An affinity biosensor system for characterisation of biomolecular interactions, *Biochem. Soc. Trans.*, 27, 335–340, 1999.

61. Salamon Z., Brown M. F., and Tollin G., Plasmon resonance spectroscopy: Probing molecular interactions within membranes, *Trends Biochem.*, 4, 213–219, 1999.

62. Liedberg B. and Johansen K., Affinity biosensing based on surface plasmon resonance detection, in *Biosensors: Techniques and Protocols*, ed. K. Rogers and A. Mulchandani (Totowa, NJ: Humana Press 1998); *Meth. Biotechnol.*, 7, 31–54, 1998.

63. De Bruijn H. E., Kooyman R. P. H., and Greve J., Choice of metal and wavelength for SPR sensors: Some considerations, *Appl. Optics.*, 31, 440–442, 1992.

64. Matsubara K., Kawata S., and Minami S., Optical chemical sensor-based on surface plasmon resonance. *Appl. Optics*, 27(6), 1160–1163, 1988.

65. Matsubara K., Kawata S., and Minami S., A compact surface plasmon resonance sensor for water in process, *Appl. Spect.*, 42, 229–240, 1988.

66. Stenberg E., Persson B., Roos H., and Urbaniczky C., Quantitative determination of protein with surface plasmon resonance by using radiolabelled proteins, *J. Colloid Interfac. Sci.*, 143, 513–526, 1991.

67. Löfås S. and Johnsson B., A novel hydrogel matrix on gold surfaces in surface plasmon resonance sensor for fast and efficient covalent immobilization of ligands, *J. Chem. Socc. Chem. Commun.*, 21, 1526–1528, 1990.

68. Hashimoto S., Principles of BIACORE, in *Real-Time Analysis of Biomolecular Interactions-Applications of BIAcore* (Tokyo, Japan: Springer-Verlag, 2000).

69. Hahnefeld C., Drewianka S., and Herberg F. W., Determination of kinetic data using surface plasmon resonance biosensors, in *Methods in Molecular Medicine*, ed. J. Decker and U. Reischl (Vol. 94), 2nd ed., Chapter 19 (Totowa, NJ: Humana Press, 2004) pp. 299–320.

70. Canziani G., Klakamp S., and Myszka D. G., Kinetic screening of antibodies from crude hybridoma samples using BIAcore, *Anal. Biochem.*, 325, 301–307, 2004.

71. Ober R. J. and Ward E. S., The choice of reference cell in the analysis of kinetic data using BIAcore, *Anal. Biochem.*, 271, 70–80, 1999.

72. Markey F., Principles of surface plasmon resonance, in *Real-Time Analysis of Biomolecular Interactions—Applications of BIAcore*, ed. K. Nagata and H. Handa (Tokyo, Japan: Springer-Verlag, 2000) pp. 13–32.

73. Haasnoot W., Cazemier G., Koets M., and Van Amerongen A., Single biosensor immunoassay for the detection of five aminoglycosides in reconstituted skimmed milk, *Anal. Chim. Acta*, 488(1), 53–60, 2003.

74. Rasooly A., Principles plasmon resonance analysis of Staphylococcal Enterotoxin B in food, *J. Food Protect.*, 64, 37–43, 2001.

75. Nagata K. and Handa H. (eds.), Real-time analysis of biomolecular interactions-applications of BIAcore (Tokyo, Japan: Springer-Verlag, 2000).

76. Rogers K. R., Principles of affinity-based biosensors, in *Biosensors Techniques and Protocols*, ed. K. Rogers and A. Mulchandani (Totowa, NJ: Human Press, 1998); *Meth. Biotechnol.*, 7, 3–18, 1998.

77. Leonard P., Hearty S., Quinn J., and O'Kennedy R., A generic approach for the detection of whole *Listeria monocytogenes* cells in contaminated samples using surface plasmon resonance, *Biosens. Bioelectron.*, 19, 1331–1335, 2004.

78. Kaiser T., Gudat P., Stock W., Pappert G., Grol M., Neumeier D., and Luppa P. B., Biotinylated steroid derivatives as ligands for biospecific interaction analysis with monoclonal antibodies using immunosensor devices, *Anal. Biochem.*, 282(2), 173–185, 2000.

79. O'Sullivan C. K., Vaughan R., and Guilbault G. G., Piezoelectric immunosensorstheory and applications, *Anal. Letts.*, 32, 2353–2377, 1999.

80. Bizet K., Gabrielli C., Perrot H., and Therasse J., Validation of antibody-based recognition by piezoelectric transducers through electroacoustic admittance analysis, *Biosens. Bioelectron.*, 13(3–4), 259–269, 1998.

81. Janshoff A., Galla H. J., and Steinem C., Piezoelectric mass-sensing devices as biosensors—an alternative to optical biosensors? *Angew. Chem. Int. Ed. Engl.*, 39(22), 4004–4032, 2000.

82. Uttenthaler E., Schraml M., Mandel J., and Drost S., Ultrasensitive quartz crystal microbalance sensors for detection of M13-Phages in liquids, *Biosens. Bioelectron.*, 16(9–12), 735–743, 2001.

83. Cheskis B. J., Karathanasis S., and Lyttle C. R., Estrogen receptor ligands modulate its interaction with DNA, *J. Biol. Chem.*, 272(17), 11384–11391, 1997.

84. Gerdon A. E., Wright D. W., and Cliffel D. E., Quartz crystal microbalance detection of glutathione-protected nanoclusters using antibody recognition, *Anal. Chem.*, 77(1), 304–310, 2005.

85. Shen Z., Stryker G. A., Mernaugh R. L., Yu L., Yan H., and Zeng X., Single-chain fragment variable antibody piezoimmunosensors, *Anal. Chem.*, 77(3), 797–805, 2005.

86. Haab B. B., Dunham M. J., and Brown P. O., Protein microarrays for highly parallel detection and quantitation of specific proteins and antibodies in complex solutions, *Genome Biol.*, 2(2), 0004.1–0004.13, 2001.

87. Nielsen U. B., Cardone M. H., Sinskey A. J., MacBeath G., and Sorger P. K., Profiling receptor tyrosine kinase activation by using Ab microarrays, *Proc. Natl. Acad. Sci.*, 100(16), 9330–9335, 2003.

88. Nielsen U. B. and Geierstanger B. H., Multiplexed sandwich assays in microarray format, *J. Immunol. Methods.*, 290, 107–120, 2004.

89. Merkel J. S., Michaud G. A., Salcius M., Schweitzer B., and Predki P. F., Functional protein microarrays: Just how functional are they, *Curr. Opin. Biotechnol.*, 16(4), 447–452, 2005.

90. Michaud G. A., Salcius M., Zhou F., Bangham R., Bonin J., Guo H., Snyder M., Predki P. F., and Schweitzer B. I., Analyzing antibody specificity with whole proteome microarrays, *Nat. Biotechnol.*, 21(12), 1509–1512, 2003.

91. James L. C., Roversi P., and Tawfik D. S., Antibody multispecificity mediated by conformational diversity, *Science*, 299(5611), 1362–1367, 2003.

92. James L. C. and Tawfik D. S., The specificity of cross-reactivity: promiscuous antibody binding involves specific hydrogen bonds rather than nonspecific hydrophobic stickiness, *Protein Sci.*, 12(10), 2183–2193, 2003.

93. Finlay W .J. J., deVore N. C., Dobrovolskaia E. N., Gam A., Goodyear C. S., and Slater J. E., Exploiting the avian immunoglobulin system to simplify the generation of recombinant antibodies to allergenic proteins, *Clin. Exp. Allergy*, 35(8), 1040–1048, 2005.

94. Davies E. L., Smith J. S., Birkett C. R., Manser J. M., Anderson-Dear D. V., and Young J. R., Selection of specific phage-display antibodies using libraries derived from chicken immunoglobulin genes, *J. Immunol. Meth.*, 186, 125–135, 1995.

95. Yamanaka H. I., Inoue T., and Ikeda-Tanaka O., Chicken monoclonal antibody isolated by a phage display system, *J. Immunol.*, 157, 1156–1162, 1996.

96. Andris-Widhopf J., Rader C., Steinberger P., Fuller R., and Barbas C. F., 3rd., Methods for the generation of chicken monoclonal antibody fragments by phage display, *J. Immunol. Methods.*, 28, 159–181, 2000.

97. Foti M., Granucci F., Ricciardi-Castagnoli P., Spreafico A., Ackermann M., and Suter M., Rabbit monoclonal Fab derived from a phage display library, *J. Immunol. Meth.*, 213(2), 201–212, 1998.

98. Lang I. M., Barbas C. F., 3rd, and Schleef R. R., Recombinant rabbit Fab with binding activity to type-1 plasminogen activator inhibitor derived from a phage-display library against human alpha-granules, *Gene.*, 172(2), 295–298, 1996.

99. Rader C., Ritter G., Nathan S., Elia M., Gout I., Jungbluth A. A., Cohen L. S., Welt S., Old L. J., Barbas C. F. 3rd., The rabbit antibody repertoire as a novel source for the generation of therapeutic human antibodies, *J. Biol. Chem.*, 275(18), 13668–136676, 2000.

100. Li Y., Cockburn W., Kilpatrick J. B., and Whitelam G. C., High affinity scFvs from a single rabbit immunized with multiple haptens, *Biochem. Biophys. Res. Comm.*, 268, 398–404, 2000.

101. Popkov M., Jendreyko N., Gonzalez-Sapienza G., Mage R. G., Rader C., and Barbas C. F. 3rd. Human/mouse cross-reactive anti-VEGF receptor 2 recombinant antibodies selected from an immune b9 allotype rabbit antibody library, *J. Immunol. Methods.*, 288(1–2), 149–164, 2004.

102. Els Conrath K., Lauwereys M., Wyns L., and Muyldermans S., Camel single-domain antibodies as modular building units in bispecific and bivalent antibody constructs, *J. Biol. Chem.*, 276(10), 7346–7350, 2001.

103. Huang Y., Verheesen P., Roussis A., Frankhuizen W., Ginjaar I., Haldane F., Laval S., et al., Protein studies in dysferlinopathy patients using llama-derived antibody fragments selected by phage display, *Eur. J. Hum. Genet.*, 13(6), 721–730, 2005.

104. Charlton K. A., Moyle S., Porter A. J., and Harris W. J., Analysis of the diversity of a sheep antibody repertoire as revealed from a bacteriophage display library, *J. Immunol.*, 15, 164(12), 6221–6229, 2000.

105. McElhiney J., Drever M., Lawton L. A., and Porter A. J., Rapid isolation of a single-chain antibody against the cyanobacterial toxin microcystin-LR by phage display and its use in the immunoaffinity concentration of microcystins from water, *Appl. Environ. Microbiol.*, 68(11), 5288–5295, 2002.

106. Brichta J., Vesela H., and Franek M., Production of scFv recombinant fragments against 2,4-dichlorophenoxyacetic acid hapten using naïve phage library, *Vet. Med. Czech.*, 9, 237–247, 2003.

107. Hennecke F., Krebber C., and Pluckthun A., Non-repetitive single-chain Fv linkers selected by selectively infective phage (SIP) technology, *Protein Eng.*, 11(5), 405–410, 1998.

108. Holliger P., Prospero T., and Winter G., "Diabodies": Small bivalent and bispecific antibody fragments, *Proc. Nat. Acad. Sci.*, 90, 6444–6448, 1993.

109. Perisic O., Webb P. A., Holliger P., Winter G., and Williams R. L., Crystal structure of a diabody, a bivalent antibody fragment, *Struct.*, 2, 1217–1226, 1994.

110. McGuinness B. T., Walter G., FitzGerald K., Schuler P., Mahoney W., Duncan A. R., and Hoogenboom H. R., Phage diabody repertoires for selection of large numbers of bispecific antibody fragments, *Nat. Biotech.*, 14, 1149–1154, 1996.

111. Kortt A. A., Lah M., Oddie G. W., Gruen C. L., Burns J. E., Pearce L. A., Atwell J. L., McCoy A. J., Howlett G. J., Metzger D.W., and Webster R. G., and Hudson P. J. Single-chain Fv fragments of anti-neuraminidase antibody NC10 containing five- and ten-residue linkers form dimers and with zero-residue linker a trimer, Protein Eng, 10(4), 423–433, 1997.

112. Atwell J. L., Breheney K. A., Lawrence L. J., McCoy A. J., Kortt A. A., and Hudson P. J., scFv multimers of the anti-neuraminidase antibody NC10: length of the linker between VH and VL domains dictates precisely the transition between diabodies and triabodies., *Protein Eng.*, 12(7), 597–604, 1999.

113. Todorovska A., Roovers R. C., Dolezal O., Kortt A. A., Hoogenboom H. R., and Hudson P. J., Design and application of diabodies, triabodies and tetrabodies for cancer targeting, *J. Immunol. Methods.*, 248(1–2), 47–66, 2001.

114. Huang B. C., Foote L. J., Lankford T. K., Davern S. M., McKeown C. K. and Kennel S. J., A diabody that dissociates to monomer forms at low concentration: effects on binding activity and tumor targeting, *Biochem. Biophys. Res. Commun.*, 327(4), 999–1005, 2005.

115. Krebber A., Bornhauser S., Burmester J., Honegger A., Willuda J., Bosshard H. R., and Pluckthun A., Reliable cloning of functional antibody variable domains from hybridomas and spleen cell repertoires employing a reengineered phage display system, *J. Immunol. Methods.*, 201(1), 35–55, 1997.

116. Pluckthun A. and Pack P., New protein engineering approaches to multivalent and bispecific antibody fragments, *Immunotechnology*, 3(2), 83–105, 1997.

117. Raats J. M. and Hof D., Recombinant antibody expression vectors enabling double and triple immunostaining of tissue culture cells using monoclonal antibodies, *Eur. J. Cell. Biol.*, 84(4), 517–521, 2005.

118. Barbas C. F., Kang A. S., Lerner R. A., and Benkovic S. J., Assembly of combinatorial libraries on phage surfaces—the Gene—111 site, *P.N.A.S.*, 88(18), 7978–7982, 1991.

119. Hoogenboom H. R., Griffiths A. D., and Johnson K. S., Multi-subimit proteins on the surface of filamentous phage—methodologies for displaying antibody (Fab) heavy and light chains, *Nucl. Acids Res.*, 19 (15), 4133–4137, 1991.

120. Rothlisberger D., Honegger A. and Pluckthun A., Domain interactions in the Fab fragment: A comparative evaluation of the single-chain Fv and Fab format engineered with variable domains of different stability, *J. Mol. Biol.*, 347(4), 773–789, 2005.

121. Townsend S., Finlay W. J. J., Hearty S., and O'Kennedy R., Optimizing recombinant antibody function in SPR immunosensing; the influence of antibody structural format and chip surface chemistry on assay sensitivity, *Biosens. Bioelectron.*, 22(2), 268–274, 2006.

122. Hust M. and Dubel S., Mating antibody phage display with proteomics, *Trends. Biotechnol.*, 22(1), 8–14, 2004.

123. Zhou B., Wirsching P., and Janda K. D., Human antibodies against spores of the genus Bacillus: a model study for detection of and protection against anthrax and the bioterrorist threat, *Proc. Natl. Acad. Sci.*, 99(8), 5241–5246, 2002.

124. Schwarz M., Rottgen P., Takada Y., Le Gall F., Knackmuss S., Bassler N., Buttner C., Little M., Bode C., and Peter K., Single-chain antibodies for the conformation-specific blockade of activated platelet integrin alphaIIbbeta3 designed by subtractive selection from naive human phage libraries, *FASEB J.*, 18(14), 1704–1706, 2004.

125. Van Antwerp J. J. and Wittrup K. D., Fine affinity discrimination by yeast surface display and flow cytometry, *Biotechnol. Prog.*, 16(1), 31–37, 2000.

126. Daugherty P. S., Chen G., Olsen M. J., Iverson B. L., and Georgiou G., Antibody affinity maturation using bacterial surface display, *Protein Eng.*, 11(9), 825–832, 1998.

127. Daugherty P. S., Olsen M. J., Iverson B. L., and Georgiou G., Development of an optimized expression system for the screening of antibody libraries displayed on the *Escherichia coli* surface, *Protein Eng.*, 12(7), 613–621, 1999.

128. Russell S. J., Hawkins R. E., and Winter G., Retroviral vectors displaying functional antibody fragments, *Nucleic Acids Res.*, 21(5), 1081–1085, 1993.

129. Mattheakis L. C., Bhatt R. R., and Dower W. J., *In vitro* polysome display system for identifying ligands from very large reptide libraries, *P.N.A.S.*, 91(19), 9022–9026, 1994.

130. Hanes J. and Pluckthun A., In vitro selection and evolution of functional proteins by using ribosome display, *Proc. Natl. Acad. Sci.*, 94(10), 4937–4342, 1997.

131. Chen G., Hayhurst A., Thomas J. G., Harvey B. R., Iverson B. L., and Georgiou G., Isolation of high-affinity ligand-binding proteins by periplasmic expression with cytometric screening (PECS), *Nature Biotechnol.* 19(6), 537–542, 2001.

132. Harvey B. R., Georgiou G., Hayhurst A., Jeong K. J., Iverson B. L., and Rogers G. K., Anchored periplasmic expression, a versatile technology for the isolation of high-affinity antibodies from Escherichia coli-expressed libraries, *Proc. Natl. Acad. Sci.*, 101(25), 9193–9198, 2004.

133. Vikholm I., Self-assembly of antibody fragments and polymers onto gold for immunosensing, *Sens. Actuators B*, 106, 311–316, 2004.

134. Vikholm-Lundin I. and Albers W. M., Site-directed immobilization of antibody fragments for detection of C-reactive protein, *Biosens. Bioelectron.*, 21(7), 1141–1148, 2006.

135. Karyakin A. A., Presnova G. V., Rubtsova M. Y., and Egorov A. M., Oriented immobilization of antibodies onto the gold surfaces via their native thiol groups, *Anal. Chem.*, 72, 3805–3811, 2000.

136. Steinhauer C., Wingren C., Hager A. C., and Borrebaeck C. A., Single framework recombinant antibody fragments designed for protein chip applications, *Biotechniques.*, *Suppl*: 38–45, 2002.

137. Wingren C., Steinhauer C., Ingvarsson J., Persson E., Larsson K., and Borrebaeck C. A. K., Microarrays based on affinity-tagged single-chain Fv antibodies: Sensitive detection of analyte in complex proteomes, *Proteomics*, 5, 1281–1291, 2005.

138. Peluso P., Wilson D. S., Do D., Tran H., Venkatasubbaiah M., Quincy D., Heidecker B., Poindexter K., Tolani N., Phelan M., Witte K., Jung L. S., Wagner P., and Nock S., Optimizing antibody immobilization strategies for the construction of protein microarrays, *Anal. Biochem.*, 312, 113–124, 2003.

139. Lu B., Smyth M. R., and O'Kennedy R., Oriented immobilization of antibodies and its applications in immunoassays and immunosensors, *Analyst*, 121, 29R–32R, 1996.

140. Angenedt P., Glökler J., Murphy D., Lehrach H., and Cahill D. J., Towards optimized antibody microarrays: a comparison of current microarray support materials, *Anal. Biochem.*, 309, 253–260, 2002.

141. Angenedt P., Glökler J., Sobek J., Lehrach H., and Cahill D. J., Next generation of protein microarray support materials: Evaluation for protein and antibody microarray applications, *J. Chromatog. A*, 1009, 97–104, 2003.

142. Pavlickova P., Schneider E. M., and Hug H., Advanced in recombinant antibody microarrays, *Clin. Chim. Acta.*, 343, 17–35, 2004.

143. Kim J. K., Shin D. S., Chung W. J., Jang K. H., Lee K. N., Kim Y. K., and Lee Y. S., Effects of polymer grafting on a glass surface for protein chip applications, *Colloids. Surf. B*, 33, 67–75, 2004.

144. Lal S. P., Christopherson R. I., and dos Remedios C. G., Antibody arrays: an embryonic but rapidly growing technology, *Drug. Discov. Today.*, 7, S143–S149, 2002.

145. Elia G., Silacci M., Scheurer S., Scheuermann J., and Neri D., Affinity-capture reagents for protein arrays, *Trends. Biotechnol.*, 20, S19–S22, 2002.

146. Charles P. T., Goldman E. R., Rangasammy J. G., Schauer C. L., Chen M. S., and Taitt C. R., Fabrication and characterization of 3D hydrogel microarrays to measure antigenicity and antibody functionality for biosensor applications, *Biosens. Bioelectron.*, 20(4), 753–764, 2004.

147. Wacker R., Schröder H., and Niemeyer C. M., Performance of antibody microarrays fabricated by either DNA-directed immobilization, direct spotting, or streptavidin-biotin attachment: a comparative study, *Anal. Biochem.*, 330, 281–287, 2004.

148. Neubert H., Jacoby E. S., Bansal S. S., Iles R. K., Cowan D. A., and Kicman A. T., Enhanced affinity captures MALDI-TOF MS: Orientation of an immunoglobulin G using recombinant protein G, *Anal. Chem.*, 74(15), 3677–3683, 2002.

149. Wang H., Liu Y., Yang Y., Deng T., Shen G., and Yu R., A protein A-based orientation-controlled immobilization strategy for antibodies using nanometer-sized gold particles and plasma-polymerized film, *Anal. Biochem.*, 324(2), 219–226, 2004.

150. Oh B. K., Lee W., Chun B. S., Bae Y. M., Lee W. H., and Choi J. W., The fabrication of protein chip based on surface plasmon resonance for detection of pathogens, *Biosens. Bioelectron.*, 20(9), 1847–1850, 2005.

151. Oh S. W., Moon J. D., Lim H. J., Park S. Y., Kim T., Park J., Han M. H., Snyder M., and Choi E. Y., Calixarene derivative as a tool for highly sensitive detection and oriented immobilization of proteins in a microarray format through noncovalent molecular interaction, *FASEB J.*, 19(10), 1335–1337, 2005.

152. Hirlekar Schmid A., Stanca S. E., Thakur M. S., Ravindranathan Thampi K., and Raman Suri C., Site-directed antibody immobilization on gold substrate for surface plasmon resonance sensors, *Sens. Actuators B*, 113(1), 297–303, 2006.

153. Blank K., Lindner P., Diefenbach B., and Pluckthun A., Self-immobilizing recombinant antibody fragments for immunoaffinity chromatography: generic, parallel, and scalable protein purification, *Protein Expr Purif.*, 24(2), 313–322, 2002.

154. Song E., Zhu P., Lee S. K., Chowdhury D., Kussman S., Dykxhoorn D. M., Feng Y., Palliser D., Weiner D. B., Shankar P., Marasco W. A., and Lieberman J., Antibody-mediated *in vivo* delivery of small interfering RNAs via cell-surface receptors, *Nat. Biotechnol.*, 23(6), 709–717, 2005.

155. Diamond D. and Nolan K., Calixarenes: Designer ligands for chemical sensors, *Anal. Chem.*, 73 (1), 22A–29A, 2001.

156. Lee Y., Lee E. K., Cho Y. W, Matsui T., Kang I., Kim Y., and Hi-Han M., Proteochip: A highly sensitive protein microarray prepared by a novel method of protein immobilization for application of protein-protein interaction studies, *Proteomics*, 3, 2289–2304, 2003.

157. Lee H. Y., Jung H. S., Fujikawa K., Park J. W., Kim J. M., Yukimasa T., Sugihara H., and Kawai T., New antibody immobilization method via functional liposome layer for specific protein assays, *Biosens. Bioelectron.*, 21(5), 833–838, 2005.

8 Electronic Nose Applications in Medical Diagnose

Corrado Di Natale, Eugenio Martinelli,
Marco Santonico, Arnaldo D'Amico
Department of Electronic Engineering
University of Rome "Tor Vergata"
Via del Politecnico, Roma, Italy

Giorgio Pennazza
Faculty of Engineering, University
"Campus Bio-Medico di Roma"
Via Alvaro del Portillo, Italy

Roberto Paolesse
Department of Chemical Science and Technology
University of Rome "Tor Vergata"
Via della Richerca Scientifica, Italy

Claudio Roscioni
Azienda Ospedaliera S. Camillo-Forlanini
Via Portuense, Italy

CONTENTS

8.1 INTRODUCTION

Noninvasive techniques for early diagnosis are of great importance in health-care strategies. The importance of prevention is crucial in increasing the number of successful treatment of diseases. On the other hand, prevention in order to be accepted by population necessitates noninvasive procedures. Among the many different approaches to diagnose, chemical analysis is an optimal candidate to satisfy both prevention and noninvasiveness [1]. The importance of chemistry in medical diagnoses was efficiently stated by Jellum et al. [2], "If one is able to identify and determine the concentration of all compounds inside the human body, including high molecular weight as well as low molecular weight substances, one would probably find that almost every known disease would result in characteristic changes of the biochemical composition of the cells and the body fluids." This proposition is only partially exploited by the current clinical chemistry where only the composition of human fluids (bloods and urines) is analyzed, and correlation between disease and chemical compounds are available for a very large class of pathologies. In addition, it is clear that the chemical composition of either a solid or a liquid is reflected in its volatile part. This process is mediated by the physical and chemical properties of each molecule. As a result, the volatile organic compounds (VOCs) found in the air surrounding living beings may contain meaningful information about the internal chemistry of the body and then it can provide a means for the identification of diseases.

The perception of volatile compounds through the human sense of smell is called odor. Odor was actively used in the past for medical diagnosis, and relationships between perceived odor and diseases were found for many kinds of pathologies [3]. In modern times, analytical chemistry developed instruments, such as gas chromatography (GC) and/or GC linked with mass spectrometry (GC-MS), that allow for a separation and identification of compounds in complex mixtures. The application of these instruments enlighted the human perceptions identifying, in some cases, the molecules responsible of typical odors occurring in specific diseases. On the other hand, analytical chemistry pointed out, as potential indicators of disease, molecules that are not perceived by the smell. Very promising results of these studies were those concerned with the identification of VOCs produced by microorganisms that colonize the human body. The positive results obtained with in vitro cultures were soon complemented with measurements on human substrates such as urine and blood. Analytical chemistry studies arose the interest in the electronic nose community and rather soon medical diagnosis started to be considered as possible application field of electronic nose. To this regard, the advent of a machine able to record human odor for diagnostics scope was devised since the sixties [4]. The history of artificial olfaction began in the eighties when the absence of selectivity, one of the major drawbacks of chemical sensors, was taken into consideration as the basis for a novel instrument able to provide global information about samples. This qualitative approach to the global properties of a sample is very similar to the strategy of natural olfaction processes with odorants [5]. These instruments are basically arrays on nonselective sensors. Sensors are characterized by a wide spectrum of sensitivity to many odorants, with a large overlap of response toward several compounds. This fundamental characteristic of artificial sensors is similar to that found in natural olfaction receptors [6]. This similarity is the basis on which artificial olfaction systems are developed. The sensor's response is not univocally correlated with the concentration of a single compound, but rather it is a sort of combination of the whole chemical information contained in a sample. The performance of natural olfaction at molecular discrimination and recognition results from the complex sensory signal processing carried out in the olfactory bulb and cortex. In the same way, most of the features of the artificial olfaction are revealed after a proper application of multicomponent data analysis ranging from classical statistics to chemometrics and neural networks [7]. The development of electronic noses has become soon a well-consolidated field of research, and several examples of these arrays have been widely reported in the literature and different instruments are also commercially available [8].

It has to be remarked that in spite of the widely accepted term "electronic nose," current devices are still far from the structure and functions of natural olfaction sense. The unique common feature between artificial and natural system is that both are largely based on arrays of nonselective sensors. The concept underlying electronic nose systems has been demonstrated to be independent on the particular sensor mechanism; indeed during the last two decades almost all the available sensor technologies have been utilized as electronic noses. Clearly, all these sensors are very different from the natural receptors. These dissimilarities make the perception of electronic nose very different from that of natural olfaction, so that the instrumental perception of the composition of air cannot be called odor measurement because odor is the sensation of smell as perceived by human olfaction. Nonetheless, the term odor analysis with electronic noses is now largely adopted, but it is important to keep in mind, especially in medical applications, that the electronic nose measurement may be very distant from the human perception.

The application of electronic noses to medical issues may be divided into two main areas related to measures in vitro and in vivo. In vitro measures were addressed to the identification of bacteria strains in cultures. In both cases, the identification of foreign microorganisms as source of infections is of great interest. The identification of bacteria in cultures has been demonstrated since several years [9,10]. Although, microorganisms' identification has a number of different applications including environment and food industry, a great interest about bacteria is the possibility to identify the pathogenic bacteria strain either in the environment or, for diagnostic scopes, in living tissues. Positive results were obtained to this regard about the identification of bacteria responsible of gynecological [11], stomach (*Helicobacter pilory*) [12], and ophthalmic infections [13]. These investigations triggered the possibility to detect infections in humans. The rapid identification of viral or bacterial origin of pathology is of great importance for a rapid therapy and, most of all, to avoid the massive use of antibiotics. To this regard, it is important to mention the results obtained in the case of pneumonia diagnosis [14]. It is interesting to indicate, as a further support to the assumed connection between odor and diagnoses, the feasible use of trained dogs as "diagnostic instruments." The perception of pathologies by dogs was anecdotically reported in the past, and in the last couple of years some result appeared in literature concerned with the detection of cancers of skin [15], bladder [16], lung, and breast [17]. These experiences point to the fact that some diseases are associated to particular classes of volatile compounds. Nonetheless, since dogs are rather evolved animals and their perception is likely based on more than the olfaction, it is not completely clear that, in these cases, odor is a sufficient source of information for diagnosis. In this chapter, the application of electronic noses to medical diagnosis is illustrated with particular emphasis to in vivo measurements of skin and breath odors. The diagnostic potentialities of these techniques in the case of lung cancer and schizophrenia are discussed in the following sections.

8.2 THE CASE OF BREATH ANALYSIS

Since several organs contribute to the composition of breath, it is expected to be rich in information about many biochemical processes and their alteration due to pathologies [18]. Modern breath analysis started in the seventies when Pauling et al. found more than 200 different compounds by analyzing breath with GC [19]. In the following decades, a great interest arose about the physiological meaning of volatile substances in breath and some correlations between specific compounds and pathologies emerged [20]. The fundamental components of breath are a mixture of nitrogen, oxygen, carbon dioxide, water, and inert gases. Beside these there is a long list of compounds occurring at concentrations in nmol/l–pmol/l (ppbv–pptv) range. Among these compounds those indicating the presence of particular pathologies have to be searched [21]. Volatile compounds may be generated in the body or may even be absorbed from the environment. These two groups of substances, exogenous and endogenous ones, have an informative content useful to be investigated for different scopes. Exogenous molecules, especially halogenated organic compounds, may be analyzed for environmental or expositional issues to determine compound uptake and elimination into the body. But to monitor metabolic or pathological processes endogenous substances are most important. For some of these compounds there are hypothesis about the production mechanisms leading to a clear and assessed relationship with the pathology. In other cases, different abundances are only observed and no clear explanation of their connection with diseases is available at the moment. On the basis of these studies lists VOCs or sets of VOCs seem to have predictive value even in complex diseases or disease states, such as liver diseases [22], heart failures [23], breast cancer [24], and schizophrenia [25]. The following list of endogenous VOCs with relative diseases is not exhaustive. The treatment is organized in a list of the main biomarkers for which either theoric or experimental connections with diseases have been individuated.

8.2.1 ETHANE AND PENTANE

A close correlation between clinical conditions with high peroxidative activity and the exhalation of ethane and pentane has been demonstrated [26]. There are still some problems in the interpretation of the hydrocarbon breath test for ethane and pentane because there are other potential sources of hydrocarbons in the body, such as protein oxidation and bacterial metabolism. Nonetheless, these secondary sources seem not to interfere with the breath test analysis [27]. Since hydrocarbons are stable end products of lipid peroxidation and they show low solubility in blood, they can be excreted into breath within minutes of their formation in tissues providing a direct evidence of the progress of oxidative damages [28].

8.2.2 ISOPRENE

Isoprene is known to be derived from the cholesterol synthesis pathway [29]. Nonetheless, a fraction of isoprene in breath may be of bacterial origin and may be indicative of oxidative damage of the fluid lining of the lung [30] and in more complex diseases such as cystic fibrosis [31].

8.2.3 ACETONE

It is one of the most abundant compounds in human breath. Acetone concentration strongly increases in patients with (uncontrolled) diabetes mellitus [32].

8.2.4 ETHANOL AND METHANOL

Ethanol is an obvious indicator of alcohol addiction, but there are potential source of endogenous short-chain alcohols in the intestinal bacterial flora [33].

8.2.5 SULPHUR COMPOUNDS

Under normal conditions, the concentrations of sulfur-containing compounds in human blood and breath are very low, and a high concentration is likely related to liver diseases. Compounds like ethyl mercaptan, dimethylsulfide, or dimethyldisulfide are responsible for the characteristic odor in the breath of cirrhotic patients.

8.2.6 AMINES

Amines are typical products of putrefaction processes; they are also of the characteristic odor of uremic breath due to elevated concentrations of *dimethylamine* and *trimethylamine* [34].

8.2.7 AMMONIA

Ammonia is another compound abundant in the breath of uremic patients and in cases of patients with severe kidney failure [35]. The cases mentioned earlier show that large classes of pathologies are responsible of the alteration of breath composition; nonetheless, not all these diseases would take a decisive benefit from breath analysis because either other efficient techniques are still in use (e.g., for the diabetes) or the alteration of breath occurs when other symptoms are clearly evident and the disease has been diagnosed. In other cases, breath analysis may cover the initial phases of the disease when symptoms are seldom manifested and they still do not completely justify the application of invasive diagnostic tools. This situation becomes very important in case of cancers where early detection dramatically reduces the fatalities. For this reason, the correlation between lung cancer and breath analysis has been actively studied in the last two decades.

8.2.8 Lung Cancer

Since the eighties, a number of compounds were indicated as peculiar of the breath of lung cancer affected people [36,37]. These are either normally absent or, in case of disease, are present at anomalous concentrations. The main part of these compounds is listed in Table 8.1. A wide investigation about the possibilities to identify lung cancer from the analysis of the breath was carried out by Phillips, who pioneered the studies on breath analysis. Phillips also defined an optimized breath collecting apparatus (BCA), which allowed collecting 10 L of breath during a 5-minute continuous breathing of the patient [38]. In one of these studies [39], investigating the breath of about 60 patients affected by lung cancer, the application of pattern recognition methods to GC data provided a correct diagnosis in 71% of the cases.

TABLE 8.1
The Main Part of Listed Compounds

Acids	Carbonyls	Alcohol	Steroids
n-Hexanoic	γ-C8-Lactone	Phenol	17-Oxo-5a-androsten-3-yl sulfate
2-Methylhexanoic	γ-C9-Lactone	Tetradecanol	Cholesterol
3-Methylhexanoic	γ-C10-Lactone	n-Hexadecanol	Squalene
4-Ethylpentanoic			5α-Androst-16-en-3a-ol
(Z)-3-Methyl-2-hexanoic			5α-Androst-16-en-3b-ol
2-Ethylhexanoic			5α-Androst-16-en-3-one
n-Ethylheptanoic			
2-Methylheptanoic			
(E)-3-Methyl-2-hexanoic			
n-Octanoic			
2-Methyloctanoic			
4-Ethylheptanoic			
n-Nonanoic			
2-Methylnonanoic			
4-Ethyloctanoic			
n-Decanoic			
2-Methyldecanoic			
4-Ethylnonanoic			
9-Decenoic			
n-Undecanoic			
4-Ethyldecanoic			

About 67 possible indicators were found in the breath of patients. Among them only 22 were determinant for the correct identification of the disease. In particular, the differences with normal breath were not related to novel compounds but rather to a different concentration of some compounds that are present also in the normal breath. Among these compounds isoprene, benzene, and four derivates of benzene (o-toluidine and aniline among the others) were found. It is worth to note that the presence of benzene in human breath is still not explained.

In conclusion, no individual compound can be considered as a marker, but the disease identification becomes possible when many compounds are considered and pattern recognition is applied. This condition meets very well with the characteristics of electronic noses because in these instruments the response of each sensor is not univocally correlated with the concentration of a single compound but it is rather a sort of combination of all the chemical information contained in a sample. In the following section, the results of an investigation aimed at studying the possibility to identify lung cancer affected individuals from the analysis of their breath are illustrated [40]. Experiments were carried out with the electronic nose developed at the University of Rome Tor Vergata [41]. This instrument is based on eighth Quartz microbalances coated by different metalloporphyrins [42]. A total number of 42 volunteers, affected by various forms of lung cancer, have been recruited at the C. Forlanini Hospital in Rome. Thirty-five of them were hospitalized waiting for a surgical treatment. Nine patients have been checked after a surgical removal of the tumor mass from the lung. Two patients were measured before and after the surgical operation. Eighteen volunteers have been recruited among the medical and nurse staff of the hospital, as reference. Each subject was required to follow the same diet and the same procedure for mouth hygiene. Measurements have been performed in the morning before any food intake. Individuals were required to deeply breathe in a sterile bag (volume of about 4 L). Breath samples were immediately analyzed with the electronic nose. All measurements were performed on-site. Each subject was measured twice, and the average measurement was used in the data analysis. In order to minimize the influence of possible instrumental drift, the measurement sequence was randomized. The whole experiment lasted five weeks. Postsurgical patients were checked about 1 month after the operation, in the occasion of a periodical control at the hospital site. Figures 8.1 and 8.2 show the main steps of the breath analysis procedure: the breath collection and the electronic nose measurement.

The data analysis presented here has been performed by partial least squares-discriminant analysis (PLS-DA). Partial least squares (PLS) allows for a straight calculation of the discriminant analysis avoiding the drawbacks due to sensor correlation. Furthermore, PLS provides a decomposition of the sensors data in latent variables that can be plotted to provide a visual representation of the classification properties [43]. Three different classes were considered (lung cancer diseased, reference group, and post-surgery patients). In this study we were concerned with testing the capability of the electronic nose to correctly classify the groups of subjects. The simplest technique that can be used for the scope is the linear discriminant analysis. As in any supervised classification technique, classes have to be chosen a priori. The natural choice for the samples in this experiment consisted of three classes including the patients with lung cancer, the controls, and the patients after the surgery.

FIGURE 8.1   The first step of breath analysis for lung cancer detection. Patient is asked to breath in a double compartment bag. The first part of the breath is removed and the second part is then used for analysis.

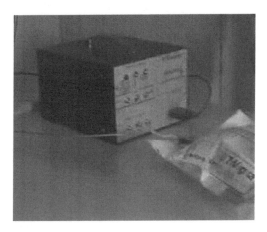

FIGURE 8.2   The bag with the collected breath is connected to the inlet of the electronic nose. In picture the latest version of the electronic nose developed at the University of Rome Tor Vergata is shown.

With this classification scheme a PLS-DA model was built. The best fitting model included five latent variables. Figure 8.3 shows the plot of the first two latent variables of the PLS-DA model. In the plot, about 92% of the total variance of the data is represented. As it can be seen, a clear separation between the data related to patients with lung cancer and the other samples is observed. On the other hand, the samples related to postsurgery and healthy reference show some overlap. A numerical evaluation of the classification properties can be obtained by considering the cross-validation of the PLS-DA method according to a "leave-one-out" technique [44].

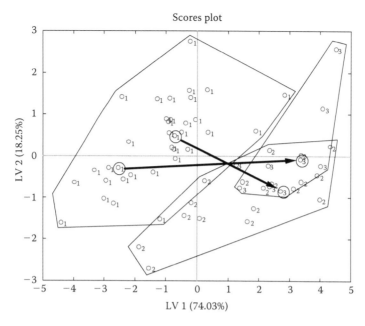

FIGURE 8.3   PLS-DA score plot of the first two latent variables (74% and 18% of explained variance respectively). Classes are labeled as—(1) lung cancer, (2) controls, and (3) postsurgery. Arrows indicate the two patients measured before and after the surgical treatment.

The results indicated that 100% of the lung cancer patients were correctly classified, 94% of the controls were correctly identified, and 6% of them have been classified as belonging to the postsurgery group. Concerning the samples of the postsurgery group, 44% have been classified as belonging to an autonomous class, while 56% as been classified as healthy controls. It is worth mentioning that the data related to the two patients were measured twice, before and after surgery. In Figure 8.3, the migration of data points from the class of lung cancer diseases to the class of controls is highlighted for the two cases measured before and after surgery.

8.3   THE CASE OF SKIN ANALYSIS

Human skin is known to be colonized by a huge number of bacteria that live as commensals on the surface and within the follicles. It is possible to describe the basic pattern of colonization of a healthy human skin. Variations of this pattern may be observed: dry skin supports a low level of colonization, while moist areas provided with sebaceous and apocrine glands are the most heavily populated. The resident aerobic flora consists of gram-positive cocci of staphylococcus and micrococcus, and a variety of gram-positive rods, mainly corynebacterium. The main anaerobic residents are propionibacteria, which are localized in the follicles of the sebaceous glands of adults. The microbial flora, usually localized on the skin, appears to have several functions, the most important of which is probably the defense against pathogenic bacterial and micotic infections. In an adult subject, skin living microorganisms may be mainly observed, at different concentrations, in the following sites: the nasal vestibule, the external auditory meatus, the axilla, the perineum and the groin, the scalp, the face, and the limbs. The number of microorganisms changes with age, sex, and race. Sweat is sterile and mostly odorless by its own when secreted. The main role of the bacteria in the odor formation appears to be the breakage of the precursor–odorant complex and the cleavage of the covalent bonds holding the acid molecules to the precursors [45]. As an example, axillary odor is a distinctive malodorous scent of adults that is generated when gram-positive microorganisms interact with the apocrine sweat. Recently, numerous studies demonstrated that the characteristic human axillary odors consisted of C6 to C11 normal, branched and unsaturated aliphatic acids, alcohol, carbonyls, and some steroids. Among the steroids, 5a-androst-16-en-3-one is supposed to represent a male pheromone. This compound results from the metabolism of coryneform bacteria that are reactive especially in axillas in men while in other mammalians it acts as a pheromone, for example, pigs, boars [46]. In addition, since different levels of 5a-androst-16-en-3-one were found to be significantly higher in men than in women, questions arose on whether, how and how much females might perceive it. Concerning the role of skin headspace composition in pathology identification, it is widely accepted that some diseases may be associated with skin odor alterations [3]. In general, since the body odor results from the combined action of skin glands and bacterial populations, it is expected to be sensitive to a large number of biochemical processes and to any alteration of them induced by pathological states.

In the following section the case of schizophrenia is discussed and an experiment aimed at identifying patients according to the chemical profile of the skin headspace is illustrated.

8.3.1 SCHIZOPHRENIA

Schizophrenia is generally defined as a complex pathology not immediately connected to specific genetic and environmental causes. The observation of a number of simultaneous behavioral, emotional, and cognitive disturbances is the adopted protocol for the diagnosis of this disease [47]. For this reason, the introduction of objective diagnostic tools for schizophrenia is highly recommended. Categorical systems of classification of mental disorders provide clear descriptions of diagnostic classes to enable clinicians and investigators to diagnose, communicate about, study, and treat people with mental disorders. Nonetheless, most of the reported clinical presentations are not clearly classified according to the available etiopathogenic or pathophysiological criteria. Since a number of psychotic symptoms, present in both schizophrenia and other mental disorders (i.e., bipolar disorders, personality disorders), are found as continuously distributed phenomena without clear boundaries, the adoption of a dimensional system able to classify clinical presentations on quantification of attributes rather than on the assignment to categories seems more promising. Nonetheless, this is a common problem whenever decisions have to be taken on the basis of qualitative observations. In recent years, genetic and environmental influences on schizophrenia were studied, putting in evidence that beside any neuronal, neuroanatomic, and neurochemical factors, schizophrenia has familial and genetic components and also that environmental factors together play a significant role in the pathophysiology of the disease. From the biochemical point of view, some hypotheses connect symptoms of schizophrenia to a functional hyperactivity of the cerebral dopaminergic system. High levels of dopamine may result from excessive biosynthesis of the amino acid tyrosine, the precursor of dopamine. Dopamine growth was linked to a faulty gene that codes for the enzyme dopamine-*b*-hydroxylase, which converts dopamine to norepinephrine. As a partial demonstration, the block of this enzyme with the drug disulfiram was demonstrated long time ago [48]. In alcoholics who overdosed on disulfiram, the treatment resulted in symptoms indistinguishable from schizophrenia.

In the late sixties, the sweat of schizophrenics of an odorous substance (known as trans-3-methyl-2-hexenoic acid, a metabolic product of 6-hydroxydopamine) [49,50] supported the dopamine hypothesis. After that it was argued that since trans-3-methyl-2-hexenoic acid is a metabolic product of 6-hydroxydopamine its presence in schizophrenic patients could be explained through an auto-oxidization mechanism of dopamine excess to 6-hydroxydopamine [51]. As a further consequence, this mechanism also causes degeneration of peripheral sympathetic nerve terminals, resulting in a marked and long-lasting depletion of norepinephrine. Further evidence supporting 6-hydroxydopamine as a neural degenerative agent of noradrenergic nerve endings came from the known hallucinogenic activity in humans of phenethylamine derivatives with the same 2,4,5-substitution pattern of 6-hydroxydopamine. A peculiar skin odor associated with schizophrenia was observed in some subjects several years ago [52]. As mentioned before, the off-odor was related to the presence of as trans-3-methyl-2-hexenoic acid, and this compound was proposed as a chemical marker for the diagnosis of schizophrenia. This assumption was not confirmed by later gas chromatography-mass spectrometry (GC-MS) investigation because the studies revealed a large variability of the concentration among individuals and also to the fact that no absolute relation between the presence of this compound and the occurrence of the disease was observed. However, the original idea was not completely abandoned, and more recently, anomalies in the chemical composition of the expired breath individuals were also reported [21]. Recently, a number of patients of a therapy unit in Rome were involved in a measurement campaign aimed at identifying schizophrenic patients from the analysis of the volatile component present in their sweat [53]. Three groups of individuals were recruited according to the following classes: schizophrenic, other mental disorders, and a control group. Mental disorders were diagnosed according to the fourth edition of *Diagnostic and Statistical Manual of Mental Disorders*. The body odor was sampled on the upper side of the forearms. The sampling was performed with a cotton compress applied for 30 minutes on the skin [54]. One half of each compress was then analyzed with the electronic nose and the other one with the GC-MS. In order to remove the influence of detergents and perfumes, forearms were washed no less than 2 hours before the sampling with the same nonperfumed and neutral soap. After washing, forearms were dried with sterile cotton and kept in air, without any contact with clothes. Staff at the therapy unit supervised the correct fulfillment of the procedure. The analysis was carried out by GC and the electronic nose developed at the University of Rome Tor Vergata. Results showed that the chemical composition of the odor emanating from the skin of individuals has a certain relation with the occurrence of mental disorders and schizophrenia in particular. GC-MS analysis revealed that trans-3-methyl-2-hexenoic acid is not a reliable indicator because it is not always present in the odor of schizophrenic subject, and sometimes it composes the body odor of healthy patients. Nonetheless, when it is present, its concentration is higher is schizophrenic than in other individuals. On the other hand, GC-MS spectra show a more rich composition of the odor of schizophrenic patients. Gas sensors exhibited higher responses to schizophrenic patients with respect to the others confirming the trend evidenced by GC-MS. Both GC-MS and sensor array data were analyzed with a classification algorithm (PLS-DA) in order to ascertain the possibility to discriminate, from these measurements, patients affected by schizophrenia from either those affected by different mental disorders or the population of healthy references. Identification performance was about 80% for both GC-MS and sensor array.

8.4   CONCLUSIONS

This chapter reports a survey about medical applications of electronic noses in particular for in vivo applications. The cases of breath and skin analysis have been discussed evidencing the most documented cases of correlations between diseases and alteration in odor composition. For each case, an example taken from the experience of the authors has been illustrated. Results obtained so far demonstrate that since the composition of the volatile part of the body contains some information about the chemical composition of the body, it is correlated with the health status. In particular, some diseases introduce strong deviations from the normal composition of the volatiles. More studies are necessary both from the instrumental point of view to improve the characteristics of sensors and from the physiological point of view to understand which modifications of chemical composition are directly related to the diseases and which are the mechanisms producing these correlations. Some consideration about the electronic nose is necessary; these can be considered as inductive instruments. Namely, their ability to correctly identify odors is based on the accumulated experience. Then, in order to eliminate any possible suspect of false correlations it is necessary to collect a large number of cases. This task may be easily accomplished in many applications (e.g., food analysis) where large quantities of samples are available; in the case of medical applications ethical and practical issues limit the collection of cases. As a consequence there is a limitation in the generalization of the achieved results. The currently available results at the moment only indicate that for their relative simplicity of use arrays of cross-selective gas sensors (electronic noses) can be considered as a viable technique to be pursued for the development of future diagnostic instruments. To achieve these goals a major effort is necessary extending the experiments from local to global cases. As a future perspective, it is worth to mention the possibility that the electronic nose can be exploited in a telemedicine context. Indeed, the main objective of electronic noses, in their attempt of mimicking natural olfaction, is the translation of the olfaction perception in electronic signals that can be recorded, analyzed, stored, and transmitted. Each electronic nose, based on a given sensor technology and with a proper measurement methodology, accumulates with time its own olfaction memory. Once the sensor signals are compared with the olfaction memory, an odor can be recognized. Of course, sensor signals and olfaction memory are not required to be present in the same place. Computer technology allows a sort of delocalization of sensor perception (where the odor interacts "physically" with the sensor array) and olfaction memory (where the signals are compared with the memory and then identified). In this way an odor occurring in one place and locally perceived by a sensor array may be recognized using an olfaction memory physically resident on a remotely located computer. In the future, this opportunity may be of great benefit to screen people living in sparse areas.

ACKNOWLEDGMENTS

This paper is gratefully dedicated to Fr. Alessandro Mantini, OFM who in years 1994–1998 pioneered these studies at the University of Rome Tor Vergata.

REFERENCES

1. Turner A. P. F. and Magan N., Electronic noses and disease diagnostic *Nature review, Microbiology*, 2, 161–166, 2004.
2. Jellum E., Stokke O., and Eldjam L., Application of gas chromatography, mass spectrometry and computer methods in clinical biochemistry, *Anal. Chem.*, 46, 1099–1166, 1973.
3. Smith M., Smith L. G., and Levinson B., The use of smell in differential diagnosis, *Lancet*, 2, 1452–1453, 1982.
4. Roscioni C. and DeRitis G., On the possibility of using odors as diagnostic test of disease (in Italian), *Ann. Ist. C. Forlanini*, 28, 457–461, 1968.
5. Persaud K. and Dodds G., Analysis of discrimination mechanisms in the mammalian olfactory system using a model nose, *Nature*, 299, 352–355, 1982.
6. Malnic B., Hirono J., Sato T., and Buck L., Combinatorial receptor codes for odors, *Cell*, 96, 713–723, 1999.
7. Hines E. L., Llobet E., and Gardner J. W., Electronic noses: review of signal processing techniques, *IEE Proceedings of Circuits Devices Systems*, 146, 297–310, 1996
8. Nagle T., Schiffmann S., Gardner J., and Pearce T. (ed.), *Handbook of Machine Olfaction* (Weinheim: J. Wiley and Sons, 2003).
9. Gardner J. W., Craven M., Dow C., and Hines E. L., The prediction of bacteria type and culture growth phase by an electronic nose with a multi-layer perceptron network, *Meas. Sci. Technol.*, 9, 120–127, 1998.
10. Gibson T. D., Prosser O., Hulbert J. N., Marshall R., Corcoran P., Lowery P., Ruck-Keene E. A., and Heron S., Detection and simultaneous identification of micro-organisms from headspace samples using an electronic nose, *Sens. Actuators B*, 44, 413–422, 1997.
11. Chendiok S., Crawley B. A., Oppenheim B. A., Chadwick P. R., Higgins S., and Persaud K. C., Screening for bacterial vaginosis: A novel application of artificial nose, *J. Clin. Pathol.*, 50, 731–790, 1997.
12. Pavolu A. K., Magan N., Sharp D., Brown J., Barr H., and Turner A. P. F., An intelligent rapid odour recognition model in discrimination of Helicobacter pylori and other gastroesophageal isolates in vitro, *Biosens. Bioelectron.*, 15, 333–342, 2000.

13. Boilot P., Hines E. L., Gardner J. W., Pitt R., John S., Mitchell J., and Morgan D. W., Classification of bacteria responsible for ENT and eye infections using the cyranose system, *IEEE. Sens. J.*, 2, 247–253, 2002.

14. Hanson W. III and Steinberger H., The use of a novel electronic nose to diagnose the presence of intrapulmonary infection, *Anesthesiology*, 87, 269, 1997.

15. Pickel D., Manucy G., Walker D., Hall S., and Walker J., Evidence for canine olfactory detection of melanoma, *Appl. Anim. Behav. Sci.*, 89, 107–116, 2004.

16. Willis C., Church S., Guest C., Cook A., McCarthy N., Bransbury A., Church M., and Church J., Olfactory detection of human bladder cancer dogs: Proof of principle study, *Br. Med. J.*, 329, 712–718, 2004.

17. McCulloch M., Jezierski T., Broffman M., Hubbard A., Turner K., and Janecki T., Diagnostic accuracy of canine scent detection in early- and late-stage lung and breast cancers, *Integr. Cancer. Ther.*, 5, 1–10, 2006.

18. Miekisch W., Schubert J. K., and Noeldge-Schomburg G., Diagnostic potential of breath analysis-focus on volatile organic compounds, *Clin. Chim. Acta.*, 347, 25–39, 2004.

19. Pauling L., Robinson A. B., Teranishi R., and Cary P., Quantitative analysis of urine vapour and breath by gas—liquid partition chromatography, *Proc. Nat. Acad. Sci.*, 68, 2374–2376, 1971.

20. Phillips M., Breath test in medicine, *Sci. Am.*, 267, 74–79, 1992.

21. Phillips M., Herrera J., Krishnan S., Zain M., Greenberg J., and Cataneo R., Variation in volatile organic compounds in the breath of normal humans, *J. Chromatogr. Biomed. Appl.*, 729, 75–88, 1999.

22. Sehnert S. S., Jiang L., Burdick J. F., and Risby T. H., Breath biomarkers for detection of human liver diseases: Preliminary study, *Biomarkers*, 7, 174–187, 2002.

23. Phillips M., Cataneo R. N., Greenberg J., Grodman R., and Salazar M., Breath markers of oxidative stress in patients with unstable angina, *Heart. Dis.*, 5, 95–99, 2003.

24. Phillips M., Cataneo R., and Ditkoff B., Volatile markers of breast cancer in the breath, *Breast J.*, 9, 184–191, 2003.

25. Phillips M., Sabas M., and Greenberg J., Increased pentane and carbon disulphide in the breath of patients with schizophrenia, *J. Clin. Pathol.*, 46, 861–864, 1993.

26. Aghdassi E. and Allard J. P., Breath alkanes as a marker of oxidative stress in different clinical conditions, *Free. Radic. Biol. Med.*, 28, 880–886, 2000.

27. Kneepkens C. M., Lepage G., and Roy C. C., The potential of the hydrocarbon breath test as a measure of lipid peroxidation, *Free. Radical. Biol. Med.*, 17, 127–60, 1994.

28. Risby T. H. and Sehnert S. S., Clinical application of breath biomarkers of oxidative stress status, *Free. Radical. Biol. Med.*, 27, 1182–1192, 1999.

29. Stone B. G., Besse T. J., Duane W. C., Evans C. D., and DeMaster E. G., Effect of regulating cholesterol biosynthesis on breath isoprene excretion in men, *Lipids*, 28, 705–708, 1993.

30. Foster M. W., Jiang L., Stetkiewicz P. T., and Risby T. H., Breath isoprene: Temporal changes in respiratory output after exposure to ozone, *J. Appl. Physiol.*, 80, 706–710, 1996.

31. McGrath L. T., Patrick R., Mallon P., Dowey L., Silke B., Norwood W., and Elborn S., Breath isoprene during acute respiratory exacerbation in cystic fibrosis, *Eur. Respir. J.*, 16, 1065–1069, 2000.

32. Lebovitz H. E., Diabetic ketoacidosis, *Lancet*, 345, 767–772, 1995.

33. Cope K., Risby T., and Diehl A. M., Increased gastrointestinal ethanol production in obese mice: Implications for fatty liver disease pathogenesis, *Gastroenterology*, 119, 1340–1347, 2000.

34. Simenhoff M. L., Burke J. F., Saukkonen J. J., Ordinario A. T., and Doty R., Biochemical profile or uremic breath, *New Engl. J. Med.*, 297, 132–135, 1977.

35. Davies S., Spanel P., and Smith D., Quantitative analysis of ammonia on the breath of patients in end-stage renal failure, *Kidney Int.*, 52, 223–228, 1997.

36. O'Neil H. J., Gordon S. M., O'Neil M. H., Gibbons R. D., and Szidon J. P., A computerized classification technique for screening for the presence of breath biomarkers in lung cancer, *Clin. Chem.*, 34, 1613–1618, 1988.

37. Preti G., Labows J., Kostelic J., Aldinger S., and Daniele R., Analysis of lung air from patients with brochogenic carcinoma and controls using gas chromatography-mass spectrometer, *J. Chromatogr.*, 432, 1–11, 1988.

38. Phillips M. and Greenberg J., Methods for the collection and analysis of volatile compounds in the breath, *J. Chromatogr.*, 564, 242–249, 1991.

39. Phillips M., Gleeson K., Hughes J. M. B., Greenberg J., Cataneo R. N., Baker L., and Mc Vay W. P., Volatile organic compounds in breath as markers of lung cancer: a cross-sectional study, *Lancet*, 353, 1930–1933, 1999.

40. Di Natale C., Macagnano A., Martinelli E., Paolesse R., D'Arcangelo G., Roscioni C., Finazzi-Agrò A., and D'Amico A., Lung cancer identification by the analysis of breath by means of an array of non-selective gas sensors, *Biosens. Bioelectron.*, 18, 1209–1218, 2003.

41. D'Amico A., Di Natale C., Macagnano A., Davide F., Mantini A., Tarizzo E., Paolesse R., and Boschi T., Technology and tools for mimicking olfaction: Status of the Rome Tor Vergata Electronic Nose, *Biosens. Bioelectron.*, 13, 711–721, 1998.

42. D'Amico A., Di Natale C., Paolesse R., Mantini A., and Macagnano A., Metalloporphyrins as basic material for volatile sensitive sensors, *Sens. Actuators B*, 65, 209–215, 2000.

43. Wold S., Sjöström M., and Eriksson L., PLS-regression: A basic tool of chemometrics, *Chemom. Intell. Lab. Syst.*, 58, 109–130, 2001.

44. Fukunaga K., *Introduction to Statistical Pattern Recognition* (New York: Academic Press, 1992).

45. Zeng X., Leyden L., Lawley H., Sawano K., Nohara I., and Preti G., Analysis of characteristic odors from human male axillae, *J. Chem. Ecol.*, 17, 1469–1492, 1991.

46. Grammer K., *5-a-androst-16en-3a*-on: A male pheromone? A brief report, *Ethol. Sociobiol.*, 14, 201–208, 1993.

47. Sawa A. and Snyder S. H., Schizophrenia: Diverse approaches to a complex disease, *Science*, 296, 692–695, 2002.

48. Angst A., Psychoses in disulfiram (anatabus) treatment; review of literature and etiology, *Schweiz Med. Wochenschrift*, 46, 1304–1306, 1956.

49. Smith K., Thomspon G. F., and Koster H. D., Sweat in schizophrenic patients: Identification of the odorous substance, *Science*, 166, 398–399, 1969.

50. Gordon S. G., Smith K., Rabinowitz J. L., and Vagelos P. R., Studies of *trans*-3-methyl-2-hexenoic acid in normal and schizophrenic humans, *J. Lipid Res.*, 14, 495–503, 1973.

51. Stein L. and Wise C. D., Possible etiology of schizophrenia: Progressive damage to the noradrenergic reward system by 6-hydroxydopamine, *Science*, 171, 1032–1036, 1971.

52. Smith K. and Sines J., Demonstration of a peculiar odor in the sweat of schizophrenic patients, *Arch. Gen. Psychiat.*, 212, 184–188, 1960.

53. Di Natale C., Paolesse R., D'Arcangelo G., Comandini P., Pennazza G., Martinelli E., Rullo S., et al., Identification of schizophrenic patients by examination of body odor using gas chromatography-mass spectrometry and a cross selective gas sensor array, *Med. Sci. Monit.*, 11, 366–375, 2005.

54. Di Natale C., Macagnano A., Paolesse R., Tarizzo E., Mantini A., and D'Amico A., Human skin odor analysis by means of an electronic nose, *Sens. Actuators B*, 65, 216–219, 2000.

9 DNA Biosensor for Environmental Risk Assessment and Drugs Studies

Graziana Bagni, Marco Mascini
Department of Chemistry
University of Florence
Via della Lastruccia, Sesto Fiorentino (Firenze), Italy

CONTENTS

9.1 INTRODUCTION

A biosensor is an analytical device, which combines a biological recognition element, which confers selectivity, with a transducer, which provides sensitivity and converts the recognition event into a measurable electronic signal [1]. The biological recognition element can be an enzyme, an antibody, a nucleic acid, a microorganism, a tissue, or organelle, while the transducer can be optical, electrochemical, thermal, piezoelectric, or magnetic [2,3]. The major performance characteristics of biosensors are the minimum sample preparation, the simplicity of the apparatus, the obtaining of fast results, and the possibility for continuous readings; moreover, they are cost-effective, small, and becoming miniaturized with new technologies.

Biosensors hold great promise for the task of environmental monitoring and control [4]. For example, the specific interaction of an immobilized biological layer with target pollutants provides the basis for analytical devices for laboratory or field use. While environmental applications of biocatalytic (enzyme) and immunosensors have greatly increased during the 1990s, little attention has been given to the development of nucleic acid probes for environmental surveillance. Advances in molecular biology and biotechnology have set the stage for exciting possibilities for DNA-based environmental biosensors [5–10]. Such recognition layers could play a major role in future environmental analysis. Since the toxic action of numerous pollutants (e.g., carcinogens and mutagens) is related to their interaction with DNA, it is logical to exploit these events for designing new environmental biosensors. The modification of a transducer surface, through the immobilization of a nucleic acid recognition layer, can form the basis for new sensing devices and provide solutions to various environmental problems. For example, Palecek's group has shown that nucleic acid modified mercury drop electrodes can be used as voltammetric studies of DNA interactions with drugs and proteins based on the adsorption of the biomacromolecule at the surface [11]. Wang's laboratory has focused on the chronopotentiometric transduction of various pollutant-DNA interactions at nucleic acid modified carbon electrodes for the biosensing of toxic substances [5–7]. In the literature, some examples of electrochemical DNA biosensors for the analysis of standard solutions of many different toxic compounds and real samples of waste and surface water samples have been reported. Intercalation has been observed with planar aromatic molecules, and classical compounds are daunomycin, ethidium bromide, acridine dyes, and so forth [12,13]. Alternatively, some positively charged compounds could bind to the DNA via electrostatic interaction with the phosphate backbone [14].

In our laboratory, disposable biosensors based on the immobilization of double-stranded (dsDNA) or single-stranded DNA (ssDNA) on the surface of a carbon-based screen-printed electrodes (SPE) have been developed in recent years in conjunction with chronopotentiometric and voltammetric methods in order to measure both guanine and analyte (when it is electroactive) oxidation peaks. The measure of both peaks gave information on the effect of analyte accumulation on DNA redox properties finding good correlation between the accumulation of a class of environmental pollutants, the aromatic amines, and the blockage of guanine oxidation [8]. This biosensor was used for the rapid detection of polycyclic aromatic amines favoring their intercalation in the hydrophobic double strand of dsDNA. A study of the performance of biosensors obtained by immobilizing different sources of dsDNA or ssDNA (23 base pairs oligonucleotides and DNA from commercially available sources) was performed. The product D4522 from *calf thymus* was found to be the most suitable for the preparation of a dsDNA biosensor (higher reproducibility of the measurements and sensitivity) [8]. The use of long-chain DNA was more suitable than short oligonucleotides (20-mer) for the development of the biosensor. In addition to this, dsDNA biosensor performances were strongly influenced by the physical properties of the DNA (i.e., purity, average chain length, presence of ssDNA) [8]. The biosensor was finally tested on real samples finding interesting correlation with results of classical genotoxicity [8,9].

One of the potentially major applications of a DNA electrochemical biosensor is the testing of water, food, soil, and plant samples for the presence of pathogenic microorganisms and analytes (carcinogens, drugs, mutagenic pollutants, etc.) with binding affinities for DNA.

In the present work, the electrochemical DNA biosensor is proposed as a screening device for the rapid bioanalysis of environmental pollution and drug studies. This is a different concept from proposing this device as a toxicity test since, in the term of toxicity, more complex reactions are involved than the simple binding of a molecule to another molecule (such as the binding of toxicants to the DNA molecules). Nevertheless, each sample that causes a variation in the DNA biosensor response is a sample that could contain possible dangerous pollutants, and, thus, must be monitored more carefully. Electrochemical DNA biosensor is in line with the requirements of in situ screening measurements, since all the equipment needed for the electrochemical analysis is portable. Moreover, a DNA biosensor could be a very useful test, integrated in a panel of tests since it can give rapid and easy to evaluate information on the presence of compounds with affinity to DNA. This test is one of the most competitive in terms of analysis cost and time, with the possibility of developing a very end-user-friendly format, according to the requirements of a screening test for infield measurements.

Concerning the environmental application, this biosensor is proposed as a screening method for the rapid detection of genotoxic compounds in soil samples from an area with high ecological risk, and to investigate the genotoxic effect due to the presence of polycyclic aromatic hydrocarbon (PAH) metabolites in fish bile samples. We propose the use of this biosensor instead of common biological tests for ecological/environmental risk evaluation. In fact, tests with microorganisms (i.e., *vibrio fisherii, daphnia magna*) or with sentinel organisms, like animals (i.e., *lombricus rubellus, eisenia foetida, ciprynus carpio*) or plants (i.e., *vicia faba*) are slow (analysis performed generally in one-two weeks), not specific and expensive. Instead, the molecular tests (among them, biosensors) are fast and, hence, can be used as early warning devices in areas with ecological risk; moreover, they are more specific and have low costs.

In addition to its useful role in environmental applications, it seems straightforward to employ such biosensor in drug studies, especially in cancer chemotherapy where DNA is the main target. Hence, the biosensor is proposed as a screening tool for *in vitro* DNA-drug interaction in the analysis of some antitumoral drugs. In fact, the interaction of drugs with their target biomolecules is of paramount importance in any pharmaceutical development process. A variety of techniques from molecular biology has been used to study this interaction, however, they require various labeling strategies. In recent years, there has been a growing interest in the electrochemical investigations of the interactions between anticancer drugs and biomolecules. Electrochemistry offers great advantages over the existing devices, because it provides rapid and qualitative information on the interaction between redox active species. Compared with other methods, electrochemical techniques are characterized by simplicity and require small amount of sample, thus offering advantages over commonly used assays (both biological and chemical) [5–10,15–17].

In this work, five antitumoral drugs were analyzed; each of them belonged to a family of a different type of DNA interaction. They were *cis*-diaminedichloroplatinum(II) *cis*-[Pt(Cl)$_2$(NH$_3$)$_2$] (cisplatin) as model agent of DNA alkylants, diammine(1,1-cyclobutanedicarboxylate)platinum(II) [Pt(NH$_3$)$_2$C$_4$H$_6$C$_2$O$_4$] (carboplatinum) as DNA alkylant in N7 of guanine base, titanocene dichloride [(η^5-C$_5$H$_5$)$_2$TiCl$_2$] (titanocene) as DNA's phosphate interacting, 2,2'-bypiridinedipyridineplatinum(II) bis-tetrafluoroborate [Pt(2,2'-bipy)(n-Rpy)$_2$][BF$_4$]$_2$ n = 2,4 (hereafter platinum bipy) as DNA double helix intercalant, and imidazolium *trans*-imidazoledimethylsulfoxidetetrachloro-ruthenate [Ru(III)Cl$_4$(DMSO)(Im)][ImH] (NAMI-A) as weak DNA interacting. Different response behavior has been observed with these drugs according to different interaction with DNA.

9.2 EXPERIMENTAL

9.2.1 MATERIALS

Milli-Q grade water (18 MΩ) was used for the preparation of all solutions (Milli-Q water purification system, Millipore, U.K.). Inorganic salts for buffer preparation were obtained from Merck (Milan, Italy). *Calf thymus* dsDNA type XV was purchased from Sigma (Milan, Italy). Using sterile pipettes, aliquots containing 1000 µg/mL of DNA in milli-Q water were prepared and stored in freezer (−20°C). Acetate buffer concentration was 0.25 M (pH = 4.75) with 10 mM potassium chloride (KCl).

For the analysis of soil, the effect of three benzene, two naphthalene, and four anthracene derivatives were analyzed in this work with DNA biosensor because widely present in the sampling polluted area [18–20]: sodium benzenesulfonate, 4-chloroaniline, 3,4-dichloroaniline, 2-naphtylamine, sodium 2-naphthalenesulfonate, sodium anthraquinone 2-sulfonate monohydrate, 2-anthracencarboxylic acid, 1,2-diaminoanthraquinone, and 2-anthramine (all from Sigma-Aldrich (Milan, Italy)).

For the analysis of bile and PAH derivatives, PAH metabolite standard solutions were prepared daily and were purchased from RF-Rogaland Research-Akvamiljø (Stavanger, Norway). All the fish bile sample were purchased from RF-Rogaland Research-Akvamiljø (Stavanger, Norway) and the sample solutions were prepared daily, diluting crude bile 1:1000 with acetic buffer and 1% EtOH after an initial dilution of 1:250 with acetic buffer containing 20% Ethyl Alcohol (EtOH).

For drug analysis, cisplatin and titanocene dichloride were obtained from Sigma—Aldrich (Milan, Italy), while carboplatin [21], platinum bipy [22], and NAMI-A [23] were prepared by following published procedures.

9.2.2 APPARATUS

Electrochemical measurements were performed with a PalmSens (Palm Instruments BV, Hoten, Netherlands) interfaced to a Compaq iPAQ Pocket PC with a PalmScan 1.3 software package (Palm Instruments BV, Hoten, Netherlands) and SPE.

To perform the extraction of pollutant molecules from soil, an Autovortex mixer SA2 (Stuart Scientific, Surrey, U.K.), filters 0.45 μm [Schleicher & Schuell Italia, Legnano (MI), Italy], and an ultrasound probe model VC100 Vibra Cell [Sonics and Materials (Danbury, U.S.)] were used respectively to homogenize the solvent-soil suspensions to filter the extracts and to avoid a rapid extraction.

Fluorescence measurements were performed in a quartz cuvette on a RF-1501 spectrofluorophotometer (Shimadzu, Italy). Fixed wavelength fluorescence (FF) was measured at the excitation/emission wavelength pairs of 341/383 nm for pyrene-like derivatives and slit widths were set at 10 nm for both excitation and emission wavelengths. The signal level in solvent blank was subtracted and the FF level was expressed as relative fluorescence intensity. Calibration curves were performed with 1-hydroxypyrene standard solutions (concentration range 0.0–2.0 μg/L, DL 0.03 μg/L, $R^2 = 0.999$) as model compound of pyrene-like derivatives. The standard solutions were prepared daily in methanol 48% and all the fish bile samples were prepared by diluting crude bile 1:1600 or 1:12000 for the most polluted samples with methanol 48% [24].

9.2.3 PREPARATION OF SCREEN-PRINTED ELECTRODES

Screen-printing is becoming a simple and fast method for mass production of disposable electrochemical sensors. Single-use sensors have several advantages, such as avoidance of contamination between samples and constant sensitivity of the different printed sensors. With this technique the ink used to print electrodes can be varied easily and therefore different properties of the final sensor can be achieved. The inks can be printed on several kinds of supports like glass, ceramic, and plastic sheet. Many different types of ink are commercially available and some of them are based on noble metals (Au, Pt, Ag, etc.). The most interesting materials for printing electrochemical sensors are the carbon-based inks, because of their very low firing temperature (from room temperature to 120°C) and because they can be printed on plastic sheets. Carbon can also be mixed with different compounds (mediators, enzymes, metal catalytic particles) and therefore modified sensors and biosensors can be easily mass produced.

Since the past decade, different types of sensors and biosensors based on SPE have been prepared in our laboratory, such as a choline sensor with choline oxidase immobilized on the surface of a ruthenized-carbon screen-printed electrode as part of a system for pesticide detection [25], a screen-printed ruthenium dioxide electrode for pH measurements [26], carbon-modified electrodes with mercury for heavy metals analysis [27], and two kinds of disposable electrochemical DNA biosensors for environmental screening [8–10,15,28,29].

The biosensor used in the present work was assembled using a planar, screen-printed electrochemical cell, which consisted of graphite working electrode, a graphite counter electrode, and a silver pseudo-reference electrode. The inks consisted of finely divided particles of different materials in a blend of thermoplastic resins (silver ink was Electrodag PF-410 and graphite ink was Electrodag 423 SS, both from Acheson Italiana, Milan, Italy, while titanium dioxide ink for insulating layer was Vinylfast 36–100 from Argon, Lodi, Italy).

The electrodes were screen printed in the laboratory using a DEK 248 screen-printing machine (DEK, Weymouth, U.K.).

The screen printing process consisted of forcing the inks through a screen onto a surface of the substrate plate (polyester foil) with a squeegee. Typical thickness of the film was around 20 μm. The open pattern in the screen defines the pattern, which will be printed on the substrate. The first layer printed consisted of the silver ink in order to obtain the conductive tracks and the pseudo-reference electrode. The second layer was the carbon ink to obtain both the auxiliary and working electrodes. After every step, the sheets were heated at 120°C for 10 minutes to achieve the polymerization of the printed films. The insulating ink was finally used to define the working electrode surface (Ø = 3 mm). A curing period of 20 minutes at 70°C was applied. The sensors were produced in sheets of 20 electrodes. Each electrode printed on the polyester flexible foil (Autostat CT5, thickness 125 μm, obtained from Autotype, Milan, Italy) is easily cut by scissors and fits to a standard electrical connector. To facilitate handling, the screen-printed electrochemical cells were placed onto a rigid polycarbonate-based support. Each electrode was used as disposable. The schemes of these three printing steps and of a SPE sheet are reported in Figure 9.1.

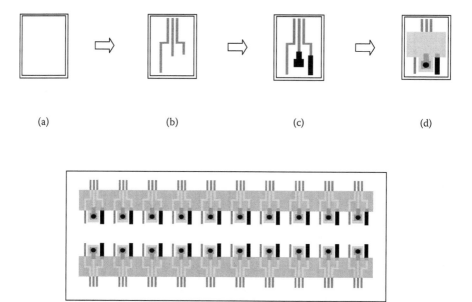

(a)   (b)   (c)   (d)

FIGURE 9.1 Scheme of screen-printing electrodes. Fabrication process of the sensor consisted of three steps including consecutive printing of the silver layer (b), carbon layer (c), and insulating layer (d) on the polyester substrate. The sensors were produced in sheets of 20 electrodes.

9.2.4 BIOSENSOR PRINCIPLE

Guanine peak was used as the transduction signal in systems detecting DNA interacting agents. As a result of interaction of double-stranded calf thymus DNA with a pollutant agent a decrease of guanine peak (measured by square-wave voltammetry) was detected (Figure 9.2). DNA modification were estimated with the value of the percentage of guanine peak height change (Signal %), which is the ratio of the guanine peak height after the interaction with a sample (S_s), and the guanine peak height after the interaction with the buffer solution (S_b): Signal % = $(S_s/S_b)*100$.

Adenine oxidation peak can also be observed and detected, but it has been chosen to use guanine peak only because it gave more reproducible signals.

Groove binding, electrostatic interactions, hydrogen and/or van der Waals bonds, and intercalation of planar condensed aromatic ring systems between adjacent base pairs are the interactions that the DNA electrochemical biosensor can detect.

The electrochemical procedure to build up the DNA biosensor included four main steps: electrochemical conditioning of the electrode surface in order to oxidize the graphite impurities and to obtain a more hydrophilic surface to avoid DNA immobilization, *calf thymus* dsDNA immobilization, interaction with the sample solution, and electrode surface interrogation.

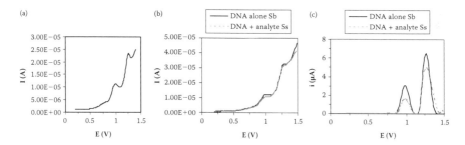

FIGURE 9.2   Redox behavior of guanine (+1.0 V vs. Ag – SPE) and adenine (+1.25 V vs. Ag – SPE) bases after a square-wave voltammetric scan carried out with graphite screen-printed working electrode. It can be noted that the acquired signal of DNA alone (a), the decrease of the DNA peaks after the interaction with an analyte solution (b), and the signals after a baseline correction (c). Electrode pretreatment: +1.6 V (vs. Ag – SPE) for 120 seconds and +1.8 V (vs. Ag – SPE) for 60 seconds in 5 ml of 0.25 M acetate buffer containing 10 mM KCl (pH = 4.75) under stirred conditions; DNA immobilization: 50 ppm Calf thymus dsDNA (in 0.25 M acetate buffer with 10 mM KCl) applying a potential of +0.5 V (vs. Ag – SPE) for 5 minute under stirred conditions; blank or sample interaction: 10 μL of the sample solutions onto the working electrode surface for 2 minutes; final measurement: square wave voltammetric scan in 0.25 M acetate buffer containing 10 mM KCl from +0.2 V to +1.40 V (E_{step} = 15 mV, $E_{amplitude}$ = 40 mV, frequency = 200 Hz).

The experiments were performed at room temperature, according to the following steps:

1. Electrode pretreatment: applying potential +1.6 V for 120 seconds and +1.8 V for 60 seconds; electrode in 5 mL of 0.25 M acetate buffer, containing 10 mM KCl (pH = 4.75), under stirred conditions.
2. DNA immobilization: 50 ppm *calf thymus* dsDNA in 0.25 M acetate buffer with 10 mM KCl, applying potential +0.5 V for 5 minutes, under stirred conditions.
3. Blank or sample interaction: 10 μL of sample solutions onto the working electrode surface for 2 minutes.
4. Measurement: a square-wave voltammetric (SWV) scan was carried out to evaluate the oxidation of guanine residues on the electrode surface. The height of the guanine peak (at +0.95 V *vs.* Ag screen-printed pseudo-reference electrode) was measured. SWV in 0.25 M acetate buffer, containing 10 mM KCl; parameters: scan from +0.2 V to +1.40 V, E_{step} = 15 mV, $E_{amplitude}$ = 40 mV, Frequency = 200 Hz.

The analysis of a sample took 11 minutes.

9.2.5   Soil Samples

Reference soil was collected in a reference site in Florence (Italy). Real soil samples were collected in the Associated National Chemical Companies (ACNA) site (Cengio, SV, Italy). ACNA is a closed organic chemical industrial factory active since 1882 with the production of explosives (nitroglycerin, dynamite, and trinitrotoluene), paints, nitric and sulphuric acids, phenols, and amines, with serious levels of contamination of soil and surface waters, where a remediation and bonification plan started in 1999. The serious environmental contamination of this area determined its inclusion in the list of national priorities for environmental reclamation. The sampling was performed from the soil layers (0–30 cm) in a specific ACNA site, called hill n°5, on March 2003. The hill was made of waste from the industry accumulated during the years and was divided into four zones: zone 1 with low contamination level; zone 2 with pseudo-reference; zone 3 with moderate pollution level, and zone 4 with high ecological risk.

The soil was dried 24 hour at r.t. [30], pulverized in a mortar, homogenized by passing a metal sieve (Ø = 0.8 mm) and stored at r.t. until extraction. The extraction was performed on 0.5 g of soil, according to an optimized procedure obtained spiking a reference soil, then extract with an ultrasound probe for 2 minutes in 10 mL of 50 mM phosphate buffer pH 7.5, followed by an equilibration of 3 minutes at r.t. and a filtration [31].

9.2.6   Fish Bile Samples

In this study, the fish species Atlantic cod (*Gadus morhua L.*) was selected since it is a widespread and common species and in addition it is important for human consumption.

The exposure experiment was performed by RF-Rogaland Research Akvamiljo laboratory (Norway).

For single PAH injection analysis, 48 individuals with an average weight of (690 ± 130) g were caught in a nonpolluted area by light trawl and returned immediately to the laboratory. The cod were acclimatized in the laboratory for one-week prior the exposure, they were maintained in 600 L tanks with flow-through seawater. They were divided into ten groups, n = 3–7, and exposed to single polycyclic aromatic compounds by intra peritoneal (i.p.) injection. The fish were sacrificed five days following exposure by a blow to the head, and the bile from each group was pooled and stored at $-80°C$ until chemical analysis. Fish were not fed during the acclimatization and exposure period.

For fish bile samples from laboratory experiments, fish from the Barents Sea were purchased from Troms Marine Yngel. They were transported in Rogaland Research Akvamiljo laboratory (Norway) in two 2000 L tanks with oxygenation, and then they were kept in quarantine for nine days before the exposures started. To obtain size distribution in all groups (six), the fish were distributed within the groups (fish weight 553 ± 93 g). The experiments lasted for 15 days and fish were not fed during the period. Exposures represented three different approaches to mime components normally present in produced water. All the six groups were held in 600 L tanks with flow-through seawater and the exposure set up was based on a continuous flow system [32] which made it possible to expose fish to relative stable concentrations of disperse crude oil or produced water for an extended period of time. The first tank contained only seawater (control group), the second one contained C_4–C_7 alkyl phenol mixture (4-tert-butylphenol, 4-n-pentylphenol, 4-n-hexylphenol, and 4-n-heptylphenol) with nominal water concentration of 2000 ng/L (expected body burden exposure concentration of 1000 µg/L), the third and the fourth contained produced water tapped at the Oseberg C field (offshore installation) and diluted with sea water in the ratio respectively of 1:200 (estimated corresponding platform distance: 200 m) and 1:1000 (distance: 1000 m), the fifth and the sixth contained respectively 0.2 mg/L crude oil (Pharmacia 56-1190-30) and 0.2 mg/L crude oil spiked with PAH mixture (with nominal concentrations of 3000 ng/L for naphthalene derivatives, 200 ng/L for phenanthrene derivatives and fluorene, 40 ng/L for dibenzothiophene, 8 ng/L for fluoranthene, 20 ng/L for chrysene derivatives, 16 ng/L for pyrene derivatives, and 1 ng/L for benzo[a]pyrene derivatives). The fish were sacrificed after the end of the exposure by a blow to the head; the bile was sampled and stored at $-80°C$.

Animal experimentation in Norway is regulated by the Ministry of Agriculture through the Norwegian Animal Welfare Act [33]. The Act states that experiments may not be performed without special permission, to be granted by a board known as the National Animal Research Authority (NARA) or persons to whom it delegates such authority. Some research projects (particularly those using fish or wild species as those ones) may require prior approval from other Ministries in addition to that obtained from the NARA. Norway has approximately 80 approved laboratory animal units; half of them are fish research stations around the cost (RF-Rogaland Research Akvamiljo is one of them).

9.2.7 STATISTICAL ANALYSIS

A one-way analysis of variance (ANOVA) was used to compare between the different sample groups. Dunnett's test was used to determine the significance of differences between the sample groups and the control (blank assay) group. When the ANOVA indicated that significant differences existed, Fisher's least-significant difference (LSD) multiple comparison test was then used. Statistical analysis was performed with Excel software. Parametric tests were preceded by tests for normal distribution per site and for homogeneity of variance between sites. A significance level of $P < 0.05$ was applied in all statistical tests.

9.3 RESULTS AND DISCUSSION

9.3.1 ENVIRONMENTAL RISK ASSESSMENT

9.3.1.1 Analysis of Soil

The soil plays a central role within ecosystems and fulfils a multitude of functions (as a habitat and with regulatory and production functions). On reaching soils, pollutants can adversely affect these functions, thus they can move to other media and cause damage to other components of the ecosystem. Soil contamination comprises either solid or liquid hazardous substances mixed with the naturally occurring soil. Usually, contaminants in the soil are physically or chemically attached to soil particles, or if they are not attached, trapped in the small spaces between soil particles. Soil contamination results when hazardous substances are either spilled or buried directly in the soil or migrate to the soil from a spill that has occurred elsewhere. Another source of soil contamination could be from water flowing near soils containing hazardous substances.

Contaminants in the soil can adversely impact the health of animals and humans when they ingest, inhale, or touch contaminated soil, or when they eat plants or animals that have been affected by soil contamination. The exposure of an organism to a carcinogen or other ecotoxic compound in soil is not related to the total concentration of that substance in soil but rather to the amount that is actually available *(bioavailability)* [34]. Most chemical extraction methods reported in literature were studied in terms of their ability to recover all the contaminants (PAH, pesticides, etc.) from a soil matrix [30,35]. In this chapter we propose the use of a fast and simple analytical extraction method to mimic the bioaccessibility determined by pollution of benzene, naphthalene and anthracene chloride, amino and sulfonate derivatives. The method, based on an ultrasonic extraction [36,37], was performed within a short time (2 minutes) and with a little amount of solvent (10 mL) [31].

The structures of the compounds analyzed are reported in Figure 9.3. Different concentrations for each compound were tested and the inhibition of the guanine oxidation peak increased with analyte concentration (Table 9.1). The results demonstrated that the compounds presented different genotoxic effect. In general, the effect was higher with increased ring number of the molecule, and also the type of the group had relevance. For example, the amino group was more toxic than a sulfonate one at the same molecular concentration. The results showed the highest toxic effect for

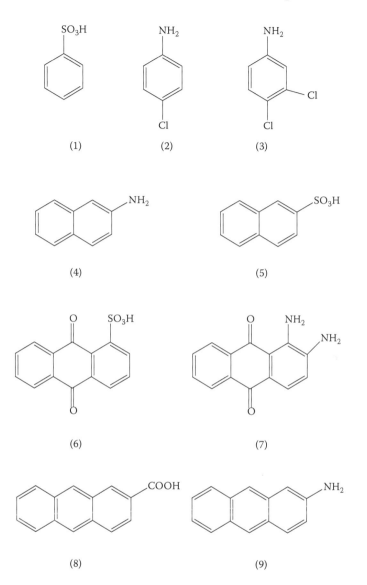

FIGURE 9.3 Structures of benzene, naphthalene, and anthracene derivatives analyzed: sodium benzenesulfonate (1), 4-chloroaniline (2), 3,4-dichloroaniline (3), 2-naphthylamine (4), sodium 2-naphthalenesulfonate (5), sodium anthraquinone 2-sulfonate monohydrate (6), 1,2-diamineanthraquinone (7), 2-anthraceencarboxilic acid (8), and 2-anthramine (9).

molecules with three rings. In fact, the naphthalene and anthracene derivatives analyzed are DNA intercalating agents [8,10,38], while the benzene derivatives are not double helix intercalant but only a weak DNA interactive (i.e., as reported in Snyder and Arnone [38], 4-chloroaniline is described as nongenotoxic and its cytotoxicity, measured as cell viability assay, is relatively low compared with other chemicals).

TABLE 9.1
DNA Biosensor Results for Benzene, Naphthalene, and Anthracene Derivatives Standard Solutions Present in ACNA Area; Experimental Conditions as in Figure 9.2

Compound	Concentration (µmol/L)	S Guanine (%)	Std. Dev. (N = 4)
Sodium benzenesulfonate	100	75	6
	50	91	8
	25	96	6
4-chloroaniline	100	78	8
	50	92	6
	25	85	10
3,4-dichloroaniline	100	43	8
	50	75	8
	25	84	6
	10	97	7
2-naphthlamine	50	54	7
	5	61	9
	1	85	12
Sodium 2-naphthalenesulfonate	150	47	8
	100	58	9
	50	79	6
Sodium anthraquinone 2-sulfonate	25	34	12
monohydrate	10	51	7
	5	55	9
	1	90	6
2-anthracenecarboxylic acid	5	42	10
	2	61	7
	1	71	5
1,2-diamineanthraquinone	5	59	10
	2	65	10
	1	79	6
2-anthramine	2.0	33	10
	1.0	47	6
	0.5	73	8

This work demonstrated that naphthalene-sulfonate derivatives with two and three aromatic rings and aromatic amines with two and three rings gave a positive result with the biosensor in a concentration range of micro and submicromolar. The EC_{50} value was 2.0 µmol/L for 2-anthracenecarboxylic acid and 1,2-diamineanthraquinone and 1.0 µmol/L for 2-anthramine.

After the extraction, ACNA soil samples were analyzed with DNA biosensor and the different zones were compared.

The comparison of the zones with the superficial sampling (maximum 5 cm in depth) did not show a significant difference between them (Figure 9.4). The

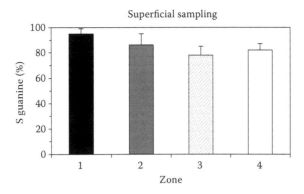

FIGURE 9.4   DNA biosensor results for superficial sampling in the four zones of hill n°5 in ACNA area. Extraction with 50 mM PBS pH 7.5, ultrasound probes 2 minutes. Electrochemical experimental conditions as in Figure 9.2.

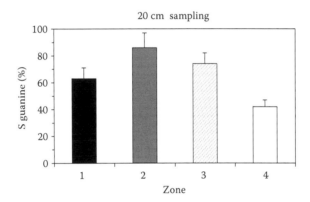

FIGURE 9.5   DNA biosensor results for 20 cm sampling in the four zones of hill n°5 in ACNA area. Extraction with 50 mM PBS pH 7.5, ultrasound probes 2 minutes. Electrochemical experimental conditions as in Figure 9.2.

comparison of the zones with 20 cm sampling depth was different, showing a trend of pollution as expected (Figure 9.5): the biosensor was able to distinguish different soil contamination sites.

A standard semiquantitative method for aromatic ring system determination as FF was used for the analysis of 2-anthramine standard solutions and some ACNA samples were tested with this methodology obtaining promising comparative results. The optimal wavelength pair, regarding sensitivity was obtained by the analysis of 0.5 µM 2-anthramine standard solutions and was 350/510 nm.

Soil samples from zone 4 were analyzed with FF as 2-anthramine-like compounds. The comparison of FF and biosensor results is reported in Figure 9.6, which shows that soil contamination increased with the sampling depth for both methods. These results confirmed the fact that this zone is the most polluted of the hill n°5.

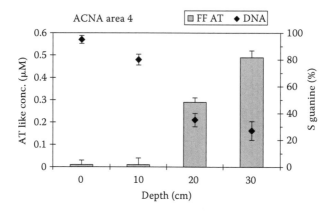

FIGURE 9.6 DNA biosensor and $FF_{350/510}$ results for different depths in zone 4 of hill n°5 in ACNA area. The samples were extracted with 50 mM PBS pH 7.5, ultrasound probe 2 minutes. Electrochemical experimental conditions as in Figure 9.2. Fixed wavelength fluorescence (FF) was measured at the excitation/emission wavelength pairs of 350/510 nm for 2-anthramine-like compounds and slit widths were set at 10 nm for both excitation and emission wavelengths. The signal level in solvent blank was subtracted and the FF level was expressed as relative fluorescence intensity. Calibration curves were performed with 2-anthramine standard solutions (concentration range 0.1–1.0 μM, $R^2 = 0.998$).

These samples were analyzed also with micronuclei and Comet tests performed by ISPESL/DIPIA laboratory (Rome, Italy). These are simple and sensitive techniques for analysis and measure of DNA damage in individual mammalian (and to some extent prokaryotic) cells [38–40]. Seed of *Vicia faba var minor* were put in ACNA soil samples and the top roots were used for the micronuclei and Comet tests in order to analyze a vegetable mutation after 5 days of exposure. The length of the top roots was also calculated in order to establish a phytotoxicity. Later, half of the roots were prepared for the micronuclei test and the other half for the Comet test. The genotoxic effect was based on the frequency of irregular anaphases and micronucleate cells. The results showed an increase in pollution from zone 1 to 4 and within the zone, with the sampling depth for all methods (Table 9.2): when there was a lower DNA signal, there was a lower primary roots length, a higher irregular anaphases or micronuclei frequency, and a higher damage class with Comet test.

Interestingly, the results obtained analyzing the ACNA soil samples with DNA biosensor were confirmed by other bioassays, hence this kind of biosensor can be very useful as a rapid screening method of analysis.

9.3.1.2 Analysis of Fish Bile

Contamination of coastal environments by chemical contaminants such as hydrocarbons (*oil spills, wastes*), pesticides (*crop treatment*), heavy metals, and various organic pollutants in dredged sediments and wastewaters is a major environmental concern. Biological effect monitoring is an important element in management

TABLE 9.2

Comparison of Biosensor Results Performed on ACNA Samples (Sampling of 2003) with Other Bioassays for Toxicity and Genotoxicity Detection

Sample Code	DNA Biosensor S Guanine (%)	Phytotoxicity Primary Roots Length (mm)	Genotoxicity Irregular Anaphase Frequency (%)	Genotoxicity Micro Nucleate Cells Frequency (%)	Comet Test Damage Class
1A	99	32.7	0.05	0.18	1
1B	87	36.7	0.06	0.20	1
1C	63	18.1	–	–	–
2A	86	35.0	0.01	0.01	1
3A	78	25.9	0.01	0.03	0
3B	60	28.2	–	–	0
3C	74	22.8	–	–	2
4A	82	21.9	0.46	0.09	–
4B	70	29.5	0.11	0.08	3
4C	42	14.5	0.25	0.12	1
4D	26	12.7	0.52	0.64	4

Notes: A—superficial sampling, B—10 cm, C— 20 cm, and D—30 cm.

and monitoring programmes and aims to assess the overall quality of the marine environment.

The oil industry is currently extending exploration and production activities into the North Sea and methods to meet demands for environmental monitoring and prediction must be developed. Several European countries are currently implementing new discharge policies and new methods are being developed to measure and predict biological effects. Internationally, a set of early warning biomarker parameters is already being recommended (e.g., by OSPAR, Oslo, and Paris Commissions) for the environmental risk evaluation of chemicals, mixtures, and effluents. The aim of this work is to study and describe the potential of certain constituents in fish sampled in the North Sea as monitoring parameters for the assessment of mixed pollution, crude oil, and produced water exposure in wild fish.

Polycyclic aromatic hydrocarbons (PAHs) are ubiquitous contaminants in the marine environment as a result of uncontrolled spills, river transport, surface runoff, and atmospheric deposition. Since several PAHs are toxic and carcinogenic, their accumulation and eventual effects in marine organisms are topics of environmental concern [24]. PAH exposure in marine organisms is often assessed by measuring the concentration of PAHs in their tissues. However, fish caught at highly polluted sites often show only trace levels of PAHs in their tissue due to their ability to metabolize these compounds. Laboratory studies have demonstrated that the presence of PAH metabolites in bile is well correlated with levels of exposure [24,41–43], as gallbladder bile is a major excretion route for PAH in fish. Bile fluid is one of the major targets since this complex biological fluid is one of the prime routes for excretion of many pollutants in fish (and other vertebrates) and since it is virtually a natural extract sample that reflects the recent ongoing exposure to a range of pollutants. Due to their liposolubility, PAHs are absorbed fairly easily by living organisms where they are accumulated or sometimes transformed into more toxic substances. In fact, after introduction into the organism, PAH are biotransformed and their metabolites are excreted into the bile and concentrated. Therefore, the detection of biliary PAH metabolites in fish may serve as an early warning parameter that indicates a recent or ongoing PAH exposure. Long-term PAH stress may led to more adverse effects such as DNA lesions (e.g., covalent bond DNA adducts) and later irreversible histopatologic effects such as DNA lesions (e.g., PAH induced tumors). Because of this, the determination of PAH metabolite levels in fish bile has been proposed as a biomarker of PAH exposure by international bodies such as OSPAR.

Possible damage to DNA molecules is of special concern and should always be considered as an aspect when environmental risk is to be evaluated. Nowadays there is a range of methods developed to detect the changes in the DNA structure. One type of DNA damage is the formation of DNA adducts, which are bonds between certain contaminant compounds and the DNA molecule. Such adducts are shown to be involved during the development of cancer in various organisms. In laboratory and field studies, DNA adducts formation in fish had been identified as a sensitive biomarker of exposure to aromatic genotoxicant [44,45]. Environmental mutagens, such as PAHs and heterocyclic amines, are known to bind nucleotides, resulting in the formation of DNA adducts [46]. Reactive oxygen species generated by pollutants also induce the formation of DNA adducts. It is known that larger PAH molecules with bay and fjord region structural elements are strong inducers of DNA adducts [46]. Since DNA adducts analyses are expensive, technically complicated, and time consuming, there is a need for a fast screening tool that can be used in environmental surveys as a first step assay in examination of genotoxic effects in fish. For such purpose, DNA-based biosensors, which are able to detect the presence of compounds with affinity for DNA measuring their effect on the guanine base, are proposed in this work as a screening device for the rapid detection of genotoxic compounds in fish bile samples as a biomarker of exposure at contaminated sites. The use of this kind of biosensor on fish bile analysis was already successfully applied [28,29]; in this work the application on two new kinds of exposures designs was tested. The aim of this study was to analyze fish for genotoxicity by examination of an acute toxicity exposure induced in a laboratory via injection of a single PAH and of a chronic oil exposure induced in a laboratory.

Preliminary studies with PAH metabolites standard solutions were performed using naphthalene, phenanthrene, chrysene, pyrene, and benzo[a]pyrene hydroxy or dihydroxy as model compounds. The structures of the compounds analyzed are reported in Figure 9.7.

These standards were measured as preliminary analysis in order to establish their behavior with the biosensor being the PAH derivatives present in animals after their metabolism. For all of them, the effect of increasing concentrations in acetate buffer solutions with 10% ethanol or acetone was investigated. In Figure 9.8, the comparison between one metabolite for each PAH class is reported. S% values decreased with increasing standard concentration and increasing number of the condensed aromatic rings according to the toxicity reported in several papers [47–49]. In Ribeiro

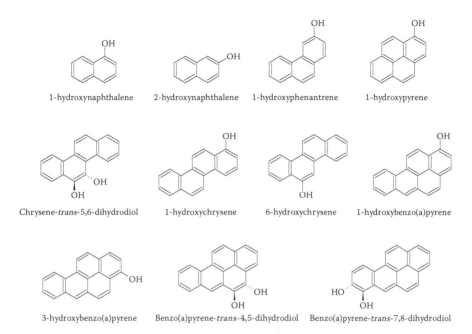

1-hydroxynaphthalene   2-hydroxynaphthalene   1-hydroxyphenantrene   1-hydroxypyrene

Chrysene-*trans*-5,6-dihydrodiol   1-hydroxychrysene   6-hydroxychrysene   1-hydroxybenzo(a)pyrene

3-hydroxybenzo(a)pyrene   Benzo(a)pyrene-*trans*-4,5-dihydrodiol   Benzo(a)pyrene-*trans*-7,8-dihydrodiol

FIGURE 9.7   Structures of the analyzed PAH metabolites.

and Ferreira [49], a new model for the photoxicity of PAH is reported, in comparison with previous ones (i.e., Ferreira [48]): these methods classified the PAH according to their phototoxicity, and generally naphthalene, phenantrene, and fluorene are in the same class (low toxic), chrysene and fluoranthene in another one (moderately toxic), pyrene and benzo[a]pyrene in the last one (extremely toxic).

An exception in biosensor results is chrysene which, even though has four rings, exerted less effect than phenantrene (three rings): this behavior could be explained by the different positions of the fused rings that may permit a less or stronger interaction with DNA. As shown in Figure 9.8, the curves decreased almost linearly until they reached a plateau caused by a saturation phenomenon. The saturation phenomenon was already studied with this kind of biosensor with others DNA interacting agents [6,8].

In Table 9.3, the detection limits, calculated using the formula recommended by IUPAC as the concentration of analyte that gives a signal at least three times higher than the blank, and the linear range for each metabolite are reported.

A simulation experiment of acute exposure to PAH was performed in fish. A single PAH compound was injected intraperitoneally in fish in different doses depending on the ring number of the compound and the fish weight. The PAH derivatives potentially present in fish after 5 days of metabolism are PAH hydroxy, dihydroxy, glucuronic, and sulphate.

Biosensor analysis on diluted bile showed that the PAHs genotoxic effects were statistically different (Figure 9.9) according to the analysis of variance (ANOVA) and Duncan's test, which was used to determine the significance of differences in mean values within each group.

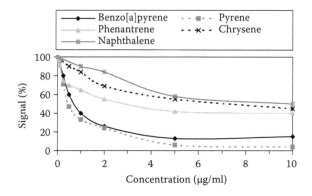

FIGURE 9.8 Comparison of the DNA biosensor response for one metabolite standard solutions for each PAH class. Average values and standard deviations were calculated over four results but error bars are omitted for clarity. Experimental conditions as in Figure 9.2.

TABLE 9.3
Features of PAH Metabolites Standard Solutions

Pah Metabolite	Detection Limit μg/mL	Linear Range μg/mL	R^2
1-hydroxynaphtalene	0.09	0–10.0	0.890
2-hydroxynaphtalene	0.1	0–5.0	0.993
1-hydroxyphenantrene	0.01	0–0.5	0.897
1-hydroxypyrene	0.006	0–1.0	0.987
Chrysene-*trans*-5,6-dihydrodiol	0.2	0–10.0	0.965
1-hydroxychrysene	0.1	0–10.0	0.985
6-hydroxychrysene	0.01	0–2.0	0.962
1-hydroxy-benzo [a] pyrene	0.2	0–10.0	0.946
3-hydroxy-benzo [a] pyrene	0.03	0–2.0	0.997
Benzo [a] pyrene-*trans*-4,5-dihydrodiol	0.01	0–1.0	0.966
Benzo [a] pyrene-*trans*-7,8-dihydrodiol	0.03	0–1.0	0.998

Notes: Experimental conditions as in Figure 9.2.

This different carcinogenic power could be due to the ring number of the compound, as observed with PAH metabolites standard solutions, the molecular conformation, and the presence of heteroatom. They were divided into two groups depending on the S % level. The compounds of the first group (naphthalene, phenantrene, and fluorene) produced S % levels around (87 ± 1) %. The second group includes fluoranthene, chrysene, pyrene, and benzo(a)pyrene and showed higher genotoxic effects with S % levels around (79 ± 4) %. The dibenzothiophene effect could not be differentiated from the control sample; the presence of a heteroatom did not facilitate interaction with dsDNA. Bile samples from fish artificially exposed to different class

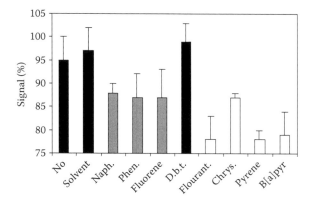

FIGURE 9.9   DNA biosensor response (S %) obtained for bile analysis of fish injected with a single PAH. Experimental conditions as in Figure 9.2. Legend: No (fish not exposed), Solvent (solvent carrier control), Napht. (Naphthalene), Phen. (Phenantrene), D.b.t. (Dibenzotiophene), Fluorant. (Fluorantene), Chrys. (Chrysene), B[a]pyr (benzo[a]pyrene).

of pollutants were tested with the DNA biosensor. The chronic exposure simulation experiments showed that the biosensor was able to distinguish between the six exposure groups.

ANOVA was used to compare between the different sample groups: the control group was different from all the others groups. When the two groups with the exposure to produce water at two different dilutions (groups 3 and 4) were compared, the results showed that they were statistically different. When the two groups between the exposure to crude oil alone and crude oil with an addition of standard solutions of PAH and AP (groups 5 and 6) were compared, the analysis showed a difference between them and the group 6 showed the lowest value of DNA signal (highest genotoxicity). This behavior was confirmed with the fluorescence analysis, which gave the pyrene-like equivalents content of the groups, and with the micronuclei assay (MN) (Table 9.4). The obtained results confirm the analysis performed with the biosensor: control group of cod showed the higher DNA signal, the lower MN incidence (kidneys) and the lower pyrene-like molecules concentration, corresponding to absence or minimal amount of pollution. The higher DNA signal and the lower MN incidence (kidneys) among the exposed groups were found in group 4 (cod exposed to produced water 1:1000) corresponding to a minimal amount of pollution, while one of the lower DNA signal, the higher MN incidence (kidneys) and the higher pyrene-like molecules concentration among the exposed groups was found in groups 3 and 6 (cod exposed to produced water 1:200 and to oil + PAH + AP) corresponding to a high amount of pollution.

The variability of a single group can be explained by interspecies variability. The results showing lower DNA signals in fish exposed to crude oil or produced water indicate that in fish the biosensor may be used for the detection of genotoxic pollution in a marine environment as a rapid alternative monitoring device.

TABLE 9.4

Average Biosensor, Fluorescence, and Micronuclei Frequency Response for Each Samples Group

Group	Type of Exposure	Individual Fish Number	Biosensor DNA Signal %[a]	Fluorescence $FF_{341/383}$ Pyrene-Like Conc. (mg/L)[a]	Micronuclei Frequency (number of micronuclei per 1000 cells scored)[b]	
1	Control	13	71 ± 7	1.0 ± 0.4	0.18 ± 0.04	0.13 ± 0.03
2	C_4–C_7 high	10	62 ± 3	1.8 ± 0.3	0.58 ± 0.09	0.48 ± 0.07
3	PW 1:200	14	56 ±8	5.6 ± 1.9	0.65 ± 0.10	0.48 ± 0.07
4	PW 1:1000	16	64 ± 8	2.6 ± 0.9	0.32 ± 0.06	0.20 ± 0.05
5	Oil	14	58 ± 9	5.1 ± 1.5	0.51 ± 0.04	0.27 ± 0.06
6	Oil + PAH + AP	14	51 ± 9	21.2 ± 5.1	0.53 ± 0.10	0.29 ± 0.10

Notes: [a] Performed in fish bile samples.
     [b] Performed in fish kidney (left) and in liver (right) samples.

9.3.1.3 Drug Analysis

The biosensor was used to study the interaction between a panel of antiproliferative metallodrugs and dsDNA as a model of the analogous interaction occurring in solution. The structures of the drugs analyzed are reported in Figure 9.10.

To assess the ability of the compounds to bind DNA, calibration curves were performed in 5 and 100 mM NaCl (concentrations that roughly mimic the NaCl content in cytoplasm and blood, respectively) and sometimes in 5 and 100 mM $NaClO_4$. Each concentration of each complex was in contact with the biosensor for 2 minutes.

The hydrolysis of metallodrugs is often the key step in their activation, but the timescale is extremely variable, ranging from tenths of a second for titanocene dichloride, to hours for cisplatin, to days for carboplatin. We found on SPE that the interaction of cisplatin with DNA increases with the solution ageing time, especially at low concentrations of chlorides [50]. Each complex was therefore preactivated for 60 minutes before testing interaction with DNA, in order to permit the transformation into the active aqua species, for the complexes needing this incubation time, and to compare the behavior at the same solution ageing time, for the complexes with a different rate of hydrolysis.

The first drug analyzed with the biosensors was cisplatin, which is one of the most potent and effective antitumor agents discovered in the past century [51]. The first clinical trials of cisplatin were carried out in 1971, and by 1983, cisplatin was the United States' biggest selling antitumor drug. Cisplatin is still one of the most widely used antitumor drugs because it is one of the most effective drugs for treating testicular, ovarian, bladder, and neck cancers. The mechanism of action of cisplatin with DNA is well known [52] and is reported in Figure 9.11.

Cationic platinum complexes, such as $[Pt(NH_3)_2(OH_2)Cl]^+$, are formed when a water molecule attacks the platinum metal center, thus eliminating a chloride ion. After losing two Cl^- ions, hydrolyzed cisplatin reacts with DNA, forming coordinative

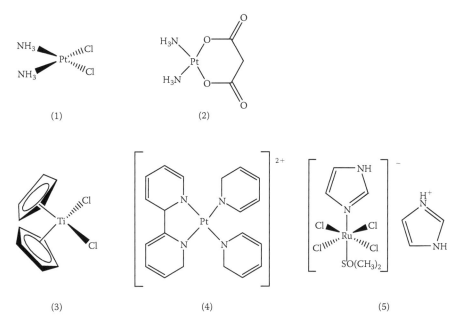

FIGURE 9.10 Structures of the metallodrugs analyzed: cisplatin (1); carboplatin (2); titanocene (3); platinum bipy (4); NAMI-A (5).

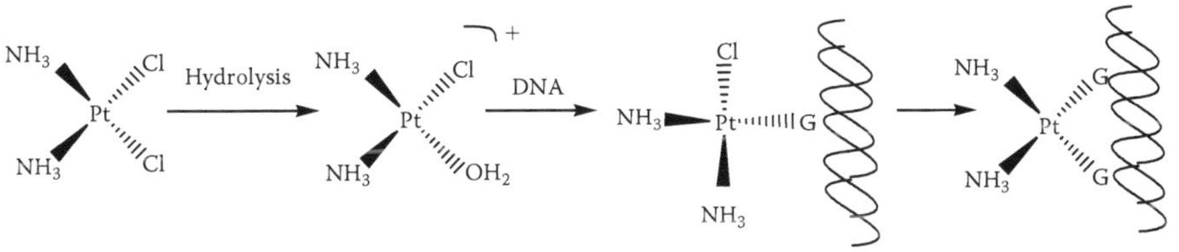

FIGURE 9.11 Mechanism of action and binding of cisplatin with DNA molecule.

bonds to nitrogen atoms of the nucleobases: the metal coordinates with N7 atoms of two adjacent guanine bases on DNA. The analysis with DNA biosensor in 5 and 100 mM NaCl has shown that the behavior of cisplatin strictly depends on the concentration of the NaCl and on the ageing time of the solution: high concentrations of chlorides inhibits the equation of cisplatin and, hence, its interaction with DNA [50] (Figures 9.12 and 9.13).

Modified versions of cisplatin have been synthesized over the past 30 years. Only a few of these have shown sufficient promise to be tested in human clinical trials. Of note is carboplatin, which has lower kidney toxicity and neurotoxicity and reduced nausea and vomiting, than cisplatin [53]. Analyzing carboplatin with the biosensor, the DNA signal % values decreased as carboplatin concentration increased (Figure 9.12).

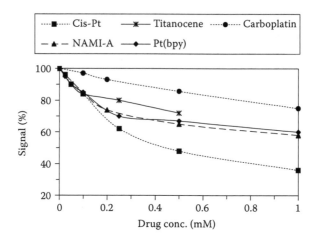

FIGURE 9.12 Comparison of guanine signal changes % obtained from the different drugs in 5 mM NaCl solution with 2 minutes of interaction with DNA biosensor. Average values and standard deviations were based on 3 or more results and are omitted for clarity (basically for all experiments about ± 4). Experimental conditions as described in Figure 9.2.

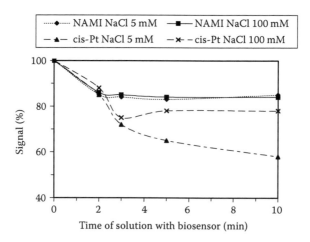

FIGURE 9.13 S % versus solution ageing time for 0.1 mM cisplatin or 0.1 mM NAMI-A in 5 mM or 100 mM NaCl solutions. Error bars (for all experiments about ± 4) are omitted for clarity. Experimental conditions as described in Figure 9.2.

S % values were lower than cisplatin and there were no differences between the two concentrations of NaCl. A slight difference was noted between the two types of electrolytes, showing a less strength of the compounds with perchlorate (data not shown). The slight decrease of the guanine oxidation signal in comparison with cisplatin is caused by the presence of chloride ions. This can be explained by a moderately efficient exchange between the 1,1-cyclobutanedicarboxylate and chlorides producing cisplatin that in turn undergoes activation by equation [54]. A similar exchange is more difficult for perchlorates being weaker Lewis bases. It has been also studied that in vitro experiments carboplatin requires a higher concentration and longer incubation times to reach the same results as cisplatin [55]. The time of stability of carboplatin was also tested, and the concentration 0.5 mM showed a small decrease in time due to the slow hydrolysis of the complex (i.e., from S % 94 to 83 after 500 minutes).

The side effects of these metal compounds and the narrow range of tumors in which they are efficient have developed new routes and many researchers have adopted a different approach. Thus, sterically hindered complexes, *trans* complexes, multinuclear platinum complexes, complexes with biologically active carrier ligands, and water-soluble complexes have been studied. In general, platinum (II)- based drugs possess high affinity for nucleotides. Platinum bipy complex has no leaving groups, therefore is devoid of any alkylating properties. Indeed, platinum bipy is able to intercalate DNA [22]. With DNA biosensor a decrease of S % was observed, but lower than cisplatin (Figure 9.12). Difficulty in the oxidation of guanine can be caused by depletion of electron density on nitrogen atom in 7-position of G (N7-G) due to the efficient and extensive delocalization throughout the DNA chain [56]. This electronic communication renders the guanine oxidation sensitive to the double positive charge localized on the Pt(II) center. Interestingly, S % value is practically time-independent since the complex does not need dissociation of inner-sphere ligands to be activated for the intercalation.

Platinum is not the only metal used in anticancer drugs. In fact, as early as 1979 titanocene had been recognized as being active against certain breast and gastrointestinal carcinomas, and is now in late-stage clinical trials along with a number of related complexes, although liver damage has been shown to be the dose-limiting side effect. Titanocene comes from the family of bis(cyclopentadienyl) metal(IV) diacido complexes, which are "bent sandwich" compounds. Interaction studies carried out using UV-vis and fluorescence spectroscopy indicated that the binding of titanocene with calf thymus DNA is dominated by phosphate coordination [57]. Obviously, the ionic interaction between Ti cation and external phosphate backbone produce a minor effect on the oxidation of G with respect to the direct coordination of its N7. Efforts to identify the biologically active species have been largely unsuccessful due to both the poor solubility of titanocene in water and the hydrolysis of the cyclopentadienyl rings, a reaction which is accelerated at high pH and results in precipitation of uncharacterized hydrolysis products [58]. Because of its poor solubility in water, we employed a 5 mM NaCl/ethanol solution 90/10 v/v; nevertheless, the maximum concentration of complex achieved was about 0.5 mM. Titanocene unambiguously showed a lower degree of interaction with SPE adsorbed dsDNA than cisplatin (Figure 9.12) as expected [57].

Ruthenium(III) complexes are increasingly attracting the interest of researchers due to their promising pharmacological properties as antitumor and antimetastatic agents [59,60]. The first representative of ruthenium(III) complexes presently undergoing phase I clinical trials is NAMI-A, which possesses outstanding antimetastatic properties [60]. The mechanism of ruthenium(III) complexes diverges profoundly from that of clinically established anticancer platinum(II) complexes [61,62]: degradation (hydrolysis, reduction) in the blood accompanied by reaction with serum proteins and other biological targets; transport into the cell due to transferrin receptor-mediated endocytosis (transferrin pathway) and enhanced permeability and retention (EPR) effect (albumin pathway); release of the complex caused by the influence of pH, ATP, and citrate; reduction in the hypoxic cell; and binding to nucleic acids [63]. In general, DNA binding is weaker for NAMI-A than for similar platinum complexes [50,61]. On the other hand, tight binding of NAMI-A to proteins has been described [50,64] and it is likely to conceive that binding to specific proteins may represent the molecular basis for its peculiar biological activity. Gallori et al. [61] showed that NAMI-A interacts with DNA at concentrations significantly higher than those at which cisplatin produces similar effects. For NAMI-A, S % value decreased as concentration increased (Figure 9.12). On the contrary, the kind and concentration of the supporting electrolyte play minor roles. In fact, NaCl, that should exert mass effect, and NaClO$_4$, which produces the noncoordinating perchlorate anion, gave similar results. This was quite surprising since the activation of NAMI-A has been supposed to involve chloride dissociation, followed by equation reactions. Very recently it has been reported that trans-[Ru$^{(III)}$Cl$_4$(DMSO-S)$_2$]$^-$ complex reacted with protonated G base coordinating it in the axial direction without detaching the four equatorial chlorides and producing trans-[Ru$^{(III)}$Cl$_4$(GH)(DMSO-S)]$^-$ (GH = protonated guanine) [65]. Since for steric reason the nucleobase is coordinated through a single N atom (in the actual case N9, N7 being protonated), the resulting interaction is certainly weaker than that corresponding to that involved in bifunctional intrastrand cross-linking, and indeed, we found that the change from 5 to 100 mM of NaCl causes a slight variation in S % (Figure 9.13). Figure 9.13 shows the trend of S % values for 0.1 mM cisplatin and 0.1 mM NAMI-A in 5 and 100 mM NaCl. Cisplatin shows lower S % values in comparison with NAMI-A, especially at low (5 mM) chloride concentration, and in this case the hydrolysis (that produces the electrophilic aquo-complex able to cross-link the DNA) proceeds for several minutes, as testified by a further decrease in S %.

Finally, we can conclude that different response behavior has been observed according to the different interaction of the drug with DNA. Cisplatin was taken as a model complex because its reaction with DNA is as well known as N7 guanine alkylant. As reported in literature, NAMI-A prefers to link to proteins instead of nucleobases [50], therefore S % values are higher than cisplatin. The binding of titanocene with DNA is due to bonds with phosphates [57] and obviously the ionic interaction between Ti cation and external phosphate backbone produce a minor effect on the oxidation of G with respect to the direct coordination of its N7, therefore S% values are higher than cisplatin. For carboplatin, S% values were higher than cisplatin, but this confirm the fact that in vitro experiments, carboplatin requires a higher concentration and longer incubation times to reach the same results as cisplatin [55].

This study of antiproliferative metallodrugs offers some interesting, albeit qualitative, information concerning (a) the reactivity of the metal complex when acting as a prodrug being activated by equation, (b) the possible mass effect of anions acting as a ligand (i.e., chlorides), (c) the intrinsic affinity of the electrophilic agent to each nucleophilic component of the DNA, and (d) the strength of perturbation caused directly to the DNA chain by such metallodrugs on the electron density of N7-G that is quantifiable in such measurements.

9.4 CONCLUSIONS

In this chapter, the electrochemical DNA biosensor is proposed as a screening device for the rapid bioanalysis of environmental pollution and drug studies. Interestingly, agreement in terms of the amount of pollution indicated was found between biosensor results and other methodologies. Therefore, this kind of biosensor may represent an easy and fast way of analysis of polluted areas, especially for infield experiments. A total time of 11 minutes is necessary to perform the measurement of a sample; therefore, such biosensor could be useful as early warning device in areas subject to ecological risk or in the qualitative analysis of different drugs for a rapid monitoring tool.

ACKNOWLEDGMENTS

The authors are grateful for the financial support, sample providing, and travel expenses within this research provided by the NFR-PROOF projects 153898/40, 164427/S40, and TOTAL E&P Norge A/S.

REFERENCES

1. Mascini M., Editorial, *Bioelectrochemistry*, 67, 1–129, 2005.
2. IUPAC, *Compendium of Chemical Terminology, The Gold Book*, ed. A. D. McNaught and A. Wilkinson, 2nd ed. (Cambridge: Blackwell Science, 1997) 148.
3. Thevenot D. R. et al., Electrochemical biosensors: Recommended definitions and classification, *Biosens. Bioelectron.*, 16, 121–131, 2001.
4. Bilitewski U. and Turner A. P. F., *Biosensors for Environmental Monitoring* (Amsterdam: Harwood Academic, 2000) 1.
5. Wang J. et al., DNA electrochemical biosensors for environmental monitoring. A review, *Anal. Chim. Acta*, 347, 1–8, 1997.
6. Wang J. et al., DNA-modified electrode for the detection of aromatic amines, *Anal. Chem.*, 68, 4365–4369, 1996.
7. Wang J. et al., DNA biosensor for the detection of hydrazines, *Anal. Chem.*, 68, 2251–2254, 1996.
8. Chiti G., Marrazza G., and Mascini M., Electrochemical DNA biosensor for environmental monitoring, *Anal. Chim. Acta*, 427, 155–164, 2001.
9. Lucarelli F. et al., Electrochemical DNA biosensor for analysis of wastewater samples, *Bioelectrochemistry*, 58, 113–118, 2002.
10. Lucarelli F. et al., Electrochemical DNA biosensor as a screening tool for the detection of toxicants in water and wastewater samples, *Talanta*, 56, 949–957, 2002.
11. Palecek E. and Fojta M., Detecting DNA hybridization and damage, *Anal. Chem.*, 73, 74A–83A, 2001.
12. Wang J. et al., Interaction of antitumour drug daunomycin with DNA in solution and at the surface, *Bioelectroch. Bioener.*, 45, 33–40, 1998.
13. Hashimoto K., Ito K., and Ishimori Y., Novel DNA sensor for electrochemical gene detection, *Anal. Chim. Acta*, 286, 219–224, 1994.
14. Carter M. T., Rodriguez M., and Bard A. J., Voltammetric studies of interaction of metals chelates with DNA: tris-chelated complexes of Co(III) and Fe(III) with 1,10-phenantroline and 2,2′-bipyridine, *J. Am. Chem. Soc.*, 111, 8901, 1989.
15. Lucarelli F., Marrazza G, Turner AP, Mascini M., Carbon and gold electrodes as electrochemical transducers for DNA hybridisation sensors, *Biosens. Bioelectron.*, 19, 515–530, 2004.

16. Wang J., Cai X., Rivas G., Shiraishi H., and Dontha N., Nucleic-acid immobilisation, recognition and detection at chronopotentiometric DNA chips, *Biosens. Bioelectron.*, 12(7), 587–599, 1997.

17. Erdem A. and Ozsoz M., Electrochemical DNA biosensors based on DNA-drug interactions, *Electroanal.*, 14, 965–974, 2002.

18. Esposito A., Del Borghi A., and Vegliò F., Investigation of naphthalene sulphonate compounds sorption in a soil artificially contaminated using batch and column assays, *Waste Manage.*, 22, 937–943, 2002.

19. Conte P., Agretto A., Spaccini R., and Piccolo A., Soil remediation: humic acids as natural surfactants in the washing of highly contaminated soils, *Environ. Pollut.*, 135, 515–522, 2005.

20. Avidano L., Elisa G., Paolo C. G., Elisabetta C., Characterisation of soil health in an Italian polluted site by using microorganisms as bioindicators, *Appl. Soil Ecol.*, 30, 21–33, 2005.

21. Harrison R. C., McAuliffe C. A., and Zaki A. M., An efficient route for the preparation of highly soluble platinum (II) antitumour agents, *Inorg. Chim. Acta*, 46, L15–L16, 1980.

22. Cusumano M., Di Pietro M. L., and Giannetto A., Relationship between binding affinity for calf-thymus DNA of $[Pt(2,2'-bpy)(n-Rpy)_2]^{2+}$ (n = 2,4) and basicity of coordinated pyridine, *Chem. Commun.*, 2527–2528, 1996.

23. Mestroni G., Alessio E., and Sava G., New salt of anionic complexes of Ru (III) as antimetastatic and antineoplastic agents, Int. Pat. WO 98/00431, 1998.

24. Beyer J., Aas E., Borgenvik H. K., and Ravn P., Bioavailability of PAH in effluent water from an aluminium works evaluated by transplant caging and biliary fluorescence measurements of Atlantic cod (*Gadus morhua*), *Mar. Environ. Res.*, 46(1–5), 233–236, 1998.

25. Cagnini A., Palchetti I., Lionti I., Mascini M., and Turner A. P. F., Disposable ruthenized screen-printed biosensors, *Sens. Actuators B*, 24, 85–89, March 1995.

26. Koncki R. and Mascini M., Screen-printed ruthenium dioxide electrodes for pH measurements, *Anal. Chim. Acta*, 351(1), 143–149 1997.

27. Palchetti I., Mascini M., Minunni M., Bilia A. R., Vincieri F. F., Disposable electrochemical sensor for rapid determination of heavy metals in herbal drugs, *J. Pharmaceut. Biomed.*, 32(2), 251–256, 2003.

28. Lucarelli F., Authier L., Bagni G., Marrazza G., Baussant T., Aas E., Mascini M., DNA biosensor investigations in fish bile for use as a biomonitoring tool, *Anal. Lett.*, 36(9), 1887–1901, 2003.

29. Bagni G., Baussant T., Jonsson G., Barsiene J., Mascini M., Electrochemical device for the detection of genotoxic compounds in fish bile samples, in *Proc. 9th Italian Conference Sensors and Microsystems* (London: World Scientific, 2005) 56.

30. Berset J. D., Ejem M., Holzer R., Lischer P., Comparison of different drying, extraction and detection techniques for the determination of priority polycyclic aromatic hydrocarbons in background contaminated soil samples, *Anal. Chim. Acta*, 383(3), 263–275, 1999.

31. Bagni G., Hernandez S., Mascini M., Sturchio E., Boccia P., Marconi S., DNA biosensor for the rapid detection of genotoxic compounds in soil samples, *Sensors*, 5(6–10), 394–410, 2005.

32. Sanni S., Øysæd K. B., Høivangli V., and Gaudebert B., A continuous flow system (CFS) for chronic exposure of aquatic organisms, *Mar. Environ. Res.*, 46(1–5), 97–101, 1998.

33. Smith A., The regulation of animal experimentation in Norway: An introduction, 1–28 1998, available online at http://oslovet.veths.no/book/Booklet.pdf.

34. Ruby M. V., Davis A., Schoof R., Eberle S., Sellstone C. M., Estimation of lead and arsenic bioavailability using a physiologically based extraction test, *Environ. Sci. Technol.*, 30(2), 422–430, 1996.

35. Gfrerer G., Serschen M., and Lankmayr E., Optimised extraction of polycyclic aromatic hydrocarbons from contaminated soil samples, *J. Biochem. Bioph. Meth.*, 53(1–3), 203–216, 2002.

36. Guerin T. F., The extraction of aged polycyclic aromatic hydrocarbons (PAH) residues from a clay soil using sonication and Soxhlet procedure: a comparative study, *J. Environ. Monit.*, 1, 63–67, 1999.

37. Song Y. F., Jing X., Fleischmann S., and Wilke B. -M., Comparative study of extraction methods for the determination of PAHs from contaminated soil and sediments, *Chemosphere*, 48, 993–1001, 2002.

38. Snyder R. D. and Arnone M. R., Putative identification of functional interactions between DNA intercalating agents and topoisomerase II using the V79 in vitro micronucleus assay, *Mutat. Res.*, 503, 21–35, 2002.

39. Fenech M., Chang W. P., Kirsch-Volders M., Holland N., Bonassi S., Zeiger E., HUman MicronNucleus project, HUMN project: detailed description of the scoring criteria for the cytokinesis-block micronucleus assay using isolated human lymphocyte cultures, *Mutat. Res.*, 534, 65–75, 2003.

40. Arkhipchuk V. V. and Garanko N. N., Using the nuclear biomarker and the micronucleus test on in vivo fish fin cells, *Ecotox. Environ. Safe.*, 62, 42–52, 2005.

41. Escartin E. and Porte C., Assessment of PAH pollution in coastal areas from the NW Mediterranean through the analysis of fish bile, *Mar. Pollut. Bull.*, 38, 1200–1206, 1999.

42. Aas E., Beyer J., and Goksøyr A., PAH in fish bile detected by fixed wavelength fluorescence, *Mar. Environ. Res.*, 46, 225–228, 1998.

43. Lin E. L. C., Cormier S. M., and Torsella J. A., Fish biliary polycyclic aromatic hydrocarbon metabolites estimated by fixed-wavelength fluorescence: comparison with HPLC-fluorescent detection, *Ecotox. Environ. Safe.*, 35, 16–23, 1996.

44. Aas E., Thierry B., Lennart B., Birgitta L., Ketil A.O., PAH metabolites in bile, cytochrome P4501A and DNA adducts as environmental risk parameters for chronic oil exposure: a laboratory experiment with Atlantic cod, *Aquat. Toxicol.*, 51, 241–258, 2000.

45. Aas E., Beyer J., Jonsson G., Reichert W. L., Andersen O. K., Evidence of uptake, biotransformation and DNA binding of polyaromatic hydrocarbons in Atlantic cod and corkwing wrasse caught in the vicinity of an aluminium works, *Mar. Environ. Res.*, 52, 213–229, 2001.

46. Jeannette K. W., Jeffrey A. M., Blobstein S. H., Beland F. A., Harvey R. G., Weinstein I. B., Nucleoside adducts from the in vitro reaction of benzo[a]pyrene-7,8-dihydrodiol 9,10-oxide or benzo[a]pyrene 4,5-oxide with nucleic acids, *Biochemistry*, 16, 932–938, 1977.

47. Barron M. G., Heintz R., and Rice S. D., Relative potency of PAHs and heterocycles as aryl hydrocarbon receptors agonists in fish, *Mar. Environ. Res.*, 58, 95–100, 2004.

48. Ferreira M. M. C., Polyciclic aromatic hydrocarbons: A QSPR study, *Chemosphere*, 44, 125–146, 2001.

49. Ribeiro F. A. L. and Ferreira M. M. C., QSAR model of the phototoxicity of polycyclic aromatic hydrocarbons, *Theochem—J. Mol. Struc.*, 719, 191–200, 2005.

50. Ravera M., Sara B., Claudio C., Donato C., Graziana B., Gianni S., Domenico O., Electrochemical measurements confirm the preferential bonding of the antimetastatic complex [ImH][RuCl4(DMSO)(Im)] (NAMIA) with proteins and the weak interaction with nucleobases, *J. Inorg. Biochem.*, 98, 984–990, 2004.

51. Guo Z. and Sadler P. J., Medicinal inorganic chemistry, in *Advances in Inorganic Chemistry*, Academic Press Ed., Elsevier Science B. V., Oxford, 2000, 49, 183.

52. Takahara P. M., Rosenzweig A.C., .Frederick C.A., Lippard S.J., Crystal structure of double-stranded DNA containing the major adduct of the anticancer drug cisplatin, *Nature*, 377, 649–652, 1995.

53. Wong E. and Giandomenico C. M., Current status of platinum-based antitumor drugs, *Chem. Rev.*, 99, 2451–2466, 1999.

54. Heudi O., Mercier-Jobard S., Cailleux A., Allain P., Mechanisms of reaction of L methionine with carboplatin and oxaliplatin in different media: A comparison with cisplatin, *Biopharm. Drug Dispos.*, 20, 107–116, 1999.

55. Go R. S. and Adjei A. A., Review of the comparative pharmacology and clinical activity of cisplatin and carboplatin, *J. Clin. Oncol.*, 17, 409–422, 1999.

56. Holmlin R. E., Dandliker P. J., and Barton J. K., Charge transfer through the DNA base stack, *Angew. Chem. Int. Ed. Engl.*, 36, 2714–2730, 1997.

57. Yang P. and Guo M., Interactions of organometallic anticancer agents with nucleotides and DNA, *Coordin. Chem. Rev.*, 186, 189–211, 1999.

58. Mokdsi G. and Harding M. M., Water soluble, hydrolytically stable derivatives of the antitumor drug titanocene dichloride and binding studies with nucleotides, *J. Organomet. Chem.*, 565, 29–35, 1998.

59. Clarke M. J., Zhu F., and Frasca D. R., Non-platinum chemotherapeutic metallopharmaceuticals, *Chem. Rev.*, 99, 2511–2534, 1999.

60. Sava G., Alessio E., Bergamo A., Mestroni G., *Topics in biological inorganic chemistry-metallopharmaceuticals*, ed. M.J Clarke and P.J Sadler, 1st ed. (Berlin: Springer, 1999) 143–170.

61. Gallori E., Vettori C., Alessio E., Vilchez F. G., Vilaplana R., Orioli P., Casini A., and Messori L., DNA as a possible target for antitumor ruthenium(III) complexes—A spectroscopic and molecular biology study of the interactions of two representative antineoplastic ruthenium(III) complexes with DNA, *Arch. Biochem. Biophys.*, 376, 156–162, 2000.

62. Malina J., Olga K., Bernhard K. A., Enzo B.V., Biophysical analysis of natural, double-helical DNA modified by anticancer heterocyclic complexes of ruthenium(III) in cell-free media, *J. Biol. Inorg. Chem.*, 6, 435–445, 2001.

63. Bergamo A., Gagliardi R., Scarcia V., Furlani A., Alessio E., Mestroni G., and Sava G., In vitro cell cycle arrest, in vivo action on solid metastasizing tumors, and host toxicity of the antimetastatic drug NAMI-A and cisplatin., *J. Pharmacol. Expl. Ther.*, 289, 559–564, 1999.

64. Messori L., Orioli P., Vullo D., Alessio E., Iengo E., A spectroscopic study of the reaction of NAMI, a novel ruthenium(III)anti-neoplastic complex, with bovine serum albumin, *Eur. J. Biochem.*, 267, 1206–1213, 2000.

65. Turel I., Pecanac M., Golobic A., Alessio E., Serli B., Bergamo A., Sava G., Solution, solid state and biological characterization of ruthenium(III)-DMSO complexes with purine base derivatives, *J. Inorg. Biochem.*, 98, 393–401, 2004.

10 Methods of Detection of Explosives

As Chemical Warfare Agents

Sagar Yelleti, Ebtisam S. Wilkins
Department of Chemical and Nuclear Engineering
Department of Biology
University of New Mexico
Albuquerque, New Mexico

Ravil A. Sitdikov, David Faguy, Ihab Seoudi
Department of Chemical and Nuclear Engineering
University of New Mexico
Albuquerque, New Mexico

CONTENTS

10.1 INTRODUCTION

Detection of explosives can potentially employ a wide variety of analytical techniques. In order to assess the feasibility of such techniques and design viable detection systems, molecular properties and spectroscopic signatures of the target molecule are needed. We have surveyed the techniques and methods currently under development for the detection of explosive [1–167].

10.2 AVAILABLE TECHNIQUES FOR EXPLOSIVE DETECTION

10.2.1 SNIFFING DOGS

At the present time, there does exist a reasonably reliable, cost-effective, user-friendly biological system for explosives' detection, namely, a trained dog and its handler [1,131]. Explosive-sniffing dogs are still a key backup system at many airports and crime scenes. In 1994, the Federal Aviation Administration maintained 103 explosive detection dog handling teams for airport searches [62,63], and this number is expected to triple over the next several years. The Bureau of Alcohol, Tobacco, and Firearms maintain a dog-handler team that has participated in numerous searches during the past several years. Other species, such as field mice and gerbils, have been suggested for this purpose, but dogs remain the "system of choice" for this application [63].

The process whereby dogs recognize and respond to odors is still not very well understood, and current research in this area is attempting to quantify such responses and improve the reliability of this already remarkable detection system [68,82]. While it may eventually be possible to combine chemical sensors, readout devices, and neural network processing schemes to create an "electro-optic dog" that will reduce or eliminate the variability of the natural system, we have a long way to go before this can be accomplished.

10.2.2 IMMUNOSENSORS

Several approaches based on antigen-antibody formation have been investigated [96] besides the more physically based approaches described above. In these systems, a monoclonal antibody is developed by sensitizing test animals to a protein incorporating the substance to be detected or a close analog. When the substance to be detected binds to the antibody, a change in a property such as optical transmittance [96] or fluorescence [136] may be detected, registering the presence of the explosive. In continuous-flow immunosensors [4–5,83,91], a flow of solution containing the analyte is exchanged with the immobilized antibody [83]. Plastic beads containing the antibodies have been previously treated with a fluorescent dye-labeled explosive analog that is released to the solution when the explosive interacts with the antibody. The displaced analog molecules are then detected downstream by fluorescence excitation/emission spectroscopy.

Although such immunosensors are capable of sub parts-per-million detection of 2, 4, 6-trinitrotoluene (TNT), cyclotrimethylenetrinitramine (RDX), pentaerythritol tetranitrate (PETN), and related substances, their principal drawback for an application such as airport security screening is the time required to complete the analysis [5,12,21–22,83]. For this reason, this technique has found its greatest use in analysis of explosive residues in soils and groundwater, rather than for real-time detection.

10.2.3 SYNTHETIC RECEPTORS: MOLECULAR RECOGNITION

One approach to high-selectivity detection of explosives at ultratrace levels that may overcome the processing time limitation of immunosensors is to combine a chemical sensor tailored to specific target molecules with a "molecular transducer" that responds nearly instantaneously to the presence of the analyte [97–100]. The chemosensors consist of a molecular recognition site tailored to the target molecule, such as a poly-rotaxane, cyclophane, or calixarene. The molecular transduction site may be based on fluorescence quenching, photoconductivity, or possibly surface-enhanced Raman scattering. Receptors for nitro-containing compounds have seen scant investigation to date, however, and much more work needs to be done in this area.

10.3 ELECTROCHEMICAL BIOSENSOR

Biosensors promise low cost, rapid, and simple-to-operate analytical tools. They, therefore, represent a broad area of emerging technology ideally suited for point-of-care analysis [56,145]. The following article provides a summary of biosensors, electrochemical detection, and their combination to yield an electrochemical affinity assay. Biosensors are analytical tools combining a biochemical recognition component with a physical transducer (Figure 10.1). The biological sensing element can be an enzyme, antibody, DNA sequence, or even microorganism. The biochemical component serves to selectively catalyze a reaction or facilitate a binding event. The selectivity of the biochemical recognition event allows for the operation of biosensors in a complex sample matrix, that is, a body fluid. The transducer converts the biochemical event into a measurable signal, thus providing the means for detecting it. Measurable events range from spectral changes, which are due to the production or consumption of an enzymatic reaction's product/substrate, to mass change upon biochemical complexation.

Enzymes are nature's catalysts. Like all catalysts, they increase the rate at which a reaction reaches equilibrium by providing a low-activation energy reaction pathway. They usually operate in approximately neutral pH at mild temperatures, generate no by-products, and are highly selective. Enzyme-catalyzed reactions can be selective for one substrate or a group of substrates. They are also stereo selective and stereo specific. These characteristics have resulted in the frequent use of enzymes in analytical applications.

Several thousand enzymes have been isolated, and several hundred are available commercially. They are classified by the reactions they catalyze. With

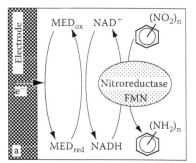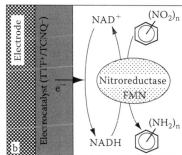

FIGURE 10.1 Generalized concept of the enzyme electrochemical biosensor for explosives: (a) with redox mediator; (b) based on catalytic reduction of NADH.

amperometric enzyme electrodes, oxidoreductase enzymes are most frequently used. Oxidoreductases catalyze the oxidation (removal of electrons) or reduction (addition of electrons) of the enzyme substrate. Since oxidoreductases are most closely associated with electrochemical processes, their turnover is easiest to observe by electrochemical detection. Enzymes, like all proteins, are made of amino acid chains folded into specific three-dimensional structures. They range in size from 10,000 to several million daltons. Besides amino acids, many enzymes also contain prosthetic groups—nicotinamide adenine dinucleotide (NADH), flavin (FAD), heme, Mg^{+2}, and Ca^{+2}—that enhance enzyme activity. With oxidoreductases, the prosthetic groups serve as temporary traps of electrons or electron vacancies.

High sensitivity, selectivity, and ability to operate in turbid solutions are advantages of electrochemical biosensors. Amperometric detection is based on measuring the oxidation or reduction of an electroactive compound at a working electrode (sensor). A potentiostat is used to apply a constant potential to the working electrode with respect to a second electrode (reference electrode). A potentiostat is a simple electronic circuit that can be constructed using a battery, two operational amplifiers, and several resistors. The applied potential is an electrochemical driving force that causes the oxidation or reduction reaction.

10.4 EXPLOSIVE DETECTION

TNT is most commonly biotransformed by reduction of the nitro groups. By the sequential addition of two electrons, a nitro group is reduced to a nitroso, a hydroxylamino, and finally to an amino group. Through successive reductions of the three nitro groups, many bacterial species sequentially transform TNT to aminodinitrotoluene, diaminonitrotoluene, and finally triaminotoluene (TAT). However, in some bacterial cultures hydroxylamino-substituted compounds persist without the production of amino-substituted compounds. Two monoamino isomers, 2-amino-4, 6-dinitrotoluene (2ADNT) and 4-amino-2,6-dinitrotoluene (4ADNT), and only one diamino isomer, 2,4-diamino-6-nitrotoluene (DANT), form in significant amounts from TNT reduction. As the electron-withdrawing nitro groups are sequentially replaced with electron-donating amino groups, the remaining nitro groups become less susceptible to reduction. Thus, TNT is reduced faster than ADNTs, which, in turn, are reduced faster than DANT. Complete anaerobic transformation of TNT to TAT appears limited by the DANT to TAT transformation.

Enzymes, which catalyze the reduction of nitro compounds using a reduced pyridine nucleotide, are termed nitroreductases and are distinguished by their sensitivity of activity to oxygen. These have been divided in Type I nitroreductases that reduce in two-electron increments and Type II nitroreductases that reduce in one-electron increments. These enzymes have other physiological roles but also exhibit activity with nitro groups. Type II nitroreductases are oxygen sensitive because the nitro anion radical formed after one-electron reduction can react with oxygen, resulting in electron transfer to oxygen yielding a superoxide radical and the original nitro group. Most nitroreductases purified from bacteria are Type I soluble flavoproteins that use nicotinamide adenine dinucleotide ($NADH \cdot H^+$) or nicotinamide adenine dinucleotide phosphate ($NADPH \cdot H^+$) as electron donors. In redox reactions, these compounds accommodate the transfer of two protons and two electrons to yield NAD⁺ and NADP⁺, respectively. Many purified enzymes and cell-free extracts exhibit activity against TNT using either $NADH \cdot H^+$ or $NADPH \cdot H^+$ as electron donors including enzymes from *Pseudomonas pseudoalcaligenes, P. aeruginosa, Bacillus* sp., *Staphylococcus* sp., *Ralstonia eutropha, Enterobacter cloacae,* and *Clostridium thermoaceticum.* The enzyme's physiological role is the coupled oxidation of $NAD(P)H \cdot H^+$ and reduction of flavin mononucleotide (FMN), but it has been shown to reduce TNT and other nitroarenes.

10.5   RESEARCH AT THE UNIVERSITY OF NEW MEXICO

The goal for the project at the University of New Mexico is to establish the feasibility of an enzymatic biosensor approach for detection of HE from the classes of nitroaromatic and nitramine explosives, primarily 2,4,6-TNT and cyclo-1,3,5-trimethylene-2,4,6-RDX, respectively, via their enzymatic reduction, and to develop a working prototype. Initially, the enzyme electrode detection system will be assembled and tested utilizing laboratory sensor approaches.

10.5.1   RESEARCH APPROACH

The explosive sensor is based on enzymatic reduction of nitroaromatic and nitramine hydrocarbons to corresponding amines or intermediates by a microbial nitroreductase enzyme [2]. Bacterial nitroreductases are highly variable, yet the genetic sequence for those that have been isolated and determined is highly conserved [2]. There are two types of nitroreductases produced by bacteria, oxygen-sensitive and oxygen-insensitive enzymes. The oxygen-sensitive enzyme is highly unstable and is inhibited by oxygen; therefore, we propose to isolate oxygen-insensitive nitroreductases.

Microorganisms grown on both TNT and RDX are screened for nitroreductase activity that is stable under aerobic conditions. We will select bacteria with demonstrated ability to convert the nitro groups of TNT or RDX to intermediates or amino groups. These organisms are cultured under the appropriate conditions in order to supply cells from which to extract the enzyme. Routine techniques are used for breaking open the cells, precipitating the proteins, separating the membrane fraction, and obtaining a crude preparation. The enzyme obtained is measured both spectroscopically and through protein assays, and perform the nitroreductase assay to demonstrate success. Bacterial nitroreductases have two low-molecular weight redox cofactors: flavine-adenine mononucleotide (FMN) and nicotinamide adenine dinucleotide (NAD) [2]. Usually FMN is strongly bound to the enzyme protein molecule, while NAD is loosely associated with the enzyme, acting as an electron/proton exchange shuttle (natural redox mediator).

There have not been any reports in the literature using nitroreductases in sensing systems to date. There is extensive experience, however, in coupling of other NAD-dependent enzymes (predominantly dehydrogenases) with electrodes for sensing purposes. The general approach is the regeneration of one of the enzyme cofactors (NAD), at the electrode, which will result in signal (usually current) generation. Direct reduction of NAD, however, takes place at a quite high over voltage. This results in numerous interference effects from redox active substances, usually present in the sample matrix. There are two possible methods of low-potential reduction of NAD: (i) a low-molecular weight mediator included in the system and (ii) modification of the electrode surface by a proper electrocatalyst to provide conditions for lowering the over voltage of NAD reduction. Both pathways have been explored in the literature. Figure 10.1 illustrates the general concept of both a mediator system based (a) and a catalytic electrode based (b) nitroreductase enzyme electrode.

10.5.2 EXPERIMENTAL

The experimental part includes the following steps:

- The design of the sensor
- Testing the sensor for different analytes (explosives)
- Obtaining the calibration curve for the sensor
- Testing for the reproducibility of the sensor
- Field testing of the sensor

10.5.3 DESIGN OF BIOSENSOR

Two kinds of sensor designs will be considered during the research. The first design is a gel-type sensor. The design is a three-electrode system, a working electrode made of platinum, a counter electrode made of platinum and a silver/silver chloride (Ag/AgCl) reference electrode. The enzyme is immobilized in a gel. About 25% glutaraldehyde is added to bovine serum albumin (BSA). To this solution the enzyme is added and the solution is stirred. The stirring is continued till the gel formation commences. When the gel formation starts, the gel is transferred to sensor body, which has the electrodes in it. The sensor body is covered with a membrane on one side, which holds the gel in the body. The side having the membrane is dipped in the solution, which is to be tested for the explosive.

The second design is a needle-type sensor. Here the enzyme is electrically deposited on the electrode surface using a conductive polymer. A conductive polymer is dissolved in phosphate buffer (we use phenylene diamine). To this solution the enzyme is added. The electrodes are dipped in the solution and a potential of +0.65 is applied for few seconds. The conductive polymer gets deposited along with the enzyme on the electrode surface. These electrodes are dipped in a solution, which is to be tested for the explosive.

10.5.4 TESTING OF SENSOR

After the sensor has been designed, the sensor is tested for a specific analyte, which is an explosive in this case. In order to work with the sensor we need to find the working potential for the given analyte. This working potential is obtained by running the cyclic voltammetry on the analyte. The given analyte is taken in a phosphate buffer and the sensor is dipped in the solution. A potential sweep of −1.0 to +1.0 is applied with a potentiostat. The change in the output current is recorded using a recorder

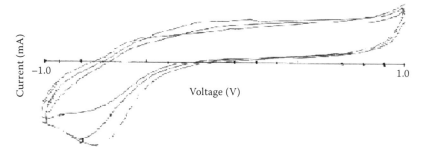

FIGURE 10.2   Cyclic voltammetry showing increase in peak with increase in concentration.

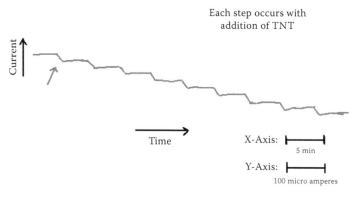

Calibration curve for TNT

FIGURE 10.3   Current versus time or current versus concentration of analyte (TNT).

for different potentials. The cyclic voltammetry curve (Figure 10.2) shows a peak in the output current at a particular voltage, which increases with the increase in the concentration of the analyte. This particular voltage is the working potential for the analyte.

Now the found working potential is applied to the sensor and the sensor is dipped in a phosphate buffer solution. When a steady current is reached, a known amount of the analyte is added to the solution and the change in the current is recorded till steady current is reached. Then the analyte concentration is increased and the steady current values are noted.

The values found for the steady current are plotted against the concentration of analyte in the solution. The plot is linear and gives us the current (steady) for a given concentration of analyte. Thus the actual concentration of the explosive in the contaminated soil or water can be found out from the calibration curve (Figure 10.3).

Now we have to test the sensor for the reproducibility. The sensor must give the same results as above on a different occasion. This test gives us the lifetime of the sensor. When the sensor tests positive in the lab, the sensor can then be tested on the fields having soils contaminated with explosives.

10.6 CONCLUSIONS

Whatever solution is chosen, its cost must not be neglected in view of its application to humanitarian demining. It is clear that no single technology has the capability to detect and recognize a mine in all situations. Costs and efficiency must be compared, hence the need for a better exchange of information between the specialists in each category.

The future avenue in sensor detection would be coming out with a sensor design, which would have less response time, economic, and highly analyte specific. An electrochemical biosensor shows promise of having a sensor, which would be portable, economic, and easy in construction and have good specificity toward a particular analyte. Till now, no work has been reported on using biosensor electrochemically for the detection of explosives. If this project comes out with some good results, the electrochemical biosensor would have a very good market for explosive detection in the near future.

REFERENCES

1. Asano K. G., Goeringer D. E., and McLuckey S. A., Parallel monitoring for multiple targeted compounds by ion trap mass spectrometry, *Anal. Chem.*, 67(17), 2739–2742, 1995.
2. Asher S. A., UV resonance Raman studies of molecular structure and dynamics: Applications in physical and biophysical chemistry, *Annu. Rev. Phys. Chem.*, 39, 537–588, 1988.
3. Alcohol, Tobacco, and Firearms, Bureau of (ATF), 1997. *Arson and Explosives Incidents Report.* Publication ATF P 3320.4 (4/97). Washington, D.C.: Bureau of Alcohol, Tobacco, and Firearms, 1995.
4. Bart J. C., Judd L. L., and Kusterbeck A. W., Environmental immunoassay for the explosive RDX using a fluorescent dye-labeled antigen and the continuous-flow immunosensor, *Sens. Actuators B*, 38–39, 411–418, 1997.
5. Bart J. C., Judd L. L., Hoffman K. E., Wilkins A. M., and Kusterbeck A. W., Application of a portable immunosensor to detect the explosives TNT and RDX in groundwater samples, *Environ. Sci. Technol.*, 31, 1505–1511, 1997.
6. Bertrand G. and Claudio B., Sensor technologies for the detection of antipersonnel mines: A survey of current research and system developments. Paper presented at the International Symposium on Measurement and Control in Robotics (ISMCR'96), Brussels, May, 9–11 1996.
7. Bruschini C. and Bertrand G., *A Survey of Current Sensor Technology Research for the Detection of Landmine, LAMI- DeTeC.* Switzerland: Lausanne, 1997.
8. Robledo L., Carrasco M., Mery D., A survey of land mine detection technology, *Int. J. Remote Sens.*, 30(9–10), 2399–2410, 2009.
9. Bertrand G. and Bruschini C., *Sensor Technologies for the Detection of Antipersonnel Mines*, A Survey of Current Research and System Developments. EPFL- LAMI DeTeC. Lausanne, Switzerland. Bibliography Page 39 GAO- 01-239 Land Mine Detection, 1996.
10. Bureau of Alcohol, Tobacco and Firearms. 1995. *Arson and Explosives Incidents Report* 4, Washington, D.C.: BATF, 1994.

11. Burchanowski C., Moler R., and Shope S., Scanned-beam x-ray source technology for photon backscatter imaging technique of mine detection: Advanced technology research, *SPIE Proc.*, 2496, 368–373, 1995.

12. Boumsellek S., Alajajian S. H., and Chutjian A., Negative-ion formation in the explosives RDX, PETN, and TNT by using the reversal electron attachment detection technique, *J. Am. Soc. Mass Spectrom.*, 3, 243–247, 1992.

13. Boumsellek S. and Chutjian A., Increased response of the reversal electron attachment detector and modeling of ion space charge effects, *Anal. Chem.*, 64(18), 2096–100, 1992.

14. McHugh C. J., Kennedy A. R., Smith W. E., and Graham D., TNT stilbene derivatives as SERRS active species, *Analyst*, 132, 986–988, 2007.

15. Carnahan B. L., Day S., Kouznetsov V., and Tarassov A., Field ion spectrometry: A new technology for cocaine and heroin detection, *SPIE Proc.*, 2937, 106–119, 1997.

16. Cerniglia C. E. and Somerville C. C., in Biodegradation of Nitroaromatic Compounds, *Env. Sci. Res. Ser.* Vol. 49 (Ed: J.C. Spain), Plenum Press, Boca Raton, FL, 99–115 (and references therein), 1995.

17. Chasan D. E. and Norwitz G., Qualitative analysis of primers, tracers, igniters, incendiaries, boosters, and delay compositions on a microscale by use of infrared spectroscopy, *Microchem. J.*, 17, 31–60, 1972.

18. Chen T. H., Kitts C. L., Cunningham D. P., and Unkefer P. J., Isolation of three hexahydro-1,3,5-trinitro-1,3,5-triazine-degrading species of the family Enterobacteriaceae from nitramine explosive-contaminated soil, *Appl. Environ. Microbiol.*, 60(12), 4608–4611, 1994.

19. Carper W. R. and Stewart J. J. P., Effects of isotopic substitution on the vibrational spectra of 2,4,6-trinitrotoluene, *Spectrochim. Act., Part A*, 43A, 1249–1255, 1987.

20. Campion A., Raman spectroscopy of molecules adsorbed on solid surfaces, *Annu. Rev. Phys. Chem.*, 36, 549–572, 1985.

21. Chutjian A., Boumsellek S., and Alajajian S. H., Negative-ion formation in the explosives RDX, PETN, and TNT by using the reversal electron attachment detection technique, *J. Am. Soc. Mass. Spectrom.*, 3(3), 243–247, 1992.

22. Chutjian A., Boumsellek S., and Alajajian S. H., Negative ion formation in the explosives RDX, PETN and TNT using reversal electron attachment detection technique, in Proc. First Int. Symp. Explo. Detect. Technol. FAA Atlantic City, NJ, November 13–15, 1991, Khan S. M. Ed., 571–583, 1991.

23. Chutjian A. and Darrach M. R., Improved portable reversal electron attachment vapor detection system for explosive detection, in Proc. 2nd Explos. Detect. Technol. Symp. Aviation Secur. Technol. Conf. FAA Atlantic City, NJ, November 12–15, Makky W., Chair, 176–180, 1996.

24. Boyars C., Compatibility (Safety) Tests for Taggants in Explosives and Reducing the Explosion Sensitivity of Ammonium Nitrate Fertilizer, Compendiu., Int. Explos. Symp. Fairfa, V., September 18–2, 1995, Treasury Dept., BATF, April 111–115, 1996.

25. Clapper M., Demirgian J., and Robitaille G., A quantitative method using FT-IR to detect explosives and selected semivolatiles in soil samples, *Spectroscop*, 10, 44–49, 1996.

26. Crane R. A., Laser optoacoustic absorption spectra for various explosive vapors, *Appl. Opt.*, 17, 2097–2102, 1978.

27. Claspy P. C., Pao Y.-H., Kwong S., and Nodov E., Laser optoacoustic detection of explosive vapors, *Appl. Opt.*, 15, 1506–1509, 1976.

28. Cheng C., Kirkbride T. E., Batchelder D. N., Lacey R. J., and Sheldon T. G., In situ detection and identification of trace explosives by Raman microscopy, *J. Forensic Sci.*, 40(1), 31–37, 1995.

29. Crellin K. C., Widmer M., and Beauchamp J. L., Chemical ionization of TNT and RDX with trimethylsilyl cation, *Anal. Chem.*, 69(6), 1092–1101, 1997.

30. Crowson A., Hiley R. W., Ingham T., McCreedy T., Pilgrim A. J., and Townshend A., Investigation into the detection of nitrated organic compounds, *Anal. Commun.*, 34, 213–216, 1997.

31. Danylewych-May L. L. and Cumming C., Explosive and taggant detection with ionscan, Adv. Anal. Detect. Explos. Proc. 4th Inter. Symp. Anal. Detect. Explos., September 7–1, 1992 Jerusale, Israel, Yinon J. Ed. Kluwer Academic Publishers Dordrecht, Holland, 385–401, 1993.

32. Davies J. P., Blackwood L. G., Davis S. G, Goodrich L. D., and Larson R. A., Design and calibration of pulsed vapor generators for 2,4,6-trinitrotoluene, cyclo-1,3,5-trimethylene-2,4,6-trinitramine, and pentaerythritol tetranitrate, *Anal Chem.*, 65, 3004–3009, 1993.

33. Davidson W. R., Stott W. R., Akery A. K., and Sleeman R., The role of mass spectrometry in the detection of explosives, in Proceedings of the First International Symposium on Explosives Detection Technology, Atlantic City, NJ, 663–671, 1991.

34. Davidson W. R., Thomson B. A., Sakuma T., Stott W. R., Akery A. K., and Sleeman R., Modifications to the ionization process to enhance the detection of explosives by API/MS/MS, in Proceedings of the First International Symposium on Explosives Detection Technology, Atlantic City, NJ, 653–662, 1991.

35. Danylewych-May L. L., Modifications to the ionization process to enhance the detection of explosives by IMS, Proc. First Int. Symp. Explos. Detect. Technol., FAA, Atlantic City NJ, Nov. 13–15, Khan S. M, Ed., 672–686, 1991.

36. Davies J. P., Hallowell S. F., and Hoglund D. E., Particle generators for the calibration and testing of narcotic and explosive vapor/particle detection systems, *SPIE Proc-Int Soc. Opt. Eng.*, 2092, 137, 1994.

37. Bohn R. S., and Werner W., Recovering of components from plastic bonded propellants Manfred A. Waste Management, 17(2–3) 175–185, 1998.

38. Eisenreich N., Neutz J., Seiler F., Hensel D., Stancl M., Tesitel J., Price R., et al., Selected papers presented at the Sixth International Symposium on Hazards, Prevention, and Mitigation of Industrial Explosions, 6th ISHPMIE, 2007, Halifax, Nova Scotia, August 27–September 1, 2006.

39. Ertl H., Breit U., Kaltschmidt H., and Oberpriller H., Determination of the HMX and RDX content in synthesized energetic material by HPLC, FT-MIR, and FT-NIR spectroscopies. New separation device that allows fast gas chromatography of large samples, *SPIE Proc.*, 2276, 58–68, 1994.

40. Fetterolf D. D. and Clark T. D., Detection of Trace Explosive Evidence by Ion Mobility, *Spectrometry*, 38(1), 1993.

41. Fetterolf D. D. and Clark, T. D., in Proc. Int. Symp. Explosive Detection Technology, 1st, Atlantic City, FAA, 689–702, 1992.

42. Fetterolf D. D. and Clark T. D., Detection of trace explosive evidence by ion mobility spectrometry, *J. Forensic Sci.*, 38(1), 28–39, 1993.

43. Fetterolf D. D., Donnelly B., and Lasswell L. D., Portable instrumentation: New weapons in the war against drugs and terrorism. *SPIE Proc-Int. Soc. Opt. Eng.*, 2092, 40–52, 1994.

44. Fetterolf D. D., Donnelly B. D., and Lasswell L. D., Portable instrumentation: New weapons in the war on drugs. Proceedings of the International Association of Forensic Sciences 13th Triennial Meeting 5, 232, 1993.

45. Fine D. H, Rounbehler D. P, Curby W. A., in Proc. Int. Symp. Explosive Detection Technology, 1st, Atlantic City, NJ, FAA, 505–517, 1992.

46. Fine D. H. and Wendel G. J., *SPIE Substance Detection Systems*, 2092, 131–136, 1993.

47. Fleischmann M., Hendra P. J., and McQuillan A. J., Raman spectra of pyridine adsorbed at a silver electrode, *Chem. Phys. Lett.*, 26(2), 163–65, 1974.

48. Freeman R. G., Grabar K. C., Allison K. J., Bright R. M., Davis J. A., Guthrie A. P., Hommer M. B., Jackson M. A., Smith P. C., and Walter D. G., Self-assembled metal colloid monolayers: An approach to SERS substrates, *Science*, 267, 1629–1632, 1995.

49. Fox S., and Hooley T., *SPIE Proc.*, 2093:195–20336, 1994.

50. Fulghum S. F. and Tilleman M. M., Interferometric calorimeter for the measurement of water-vapor absorption, *J. Opt. Soc. Am.*, B8, 2401–2413, 1991.

51. Fulghum S. 1993. Detection of Explosives Vapor at the PPT Level with a Laser Interferometric Calorimeter, Sci. Res. Lab., Inc., Somerville, MA, Rep. SRL-05-F-1993.

52. Funsten H. O. and McComas D. J., Apparatus and method for rapid detection of explosives residue from the deflagration signature thereof, 1997. US Patent No. 5638166.

53. Giam C. S., Ahmed M. S., Weller R. R. and Derrickson J., Fourier cycloctrol (FT-ICR) mass spectrometry of RDX, PETN and other explosives, in Proc., First Inst. Symp. Explo. Detect. Technol. FAA Atlantic City, NJ, Nov 13–15, Khan S. M., Ed. 687–688, 1991.

54. Giam C. S., Holliday T. L., Ahmed M. S., Reed G. E., and Zhao G., Pseudo molecular ion formation of explosives in FT-ICR–M., in Proc. SPIE-The int. Soc. for Optical Eng., 1994, 2092 Substance detection Syst Proc., October 5–8, 227–237, 1993.

55. Giam C. S., Zhao G., Holliday T. L., Reed G. E., and Mercado A., EC-FT-ICR-MS to predict thermolytic bond fission and products of explosives, in Proc. 5th Int. Symp. Anal. Detect Exlos. Washington D.C., December 4–8, 1995, Midkiff C., Ed., Department of Treasury BATF, October, 1997.

56. Giam C. S., Holliday T. L., Ahmed M. S., Reed G. E., Zhao G., *SPIE Proc.* 2092, 227–37, 1994.

57. Glish G. L., McLuckey S. A., Grant B. C., and McKown H. S., in Proc. Int. Symp. Explosive Detection Technology, first, Atlantic City, NJ, FAA, 642–652, 1992.

58. Gorton L., Persson B., Hale P. D., Boguslavsky L. I., Karan H. I., Lee H. S., Skotheim T. A., Lan H. L., and Okamoto Y., 1992, in Biosensors and Chemical Sensors, ACS Symposium Series, Vol. 487 (Eds: P.G. Edelman, J.Wang), ACS, Washington, D.C., pp. 56–83 (and references therein).

59. Grabar K. C., Freeman R. G., Hommer M. B., and Natan M. J., Preparation and characterization of Au colloid monolayers, *Anal. Chem. Anal. Chem.*, 67, 735–743, 1995.

60. Hasue K., Nakahara S., Morimoto J., Yamagami T., Okamoto Y., and Miyakawa T., Photoacoustic Spectroscopy of Energetic Materials, *Propellants, Explosives Pyrotechnics*, 20(4), 187–191, 1995.

61. Hallowell S. F., Davies J. P., and Gresham G. L., Qualitative/semiquantitative chemical characterization of the Auburn olfactometer, *SPIE Proc.-Int. Soc. Opt. Eng.*, 2276, 437–448, 1994.

62. Hallowell S. F., Screening people for illicit substances: A survey of current portal technology, *Talanta*, 54(3), 447–458, 2001.

63. Hargis P. J. Jr., *Opt. Soc. Am. Annu. Meet. Long Beac.*, CA, 125, 1997.

64. Henderson D. O., Silberman E., Chen N., and Snyder F. W., Diffuse Reflectance Fourier Transform Infrared Spectroscopy, *Appl. Spectrosc.*, 47, 528–532, 1993.

65. Hendra P., Molecular and Biomolecular Spectroscopy, *Spectrochim. Acta Part A*, 46(2), 121–122, 1990.

66. Henderson D. O., Mu R., Tung Y. S., and Huston G. C., Decomposition kinetics of EGDN on ZnO by diffuse reflectance infrared fourier transform spectroscopy, *Appl. Spectrosc.*, 49, 444–450, 1995.

67. Hewish M. and Ness L., Mine-Detection technologies, *Int. Defense Rev.*, 28, 40–45, October 1995.

68. Hnatnicky S., Selection and use of explosives detection devices to check hand-held luggage, *J. Testing Eval.*, 22, 282–285, 1994.

69. Hobbs J. R., Analysis of Semtex Explosives, Adv. Anal. Detect. Explos. Proc. 4th Inter. Symp. Anal. Detect. Explos. September 7–1., 1992 Jerusalem, Israel, Yinon J., Ed. Kluwer Academic Publishers Dordrecht, Holland 409–427, 1992.

70. Hobbs J. R., Analysis of Propellants by Pyrolysis Gas Chromatography/Mass Spectrometry, Proceedings of 5th International Symposium Anal. Detect. Explos. Washington, D.C. December 4, 1995 Midkiff C., Ed. Dept. of Treasury, BATF October 1997.

71. Hobbs J. R. and Conde E., Comparison of different techniques for the headspace of explosives, Proc. Third Symp. Anal. Detect. Explos. Mannheim, FRG July 10–13, 1989, pp. 41–1 to 41–18, Hobbs J. R. and Conde E., "Explosives Sample Analysis" Technical Report - U.S.D.O.T. December 1989 (Distribution Limited).

72. Hobbs J. R. and Conde E. P., A simple inexpensive thermal desorption method for the trace analysis of headspace vapors from explosives and organic nitro-Compounds, Proceedings of Interrernational Symposium on Forensic Aspects Trace Evidence, FBI Quantico, VA, June 24–28, 1991, 269. Avail. NTIS PB94–145877.

73. Hobbs J. R. and Conde E. P., Gas chromatographic retention indices for explosives, Adv. Anal. Detect. Explos. Proc. 4th Inter. Symp. Anal. Detect. Explos. September 7–10, 1992 Jerusalem, Israel Yinon J., Ed. Kluwer Academic Publishers Dordrecht, Holland, 153–164, 1992.

74. Hodges C. M. and Akhavan J., The use of fourier transform Raman spectroscopy in the forensic identification of illicit drugs and explosives, *Spectrochim. Act*, 46A(2), 303–330, 1990.

75. Holland P. M., Mustacich R. V., Everson J. F., Foreman W., Leone M., Sanders A. H., and Naumann W. J., Correlated column micro gas chromatography instrumentation for the vapor detection of contraband drugs in cargo containers, *SPIE Proc.*, 2276, 79–86, 1994.

76. Hofstetter T., Heijman C., Haderlein S., Andreneap C., and Enbach S., Complete reduction of TNT and other (poly)nitroaromatic compounds under iron-reducing subsurface conditions *Environ. Sci. Technol.*, 33, 1479–1487, 1999.

77. Hong T. Z., Tang C. P., Lin K., and Yinon J., eds., Proceedings of International Symposium on Explosives Detection. 4th, London, 20, 145–52, 1992.

78. Hochberg M., Baehr-Jones T., Wang G., Shearn M., Harvard K., Luo J., Chen B., Shi Z., Lawson R., Sullivan P., Jen A. K-Y., Dalton, L. R., Scherer, A. "Terahertz all-optical modulation in a silicon-polymer hybrid system", *Nat. Mater.*, 5, 703–709, 2006.

79. Iqbal Z., Suryanarayanan K., Bulusu S., and Autera J. R., Infrared and Raman Spectra of 1,3,5-trinitro–1,3,5-triazacyclohexane (RDX), Rep. AD–752899, US Army Picatinny Arsenal, Dover, NJ, 1972.

80. Jankowski P. Z., Mercado A. G., and Hallowell S. F., FAA explosive vapor/particle detection technology, *SPIE Proc.*, 1824, 13–24, 1993.

81. Janni J., Gilbert B. D., Field R. W., and Steinfeld J. I., Infrared absorption of explosive molecule vapors, *Spectrochim. Acta*, 53A, 1375–1381, 1997.

82. Steinfeld J. I., Explosives detection: A challenge for physical chemistry, *Annu. Rev. Phys. Chem.*, 49, 203–232, 1998.

83. Judd L. L., Kusterbeck A. W., Conrad D. W., Yu H., Myles H. L. Jr., and Ligler F. S., Antibody-based fluorometric assay for detection of the explosives TNT and PETN, *SPIE Proc.*, 2388, 198–204, 1995.

84. Koder R. L. and Miller A. F., Steady-state kinetic mechanism, stereospecificity, substrate and inhibitor specificity of Enterobacter cloacae nitroreductase, *Biochim. Biophys. Acta* 1387, 395–405, 1998.

85. Lovley D. R., Widman P. K., Woodward J. C., and Phillips E. J. P., Reduction of uranium by cytochrome *c*3 of *Desulfovibrio vulgaris*, *Appl. Environ. Microbiol.*, 59(11), 3572–3576, 1993.

86. Karpowicz R. J. and Brill T. B., Comparison of the molecular structure of hexahydro-1, 3,5-trinitro-s-triazine in the vapor, solution and solid phases, *J. Phys. Chem.*, 88, 348–352, 1984.

87. Kneipp K., Wang Y., Kneipp H., Perelman L. T., Itzkan I., Dasari R. R., and Feld M. S., Single Molecule Detection using Surface-Enhanced Raman Scattering. *Phys. Rev. Lett.* 78, 1667–1670, 1997.

88. Kneipp K., Wang Y., Dasari R. R., Feld M. S., Gilbert B. D., James J., and Steinfeld J. I., Near-infrared surface-enhanced Raman scattering of trinitrotoluene on colloidal gold and silver, *Spectrochim. Acta*, 51A, 2171–2175, 1995.

89. Kolla P., Gas chromatography, liquid chromatography and ion chromatography adapted to the trace analysis of explosives, *J. Chromatogr.*, 674, 309–318, 1994.

90. Kolla P., The application of analytical methods to the detection of hidden explosives and explosive devices, *Angew. Chem., Int. Ed. Engl.*, 36, 800–811, 1997.

91. Kusterbeck A.W., Judd L. L., Yu H., Myles J., Ligler F. S., Flow immunosensor detection of explosives and drugs of abuse, *Proc. SPIE – Int. Soc. Opt. Eng.*, 2092: 218–226, 1994.

92. Lacey R. J., Direct non-destructive detection and identification of contraband using Raman microscopy, *IEE Conf. Publ.*, 408, 138–141, 1995.

93. Langford M. L. and Todd J. F. J., Negative-ion Fragmentation Pathways in 2,4,6-Trinitrotoluene, *Org. Mass Spectrom.*, 28, 773–779 (1993).

94. Lawrence A.W., Goubran R. A., and Hafez H. M., Signal Improvement in Ion Mobility Spectrometry, SPIE-The International Society for Optical Engineering Symposium on Applications of Signal and Image Processing in Explosive Detection Systems, SPIE Vol. 1824, 97–108, Boston, MA, November 1992.

95. Lee H. G., Lee E. D., and Lee M. L., Proceedings of International Symposium on Explosive Detection Technology, 1st ed., ed. S. M. Khan, Atlantic City, NJ, 619–33, 1992.

96. Little B. E. and Chu S. T., Opt. *Photonics New.*, 24, 2000.

97. Marsella M. J. and Swager T. M., Designing conducting polymer-based sensors: selective ionochromic response in crown ether-containing polythiophenes, *J. Am. Chem. Soc.*, 115, 12214–12215, 1993.

98. Marsella M. J., Carroll P. J., and Swager T. M., Conducting pseudopolyrotaxanes: A chemoresistive response via molecular recognition, *J. Am. Chem. Soc.*, 116, 9347–9348, 1994.

99. Marsella M. J., Carroll P. J., and Swager T. M., Design of chemoresistive sensory materials: Polythiophene-based pseudopolyrotaxanes, *J. Am. Chem. Soc.*, 117, 9832–9841, 1995.

100. Marsella M. J., Newl R. J., Carroll P. J., and Swager T. M., Ionoresistivity as a highly sensitive sensory probe: investigations of polythiophenes functionalized with calix[4] arene-based ion receptors, *J. Am. Chem. Soc.*, 117, 9842–9848, 1995.

101. McGann W. J., Jenkins A., Ribiero K., and Napoli J., New high-efficiency ion trap mobility detection system for narcotics and explosives, *SPIE Proc.*, 2092, 64–75, 1994.

102. McGann W. J., Bradley V., Borsody A., and Lepine S., New, high-efficiency ion trap mobility detection system for narcotics and explosives, *SPIE Proc.*, 2276, 424–436, 1994.

103. McLuckey S. A., Goeringer D. E., Asano K. G., Vaidyanathan G., and Stephenson J. L. Jr., Rapid Commun., *Mass Spectrom.*, 10(3), 287–298, 1996.
104. McLuckey S. A., Goeringer D. E., and Asano K. G., High Explosives Vapor Detection by Atmospheric Sampling Glow Discharge Ionization/Tandem Mass Spectrometry, Rep. No. ORNL/TM–13166. Oak Ridge Natl. Lab., TN, 1996.
105. McLuckey S. A., Glish G. L., and Grant B.C., The simultaneous monitoring for parent ions of a specified daughter ion: A method for rapid screening applications, *Anal. Chem.*, 62, 56–61, 1990.
106. McNesby K. L. and Coffey C. S., Spectroscopic determination of impact sensitivities of explosives, *J. Phys. Chem. B.*, 101, 3097–3104, 1997.
107. Mercado A., Janni J., and Gilbert B., Image analysis of explosives fingerprint contamination using a Raman imaging spectrometer, *SPIE Proc.*, 2511, 142–152, 1995.
108. Mercado A. and Davies J. P., Quantitative assessment methodology for an infrared spectroscopic system, *SPIE Proc.*, 2092, 27–37, 1994.
109. Mercado, A.; Janni, J.; Gilbert, B. and Steinfeld, J. I., Novel Spectrometer Concepts for Explosives Detection Applications, *Proc. 2nd Explos. Detect. Technol. Symp. Aviation Secur. Technol. Conf.*, Makky, W., –Chair, FAA, Atlantic City, NJ, 91–99, Nov. 12–15,1996.
110. Moore D. S., Instrumentation for trace detection of high explosives, *Rev. Sci. Instrum.*, 75(8), 2499–2512, 2004.
111. Nacson S., Mitchner B., Legrady O., Siu T., and Nargolwalla S., A GC/ECD Approach for the Detection of Explosives and Taggants, *Proc. First Int. Symp. Explos. Detect. Technol.*, ed. S.M. Khan, FAA, Atlantic City, NJ, 714–722, Nov. 13–15, 1991.
112. Nacson S., McNelles L., Nargolwalla S., and Greenberg D., Method of Detecting Taggants in Plastic Explosives Airport Trials and Solubility of Explosives, *Proc. 2nd Explos. Detect. Technol. Symp. Aviation Secur. Technol. Conf.*, Makky, W. –Chair, FAA, Atlantic City, NJ, 38–48, Nov. 12–15, 1996.
113. Senesac L. and Thundat, T. G., Nanosensors for trace explosive detection, *Materials Today*, 11(3), 28–36, 2008.
114. Narang U., Gauger P. R., Ligler F. S., A displacement flow immunosensor for explosive detection using microcapillaries, *Anal. Chem.*, 69, 2779–2785, 1997.
115. Nelson D. D., Zahniser M. S., McManus J. B., Shorter J. H., Wormhoudt J. C., Kolb C. E., Recent Improvements in atmospheric trace gas monitoring using mid-infrared tunable diode lasers, *SPIE Proc.*, 2834, 148–159, 1996.
116. Nguyen M.-T., Jamka A. A., Cazar R. A., and Tao F.-M., Structure and stability of the nitric acid–ammonia complex in the gas phase and in water, *J. Chem. Phys.*, 106, 8710–8717, 1997.
117. Novakoff A. K., FAA bulk technology overview for explosive detection, *SPIE Proc.*, 1824, 2–12, 1992.
118. Binks P. R., Nicklin S., and Bruce N. C., Degradation of hexahydro-1,3,5-trinitro-1,3,5-triazine (RDX) by Stenotrophomonas maltophilia PB1., *Appl. Environ. Microbiol.*, 61(4), 1318–1322, April 1995.
119. Palmer D. A., Achter E. K., and Lieb D., Analysis of fast gas chromatographic signals with artificial neural systems. In. *Proc. SPIE Int. Soc. Opt. Eng.*, 1824, 109–119, 1993.
120. Pristera F., Halik M., Castelli A., Fredericks W., Analysis of explosives using infrared spectroscopy, *Anal. Chem.*, 32, 495–508, 1960.
121. Boopathy R., and Tilche A., Pelletization of biomass in anaerobic baffled reactor treating acidified wastewater, *Biores. Technol.*, 40, 101–107, 1992.

122. Riris H., Carlisle C. B., McMillen D. F., and Cooper D. E., Explosives detection with a frequency modulation spectrometer, *Appl. Opt.*, 35, 4694–4704, 1996.

123. Pinnaduwage L. A., Thundat T., Gehl A., Wilson S. D., Hedden D. L., and Lareau R. T., Desorption characteristics of uncoated silicon microcantilever surfaces for explosive and common nonexplosive vapors, *Ultramicroscopy*, 100(3), 211–216, 2004.

124. Pinnaduwage L. A., Yi D., Thundat T. G., and Hawk J. E., Non-optical explosive sensor based on two-track piezoresistive microcantilever, United States Patent and Trademark Office Pre-Grant Publication, February 2006, pat no: US20060032289.

125. Ritchie R. K., Thomson P. C., DeBono R. F., Danylewych-May L. L., and Kim L., *SPIE Proc.*, 2092, 87–93, 1994.

126. Ritchie R. K., Kuja F., Jackson R. A., Loveless A. J., and Danylewych-May L. L., *SPIE Proc.*, 2092, 76–86, 1994.

127. Riskin M., Tel-Vered R., Bourenko T., Granot E., and Willner I., Imprinting of molecular recognition sites through electropolymerization of functionalized Au nanoparticles: development of an electrochemical TNT sensor based on pi-donor-acceptor interactions, *J. Am. Chem. Soc.*, 130(30), 9726–9733, 2008.

128. Rohe T., Grunblatt E. and Eisenreich N., Proceedings of International Annual Conference Fraunhofer Institute for Chemical Technology, 85–1–85–10, 1996.

129. Rothman L. S., Gamache R. R., Tipping R. H. Rinsland C. P., Smith M. A. H., Benner D. C., Devi V. M., Flaud J.-M., Camy-Peyret C., Perrin A., Goldimn A., Massie S. T., Brown L. R., and Toth R. A., I'The HITRAN molecular database: editions of 1991 and 1992, *J. Quant. Spectrosc. Rad. Transfer.*, 48, 469–507, 1993.

130. Rothman L. S., Gamache R. R., Goldman A., Flaud J.-M., Tipping R. H., Rinsland C. P., Smith M. A. H., et al., The HITRAN molecular database: Editions of 1991 and 1992, *J. Quant. Spectrosc. Radiat. Transfer*, 48, 469–507, 1992.

131. Rouhi A. M., Land mines: Horrors begging for solutions, *Chem. Eng. News.*, 75, 14–22, 1997.

132. Rounbehler D. P., MacDonald S. J., Lieb D. P., and Fine D.H., Analysis of explosives using high speed gas chromatography with chemiluminescent detection, Proceedings of First International Symposium on Explosive Detection Technology, ed. S. M. Khan, Atlantic City, NJ, 13–15, 703–713, 1991.

133. Romera-Guereca G., Lichtenberg J., Hierlemann A., Poulikakos D., and Kang B., Explosive vaporization in microenclosures, *Exp. Therm. Fluid Sci.*, 30(8), 829–836, 2006.

134. Simpson G., Klasmeier M., Hill H., Atkinson D., Radolovich G., Lopez-Avila V., and Jones T. L., Evaluation of gas chromatography coupled with ion mobility spectrometry for monitoring vinyl chloride and other chlorinated and aromatic compounds in air samples, *J. High. Resolut. Chromatogr.*, 19, 301–312, 1996.

135. Shi G., Qu Y., Zhai Y., Liu Y., Sun Z., Yang J., and Jin L., MSU/PDDA}n LBL assembled modified sensor for electrochemical detection of ultratrace explosive nitroaromatic compounds, *Electrochem. Commun.*, 9(7), 1719–1724, 2007.

136. Shriver-Lake L. C., Breslin K. A., Golden J. P., Judd L. L., Choi J., and Ligler F. S., Fiber optic biosensor for the detection of TNT, *SPIE Proc.*, 2367, 52–58, 1995.

137. Sleeman R., Bennett G., Davidson W. R., and Fisher W., The detection of illicit drugs and explosives in real-time by tandem mass spectrometry, in Proceedings of the contraband and cargo inspection technology international symposium, Washington, D.C., 57–63, 1992.

138. Steinfeld J. I. and Wormhoudt J., Explosives detection: A challenge for physical chemistry, *Annu. Rev. Phys. Chem.*, 49, 203–232, 1998.

139. Ramos C. and Dagdigian P. J., Detection of vapors of explosives and explosive-related compounds by ultraviolet cavity ringdown spectroscopy, Applied Optics, Vol. 46, Issue 4, 620–627, 2007.

140. Spiro T. G. and Stein P., Resonance effects in vibrational scattering from complex molecules, *Annu. Rev. Phys. Chem.*, 28, 501–521, 1977.

141. Stelson A. W., Friedlander S. K., and Seinfeld J. H., A note on the equilibrium relationship between ammonia and nitric acid and particulate ammonium nitrate, *Atmos. Environ.*, 13, 369–372, 1979.

142. Stelson A. W. and Seinfeld J. H., Relative humidity and temperature dependence of the ammonium nitrate dissociation constant, *Atmos. Environ.*, 16, 983–992, 1982.

143. Strobel R. A., Noll R., and Midkiff C. R. Jr., Proc. 4th Int. Symp. Anal. Detect. Explos. in Jerusalem, Israel, Sept. 7–10, 1992.

144. Swager T. M. and Marsella M. J., Molecular recognition and chemoresistive materials, *Adv. Mater.*, 6, 595–597, 1994.

145. Rabbany S.Y., Marganski W. A., Kusterbeck A. W., and Ligler F. S., A membrane-based displacement flow immunoassay, *Biosens. Bioelectron.*, 13(9), 939–944, 1998.

146. Stott W. R., Davidson W. R., and Sleeman R., High specificity chemical detection of explosives by tandem mass spectrometry, in Proceedings of applications of signal and image processing in explosives detection systems, ed. J. M. Connelly, S. M. Cheung, Vol. 1824 (Boston, MA, 1992, 68–78).

147. Trott W. M., Renlund A. M., and Jungst R. G., Proceedings of the southwest conference on optics, *SPIE Proc.*, 540, 368–375, 1985.

148. Trott W. M., and Renlund A. M., Single-pulse Raman scattering studies of heterogeneous explosive materials, *Appl. Opt.*, 24, 1520–1525, 1985.

149. Trott W. M., and Renlund A. M., Single-pulse Raman scattering study of triaminotrinitro-benzene under shock compression, *J. Phys. Chem.*, 92, 5921–5925, 1988.

150. Tong Y. S., Mud R., Henderson D. O., and Curby W. A., Diffusion kinetics of TNT in acrylonitrile—butadiene rubber via FT-IR/ATR spectroscopy, *Appl. Spectros.*, 51, 171–177, 1997.

151. Vourvopoulos G., Methods for the detection of explosives and contraband, *Chem. Ind.*, 8, 297–300, 1994.

152. Wehlburg J. C., Jacobs J., Shope S. L., Lockwood G. J., and Selph M. M., Image restoration techniques using Compton backscatter imaging for the detection of buried landmines, *SPIE Proc.*, 2496, 336–347, 1995.

153. Whelan J. P., Kusterbeck A. W., Wemhoff G. A., Bredehorst R., and Ligler F. S., Continuous flow immunosensor for detection of explosives, *Anal. Chem.*, 65, 3561–3565, 1993.

154. Wright A. D., Jennings K. R., and Peters R., Int. Symp. Anal. Detect. Explos. 4th, London Dordrecht: Kluwer, 291–298, 1992.

155. Wormhoudt J., Shorter J. H., McManus J. B, Kebabian P. L., Zahniser M. S., Davis W. M., Cespedes E. R., and Kolb C. E., Tunable Infrared Laser Detection of Pyrolysis Products of Explosives in Soils, *Appl. Opt.*, 35(21), 3992–3997, 1996.

156. Wormhoudt J., Zahniser M. S., Nelson D. D., McManus J. B., Miake-Lye R. C., and Kolb C. E., Infrared tunable diode laser diagnostics for aircraft exhaust emissions, in Optical Techniques in Fluid, Thermal and Combustion Flows, *SPIE Proc.*, 2546, 552–561, 1995.

157. Wormhoudt J., Kebabian P. L., and Kolb C. E., Embedded infrared fiber optic absorption studies of nitramine propellant strand burning, *Combust. Flam*, 111, 73–86, 1997.

158. Wang J., Lu F., MacDonald D., Lu J., Ozsoz M. E. S., and Rogers K. R., Screen-printed voltammetric sensor for TNT, *Talanta*, 46, 1405–1412, 1998.

159. Xu Y. and Herman J. A., Detection of nitrotoluene isomers by ion cyclotron resonance mass spectrometry using ion/molecule reactions with NO+ as reagent, *Rapid. Commun. Mass. Spectrom.*, 6(7), 425–428, 1992.

160. Yinon J., Forensic applications of mass spectrometry, *Mass. Spectrom. Rev.*, 10, 179–224, 1991.

161. Yinon J., Mass Spectrometry of explosives: Nitro compounds, nitrate esters, and nitramines, *Mass. Spectrom. Rev.*, 1, 257–307, 1982.

162. Yujiri L., Hauss B., and Shoucri M., Passive milimeter wave sensors for detection of buried mines, Proceedings, Orlando, FL, 2496, 2–6, 1995.

163. Zahniser M. S., Nelson D. D., McManus J. B., and Kebabian P. L., Measurement of trace gas fluxes using tunable diode laser spectroscopy, *Philos. Trans. R. Soc. London, Ser. A*, 351, 371–382, 1995.

164. Zhou Q. and Swager T. M., Methodology for enhancing the sensitivity of fluorescent chemosensors, energy migration in conjugated polymers, *J. Am. Chem. Soc.*, 117, 7017–7018, 1995.

165. Zhou Q. and Swager T. M., Fluorescent chemosensors based on energy migration in conjugated polymers: The molecular wire approach to increased sensitivity, *J. Am. Chem. Soc.*, 117, 12593–12602, 1995.

166. Zhu S. S., Carroll P. J., and Swager T. M., Conducting polymetallorotaxanes: A supramolecular approach to transition metal ion sensors, *J. Am. Chem. Soc.*, 118, 8713–8714, 1996.

167. Zuin L., Innocenti F., Fabris D., Lunardon M., Nebbia G., Viesti G., Cinausero M., and Palomba M., Experimental optimisation of a moderated 252Cf source for land mine detection, *Nucl. Instrum. Methods Phys. Res., Sect. A*, 449(1), 416–426, 2000.

11 Detection and Identification of Organophosphorus Compounds
As Chemical Warfare Agents

Ravil A. Sitdikov, Dmitri M. Ivnitski,
Ebtisam S. Wilkins, Ihab Seoudi
Department of Chemical and Nuclear Engineering
University of New Mexico
Albuquerque, New Mexico

CONTENTS

11.1 INTRODUCTION

The rising threat of bioterrorism over the years has prompted an ever-increasing need to detect, identify, and quantify chemical and biological agents. Rapid, accurate

295

detection of such agents is critical in order to verify whether an agent has been released and an event has occurred. In this case, a counter measure has to be implemented in a timely manner in order to protect people's health and life [1–3]. The toxic effect of organophosphorus nerve agents, such as soman, sarin, tabun, and VX (o-ethyl s-diisopropylaminomethyl methylphosphonothiolate) occurs extremely quickly on persons exposed to high concentrations, with symptoms often occurring within a matter of minutes. These agents act by blocking the transmission of nerve messages [4–7]. Toxicity manifests rapidly as increased nasal and bronchal secretions, respiratory paralysis, muscular weakness, convulsion, and death. More unspecific symptoms are tiredness, slurred speech, nausea, and hallucinations [8]. The toxic effect depends on both the concentration of nerve agent inhaled and the time of exposure. All nerve agents belong chemically to the group of organophosphorous compounds (OPCs) and they produce their toxicity through the irreversible inactivation of acetylcholinesterase (AChE) [9]. If inhaled or absorbed through the skin, a single drop of nerve agent can shutdown the body's nervous system. The most powerful is VX, but all can cause death within minutes after exposure. Due to the extremely fast-acting nature of nerve agents, treatments must be administered immediately if they are to be of any significant benefit. In many nations, the armed forces have access to an autoinjector containing antidotes to nerve agents [10–18]. This device is so easy to use that a soldier can give himself or another person an injection. Unfortunately, total protection from chemical attacks is difficult, but quick detection will undoubtedly reduce the effects.

Herbicides are chemicals that are intended to kill vegetation, but many regard herbicides as chemical weapons if used for hostile purposes. Agent Orange, an herbicide, was used during the Vietnam War to destroy jungle leaves to expose enemy troops. However, many Vietnam War veterans suffered numerous health problems blamed on exposure to Agent Orange and other toxins [19–21]. Millions of tons of pesticides, herbicides, fungicides, and insecticides are widely used throughout the world each year in agriculture, industry, and medicine [22]. They are used as plant-growth regulators, defoliants, and as chemicals applied to crops [23]. The increasing use of OPCs, mineral fertilizers, pharmaceuticals, surfactants, and many other biologically active substances results in serious environmental problems. Among the OPCs, chlorine-containing pesticides can accumulate and contaminate soil, water, and food products. Pollutants of this type are found to be present in many sampled soils, ground, and wastewaters streams. Since many of them are highly toxic, OPC accumulation in living organisms can be the cause of serious diseases [24–26]. These compounds are the most important pollutants in rivers and ground water [27]. The negative impact of pesticides on nontarget organisms depends on their toxicity, their metabolites and their transport pathways through the hydrological cycle, the soil, and food chains [28–30]. Figure 11.1 shows the chemical structures of representative compounds of the principal pesticide groups used in agriculture [31].

Organophosphorous cholinesterase inhibitors affect the human body by inhibiting three enzymes. The first is butyrylcholinesterase in the plasma, the second is AChE on the red cell, and the third is AChE at cholinergic receptor sites in tissue. They inhibit these enzymes each in different ways, and therefore their effect is not the same. Even the two AChEs have different properties, even though both have a

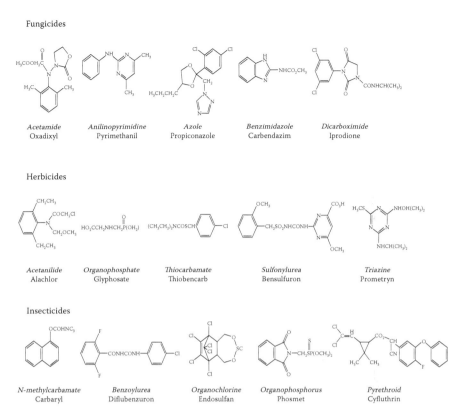

FIGURE 11.1 Chemical structures of representative compounds of pesticides used in agriculture.

high affinity for acetylcholine. The blood enzymes provide an estimate of the tissue enzyme activity. After acute exposure to a nerve agent, the erythrocyte enzyme activity most closely reflects the activity of the tissue enzyme [32,33]. There are a number of publications related to nerve agents and OPCs [34], as well as several recent books on chemical agents [35–37], with one giving a detailed summary of human studies in the United Kingdom and United States [36,38–41]. Moreover, some organochlorine pesticides such as chlorophenoxy acids and their chlorophenol derivatives contain polychlorinated dibenzodioxins and dibenzofurans, which are extremely toxic and stable compounds, as impurities. Dioxins are formed both in the course of the production of the chlorophenols and chlorophenoxy acid pesticides and during their metabolism in the environment [31].

Thus, the detection of environmentally present chemical warfare agents (CWA) requires highly sensitive, selective, and fast quantitative methods, capable of low level of pollutant detection in the battlefield and surrounding streams, waste waters, soils, plants, and food [42,43]. The most important preventive measures in this case are to rapidly determine the source of the pollutant and the magnitude of the threat using on-site measurements. The OPC assay in real samples requires the stages of extraction, clean up, and preconcentration because of the relatively low concentration of target analytes and hydrophobic nature of pesticides [44]. However, this is a complicated and time-consuming procedure. The effective testing of OPCs requires methods of analysis that meet a number of challenging criteria. Time and sensitivity of analysis are the most important limitations related to the usefulness of testing. OPCs detection methods have to be rapid and very sensitive since the presence of even extremely low levels of OPCs in the body or food may be a toxic dose. Therefore, reliable analytical methods are required for detection and identification of a variety of compounds, including not only parent OPCs but also their metabolites in complex matrices [45]. The large number of analyzed samples require the application of prescreening methods, suitable for direct field use. Fast and reliable analytical systems for determination of OPCs are strongly needed for health and environmental protection [44].

The action of many OPCs is based on their irreversible inhibition of butyrylcholinesterase activity in the plasma and AChE activity in red cells and tissues [35–41]. The OPC compounds are commonly used as pesticides, insecticides, and CWAs. Early detection of OPC neurotoxins is important for protecting air, water resources, and food supplies, in the defense against terrorist activity, and for monitoring detoxification processes [46]. Accordingly, there are growing demands for field-deployable devices for reliable on-site monitoring of OPC compounds.

Recent advances in such areas as microarray technology, microelectromechanical systems, microfluidics, and optoelectronics present new technological possibilities for producing fast, extremely sensitive, and inexpensive "smart" sensing systems for detection and identification of pesticides in the field conditions [47]. In this chapter we examine the current state of the art in detection and identification of OPC based on newly emerging sensor technologies.

11.2   CURRENT STATUS OF ANALYTICAL METHODS FOR OPC SCREENING AS CHEMICAL WARFARE AGENT

There are several different techniques employed for the detection of OPC, including conventional analytical methods; enzyme, fluorescence, and radioimmunoassay techniques; immunochemical methods; some biosensor technology, which includes optical, piezoelectric, electrochemical, microfluidic, and array-based (Lab-on-a-chip) biosensors. These techniques are described later in this chapter.

11.2.1   Conventional Analytical Methods for OPC Assay

A variety of conventional analytical methods for the detection and identification of trace levels of chemical agents are being intensively developed and used. They are gas and liquid chromatography [48,49], thin-layer chromatography [50,51], mass spectrometry [52,53], capillary electrophoresis [54], and solid-phase microextraction [55,56]. Enzyme, fluorescence, and radioimmunoassay techniques are now being seen as useful analytical tools [57–61]. Schobel et al. [58] focused on fluorescence methods for pesticide monitoring. The fluorescence methods offer a high degree of selectivity and sensitivity. Restrictions on the limits of detection due to background signals are minimized by development of solid-phase separation systems, new fluorescent probes, and new instrumentation [58]. However, these methods require extensive sample preparation, a time-consuming and complicated procedure of assay, as well as expensive and bulky equipment.

11.2.2   BIOSENSOR TECHNOLOGY

Biosensor technology is a powerful alternative to conventional analytical techniques, with high specificity and sensitivity of assay. Biosensors can be portable, small, and low-cost devices, making them highly desirable for field use. The areas for which biosensors show particular promise are clinical diagnostics, food analysis, bioprocesses, environmental monitoring, nerve agents, veterinary and agriculture. Different biosensor applications have been the subject of recent reviews [62–67]. Biosensors can also be used for detection of a broad spectrum of pesticides in complex sample matrices (blood, serum, urine, or food) with minimum sample pretreatment. There are a number of different methods of OPCs detection in References 63, 64, 68–70. Many of these methods are based on cholinesterase inhibition test. The assay is either carried out in solution or the enzyme is immobilized on the transducer surface of the sensor [71]. The latter technique provides advantages in the detection of organophosphorus pesticides in water with high sensitivity and reproducibility. A well-known example is the cholinesterase inhibitors (organophosphates and methylcarbamates), which can be measured by their inhibitory effect on cholinesterases.

Organophosphorus compounds are irreversible inhibitors of cholinesterases. The inhibitory action of organophosphates (OPs) is based on their ability to irreversibly modify the catalytic serine residue in AChEs [72]. In recent years, the use of AChEs in biosensor technology has gained enormous attention, in particular, with respect to insecticide detection [73–75]. The principle of biosensors using AChE as a biological recognition element is based on the inhibition of the enzyme's natural catalytic activity by the agent that is to be detected. The advanced understanding of the structure–function relationship of AChEs serves as the basis for developing enzyme variants, which, compared to the wild type, show increased inhibition efficiency at low insecticide concentrations and thus a higher sensitivity. Different expression systems that have been used for the production of recombinant AChEs, as well as approaches of recombinant AChEs purification that allow significantly increased sensitivity and specificity of OP and carbamate assay, are described in the reviews [76,77].

11.2.2.1   Optical Sensors for Pesticides Monitoring

Optical transducers based on porous silicon technology and nanotechnology are particularly attractive for the detection and identification of chemical agents [78–81]. Recently, a highly sensitive immunosensor using optical waveguide light mode spectroscopy (OWLS) was developed for the detection of the herbicide trifluralin [82]. Different types of immunoanalytical techniques for pesticides monitoring based on fluorescence detection are described in Reference 58. A microformat imaging sensor in combination with registration of chemiluminescence by a Charge-Coupled Device (CCD camera) was applied for the detection of the herbicide atrazine by two immunoenzyme assay formats [83]. The membrane dot blot assay allowed detection of up to 40 pg mL^{-1} of atrazine (4.5-fold enhancement). Due to the application of polyelectrolyte carriers for rapid reactants separation the membrane assay may be realized in 15 minutes. CCD-based registration of signals assures a high reproducibility of measurements; the variation coefficient is in the range 1%–7%. The effectiveness of the assays developed for controlling triazine herbicides in mineral and tap water has been demonstrated.

11.2.2.2 Piezoelectric Biosensors

The piezoelectric biosensor is based on measuring frequency changes directly related to mass change on the sensor surface. One of the main advantages of this technique is the detection of biospecific binding reactions, which allow the kinetic evaluation of affinity interactions (feature similar to the surface plasmon resonance biosensors). Besides, the piezoelectric biosensor is not costly. The limitations of this transduction method are the need for a calibration of each crystal and the possible variability when the surface is coated with an antigen or antibody [69]. Several examples, including a highly sensitive piezoelectric biosensor for detection of cholinesterase inhibitors [84,85], piezoelectric immunobiosensor for acetochlor detection [86], and a new method for the sensitive detection of cholinesterase inhibitors based on real-time monitoring using a piezoelectric biosensor [87], demonstrate practical applications of the technique to the detection of chemical agents in various media.

11.2.2.3 Electrochemical Enzyme Sensors for Pesticide Detection

A wide spectrum of organophosphate hydrolase (OPH)-based amperometric and potentiometric biosensing devices are described in literature [88,89]. Recently, amperometric OPH electrodes based on monitoring the oxidation of the p-nitrophenol product of the enzyme reaction have been described [90,91]. Improved anodic detection of p-nitrophenol is highly desired to address the high overvoltage and surface-fouling limitations associated with such transduction reaction. The majority of previous reports of sensors for pesticides including paraoxon, dichlorvos, and chlorpyrifos have utilized a second additional enzyme, choline oxidase (CHO), for the catalytic production of H_2O_2, which may then be detected amperometrically [92]. A better-designed approach is the electrochemical oxidation of thiocholine/choline [74,75,93–96]. This has already been achieved using screen-printed carbon electrodes doped with cobalt phthalocyanine, which acts as an electrocatalyst for the oxidation of thiocholine at a lowered working potential of +100 mV (versus Ag/AgCl) [74,97].

The stability of enzymes immobilized on the transducer surface is the most critical step in the design and development of these new biosensor technologies. Unfortunately, cholinesterases possess limited operational and storage stability in both solution and dry states. Drying itself also results in dehydration stress to proteins [98]. The unfolding of proteins can be prevented by using additives (e.g., sucrose) that remain in the amorphous phase with the protein and hydrogen bond to the protein in the place of water during drying [98]. Vakurov et al. [99] compare a number of methods of immobilization of AChE on carbon electrode surfaces with the aim of improving stability of OP biosensors. The simplest procedure is physical adsorption, but it provides poor sensor reproducibility and stability. The most attractive approach is enzyme immobilization via an electrochemical reduction process [74,75,99–101]. This electrochemical process includes reduction of an aryldiazonium salt resulting in the production of an aryl radical, which covalently binds to the carbon surface. The 4-nitrobenzene modified electrodes produced by this method can subsequently be electrochemically reduced to aminobenzene. There are many other methods suitable for immobilization of enzymes to amino-containing matrices [102]. Many of these methods are based on application of glutaraldehyde treatment [99,100]. Noncovalent approach based on electrostatic interaction has a number of benefits compared to covalent methods of immobilization [103]. Multipoint, noncovalent, electrostatic complex formation results in increased enzyme stability in water–organic mixtures, storage stability under dry conditions, and wet stability. Authors [104] demonstrated that polyethylene imine (PEI) modifications of the carbon electrodes significantly increase the dry and wet stability of AChE immobilized noncovalently onto PEI-modified electrodes. Two stabilizer mixtures, comprised 5% (w/v) sucrose with 1% (w/v) polygalacturonic acid (PGA) and 5% (w/v) sucrose with 0.1% (w/v) dextran sulphate were shown to effectively protect immobilized AChE against inactivation in the dry state over considerable time periods, in some cases beyond one year.

11.2.2.4 Electrochemical Immunosensors for Pesticide Assay

The development and application of immunosensors for environmental monitoring has grown steadily in recent years. Immunochemical methods provide rapid, sensitive, and cost-effective analyses for a variety of environmental contaminants. Immunoassays combine principles of immunology and chemistry into tests that are used by scientists in practically every discipline, including fields as diverse as molecular biology and environmental science. These techniques rely on biospecific interaction between an antibody as analytical reagents and target analytes (antigen) and can be used for screening major classes of pesticides in real time. Immunochemical methods of analysis are finding increasing use for determining pesticides in various samples (water, soil, food products, and biological fluids) [57,105].

The main principle of assay is the immunological reaction between an antibody immobilized on the transducer surface and target OPCs. The immunosensor takes advantage of the high selectivity, which is provided by the molecular recognition characteristic of an antibody. The immunosensors developed in the past few years for environmental analysis with particular emphasis on monitoring of OPCs levels have been summarized by Mallat et al. [106]. Different transduction elements and mechanisms of detection as well as immunosensor applications for analysis of pesticides in natural water samples in real time have been considered. The immunosensor system is based on application of solid-phase fluoroimmunoassay combined with an optical transducer chip chemically modified with an analyte derivative. Mallat et al. [106] has reviewed the analysis of atrazine, simazine, paraquat, alachlor, 2,4-dichloro-phenoxy-acetic acid (2,4-D), and isoproturon in water, while Suri et al. [107] overviewed the current status of immunosensors as well as their potential benefits and limitations for pesticide analysis. In this chapter, the basic criteria for generating specific antibodies against low-molecular-mass pesticides, which are usually nonimmunogenic in nature, are briefly discussed. This chapter also describes the fundamentals of important transducer technologies and their use in electrochemical immunosensor development for pesticides analysis [107].

A rapid immunotechnique combining separation of reactants by filtration through a porous membrane and potentiometric detection of the bound enzyme label by a pH-sensitive field-effect transistor was proposed by Plekhanova et al. [108]. The complexes to be detected are formed by the method including a homogeneous binding of immunoreactants and a polyanion carrier (polymethacrylate) followed by heterogeneous separation on a membrane incorporating an immobilized polycation (poly-N-vinyl-4-ethylpyridinium). The proposed immunoperoxidase technique for a sensitive detection of atrazine is based on the measurement of pH changes in the solution containing o-phenylenediamine, hydrogen peroxide, and ascorbic acid. Their specific detection is realized via competitive binding of free and peroxidase-labeled antigens by antibodies integrating with a (staphylococcal protein A-polyanion) conjugate. The total analysis time is 20–25 minutes, with a range of quantitative detection of 0.2–100 ng mL $(^{-1})$ for atrazine. Data scatter of replicate tests varies from 3% to 10%. Application of protein A-polyanion conjugate allows using the proposed protocol for different antigens without additional treatment of specific antiserum [108].

Electrochemical impedance spectroscopy (EIS) was evaluated for the direct determination of herbicide 2,4-D [109]. Specific antibody against 2,4-D was immobilized onto different gold electrodes. Several methods of antibody immobilization by covalent linkage to modified surface were studied [109]. Self-assembled monolayers formed using thiocompounds as cystamine, 4-aminothiophenol (ATPh), 3,3'-dithiopropionic acid di-(N-succinimidyl ester) (DTSP), and 11-mercaptoundecanoic acid (MUA) were chosen for the sensing surface activation. Three different sensor types were tested: screen-printed disc, fingerlike structures, and interdigitated array (IDA) electrodes produced by lithography. The measurements were carried out in a stationary arrangement, and the reaction between hapten and the immobilized antibody was observed online. Changes of impedance parameters were evaluated, and the optimized immobilization technique (using 4-aminothiophenol) was proposed. Impedance changes due to immunocomplex formation were evaluated, and the possibility of direct monitoring of 2,4-D binding to the antibody was demonstrated at a fixed frequency. For the strip sensor, the calibration curves were constructed in concentration range from 45-nmol l^{-1} to 0.45 mmol l^{-1} of 2,4-D.

Grennan et al. [110] describe the development of an electrochemical immunosensor for the analysis of atrazine using recombinant single-chain antibody (scAb) fragments. The sensor is based on carbon paste screen-printed electrode incorporating the conducting polymer polyaniline (PANI)/poly(vinylsulphonic acid) (PVSA). It can be used for direct mediatorless detection of specific interaction between the redox centers of antigen-labeled horseradish peroxidase (HRP) and the electrode surface. Competitive immunoassays were performed in real time using separation-free system. Analytical measurements are based on the pseudo-linear relationship between the slope of a real-time amperometric signal and the concentration of analyte in the sample. Multiple, sequential measurements of standards and samples can be performed on a single scAb-modified surface in a matter of minutes. No separation of bound and unbound species was necessary prior to detection. The system is capable of measuring atrazine to a detection limit of 0.1 ppb (0.1 µg l^{-1}). This system offers the potential for rapid, cost-effective immunosensing for the analysis of samples of environmental, medical, and pharmaceutical significance.

11.3 "LAB-ON-A-CHIP," MICROFLUIDIC AND ARRAY-BASED BIOSENSORS FOR PESTICIDE DETECTION

Miniaturization of biodetectors into a single integrated "lab-on-a-chip" system possesses great potential for environmental monitoring, which includes improved accuracy, low-power and sample consumption, disposability, and automation [111–113]. This technology adapts microfabrication techniques used in semiconductor manufacturing to convert experimental and analytical protocols into chip architectures. Chips are currently being fabricated with picoliter-sized wells and 10-microliter-sized chambers for sample preparation and detection [114–117]. The integration of microfluidic transport, total automation, and materials handling contributes to a major reduction in system retention and material transfer losses, which increases accuracy and reduces sample size requirements. However, one of the remaining barriers toward achieving true miniaturized total analysis systems is the integration of sample pretreatment for microfluidic devices. The challenge is complicated by the complexity and variation in prospective samples and analytes. There is an issue of integration and interfacing the pretreatment operation to the analysis device with which it is coupled and codependent on in terms of sample volume, time, reagent, and power consumption [118,119].

Currently, the majority of published work has concentrated on using electrokinetically driven separation schemes to separate and detect analytes of interest [120]. The electrokinetic phenomenon occurs due to the interaction of induced dipoles in the bioparticles with electric fields, and is used for movement of fluids and other materials through a network of fluid channels. In this case external pumps or valves are not needed. Precise control of fluid motion and reaction timing is achieved by changing parameters such as the current or voltage. The chip-based capillary electrophoresis system has the capacity to perform the following functions: reagent dispensing, mixing, incubation, reaction, and sample partition and analyte detection. Evidence in the experimental data produced by different organizations has shown that microchip electrophoresis is an effective process for analyzing biological agents at very high speeds and low concentrations [14,115,121]. Chip-based capillary electrophoresis technology has many benefits compared to conventional methods of analysis. For example, the chip can analyze a mixture in seconds where it would take capillary electrophoresis at least 20 minutes and gel electrophoresis 1 hour to do the same analysis. The microchip can detect a sample concentration in the range of 100 picomolar, at least two orders of magnitude greater than conventional capillary electrophoretic analysis.

Recently, results of the research on immunosensor protein chip technologies for the detection of pesticides atrazine and parathion was reported in [122]. Protein chip is a rapidly expanding technology for large-scale, high-throughput protein assay after the genechip. A competitive inhibition method for screening of poisonous chemical molecule haptens was adopted due to the aiming contaminant characteristic of small molecules. Having obtained antibodies, the derivatives of pesticide atrazine and parathion were successfully synthesized, and then the derivatives and aminopapaverine were conjugated to protein ovalbumin (OVA), used as antigens, which proved subsequently high specificity against its corresponding antibody. The detection of atrazine and papaverine was performed qualitatively and quantitatively under the optimized conditions, which includes antibody-immobilized time (2 hour), blocking buffer containing OVA, and blocking time (1 hour), sample buffer pH value (pH 8.0). The results demonstrate that fluorescence intensity increased along with the decreasing concentration of analytes. A linear trend was obtained. The limit of detection of atrazine was 0.001 and 0.01 mg/L. The possibility of a microfluidic biosensor based on an array of hydrogel-entrapped enzymes for paraoxon pesticide detection was shown by Heo and Crooks [117]. A microfluidic sensor based on an array of hydrogel-entrapped enzymes can be used to simultaneously detect different concentrations of the same analyte (glucose) or multiple analytes (glucose and galactose) in real time. The concentration of paraoxon, an acetylcholine esterase inhibitor, can be quantified using the same approach. Hydrogel micropatch arrays and microfluidic systems are easy to fabricate, and the hydrogels provide a convenient, biocompatible matrix for the enzymes. Isolation of the micropatches within different microfluidic channels eliminates the possibility of cross talk between enzymes.

A microchip for monitoring of OP nerve agents was developed by Wang et al. [115]. The assay for screening OP nerve agents based on a precolumn reaction of organophosphorus hydrolase (OPH), electrophoretic separation of the phosphonic acid products, and their contactless-conductivity detection, is described. Factors affecting the enzymatic reaction, the separation and detection processes have been assessed and optimized. The complete bioassay requires 1 minute of the OPH reaction, along with 1–2 minutes for the separation and detection of the reaction products. The response is linear, with detection limits of 5 and 3 mg/L for paraoxon and methyl parathion, respectively. Compared to conventional OPH-based biosensors, the OPH-biochip can differentiate between the individual OP substrates. The attractive behavior of the new OPH-based biochip indicates great promise for field screening of OP pesticides and nerve agents. The study also demonstrates for the first time the suitability of contactless-conductivity detection for on-chip monitoring of enzymatic reactions [115].

11.4 APPLICATION OF NANOTECHNOLOGY FOR ENVIRONMENTAL MONITORING

There is a considerable interest in the development of new microscale sensor technologies based on the application of nanoparticles for ultrasensitive detection of different agents in real time [123]. In this respect, development of new biosensing methodology based on the application of nanotechnology can provide excellent conditions for high-throughput tests that are smaller, faster, and more sensitive than conventional assays. Nanotechnology offers a wide range of applications. For example, the combination of nanoparticles with biomolecules yields new facets of bioelectronics to open new horizons of microscale sensor technology. Nanoparticles facilitate electron transfer and can be easily modified with a wide range of biomolecules and chemical ligands. Biological specificity, combined with the unique electronic properties of nanotubes, enables nanotube-based biosensors that can selectively detect target analyte in solution by using direct electronic readout, without the need for labeling. Such characteristics, together with an ease of miniaturization of sensing devices to nanoscale dimensions, make nanoparticles suitable for applications in chemical/biochemical sensing. Recently, UC Davis researchers led by Kennedy and Hammock [124] have made fluorescent nanoparticles of lanthanide oxide and europium oxide that can be coupled to biological molecules and used in antibody-based assays for pesticide residues. The nanoparticles also can be sorted magnetically. The researchers are currently investigating and carrying out these assays in microdroplets and in microchannels on etched chips [124]. A nanostructured carbon matrix has been used for the immobilization and stabilization of the Acetyl cholinesterase enzyme [125]. The carbon nano- and microparticles can provide significant enzyme stabilization, as well as the means for lowering the detection limit of the biosensor [126,127].

The nanostructured conductive carbon enzyme is immobilized by adsorption into electrodes, which also acts as the working electrode. The proposed biosensor showed very good stability under continuous operation conditions ($L_{50} > 60$ days), allowing its further use in inhibition mode. Using this biosensor, the monitoring of the organophosphorus pesticide dichlorvos at picomolar levels (1000 times lower than other systems reported so far) was achieved. The linear range of detection in flow injection system spanned six orders of magnitude (10^{-12} to 10^{-6} M). It is suggested that the ability of activated carbon to selectively concentrate the pesticide, as well as the enzyme hyperactivity within the nanopores is the reason for the decrease in the detection limit of the biosensor [20]. A disposable carbon nanotube-based biosensor was successfully developed and applied to the detection of organophosphorus (OP) pesticides and nerve agents [128]. The biosensors using AChE/CHO enzymes provided a high sensitivity, large linear range, and low-detection limits for the analysis of OP compounds [128]. Carbon nanotube technology is particularly suitable for chemical detection, as sensors suitable for sensing different analytes of interest could be configured in the form of an array to comprehensively monitor multiple analytes. Promising applications for carbon nanotube-based gas sensors include chemical leak detection or industrial chemical process monitoring (e.g., petrochemicals, hydrocarbons, nitrogen dioxide, ammonia, hydrogen, oxygen, carbon monoxide, carbon dioxide, methane, hydrogen sulfide), medical monitoring/analysis (e.g., monitoring anesthesia gases, breath gases, blood gases), biowarfare (e.g., monitoring explosives such as TNT (trinitrotoluene) or RDX (cydotrimethylene-trinitramine), or monitoring nerve agents such as GB or VX), gas alarms, and environmental pollution monitoring (including indoor air quality monitoring). Recently, chemical sensors based on individual carbon nanotubes (CNTs) have been demonstrated [129]. Altered electrical resistance of semiconducting CNTs through exposure to pesticides forms the basis for these sensors. Nanotube sensors have also exhibited fast response at room temperature to gases and a substantially higher sensitivity than existing solid-state sensors [130].

Pritchard et al. [131] describe the development of a single thiocholine enzyme-based biosensor. This biosensor is a sonochemically fabricated enzyme microelectrode array in order to impart stir-independent (convection) responses that are characteristic of microelectrodes. Microelectrode arrays with up to 2×10^5 microelectrode elements may be fabricated via the sonochemical ablation of nonconductive polymer films [132,133], which coat and thereby insulate underlying conductive surfaces [134]. Paraoxon is determined down to concentrations of 1×10^{-17} M via the use of sonochemically fabricated acetylcholine/polyaniline microelectrode array-based sensors. These sensors were fabricated via the electropolymerization of thin insulating polymer films at planar electrode surfaces and then exposed to ultrasonic vibration to ablate and thereby expose areas of underlying conductor. A conducting polymer with coentrapped AChE may then be electropolymerized at these micro-electrode cavities and together these form an enzyme microelectrode array with a population of 2×10^5 cm^{-2}. The enzymatic product thiocholine is catalyzed at the coentrapped AChE/carbon electrode and the enzymatic response is inhibited following incubation with paraoxon. Lower limits of detection on the order of 1×10^{-17} M concentration of OPCs were shown to be achievable via the use of enzyme micro-electrode arrays that are not achievable using conventional planar electrodes [135].

Carbon nanotube-based amperometric transducers have been reported to improve the amperometric biosensing of OPCs [128]. Authors have demonstrated the use of CNT for a greatly enhanced amperometric biosensing of OP compounds. The CNT-based transducer leads to a highly sensitive and stable detection of the enzymatically (OPH) liberated p-nitrophenol. Such coupling of OPH-based biorecognition and amperometric transduction on CNT transducers is advantageous over AChE-based CNT biosensors that lack specificity towards OP compounds and require addition of the substrate and an incubation period. The new OPH/CNT biosensor when coupled with handheld (battery-operated) instruments thus offer great promise for rapid field screening of OP pesticides and nerve agents in natural water resources and food supplies, thereby providing the necessary warning/alarm and related proactive action. The use of OPH amperometric biosensors for direct (including remote) measurements in untreated natural water samples has been demonstrated [136]. Such on-site applications would greatly benefit from the use of disposable OPH/CNT-coated screen-printed electrodes. Potential interferences from easily oxidizable species should be considered in such real-life applications; the Nafion coating should alleviate such electroactive interferences [136].

An amperometric biosensor for OPC pesticides based on a CNT-modified transducer and an OPH biocatalyst is described. A bilayer approach with the OPH layer on top of the CNT film was used for preparing the CNT/OPH biosensor. The CNT layer leads to a greatly improved anodic detection of the enzymatically generated p-nitrophenol product, including higher sensitivity and stability. The sensor performance was optimized with respect to the surface modification and operating conditions. Under the optimal conditions the biosensor was used to measure as low as 0.15 μM paraoxon and 0.8 μM methyl parathion with sensitivities of 25 and 6 nA/μM, respectively. Electrochemical biosensors, particularly enzyme electrodes, have greatly benefited from the ability of CNT-based transducers to promote the electron-transfer reactions of enzymatically generated species such as hydrogen peroxide [137] and from the resistance to surface fouling of such transducers. CNT-based transducers have thus been shown useful for enhancing the performance of enzyme electrodes for monitoring glucose, [138] but have not been combined with OPH-based biosensors. The accelerated oxidation of hydrogen peroxide at CNT transducers has been exploited recently for improved AChE-based inhibition biosensing of OP compounds [139]. The ability of CNT-modified electrodes to promote the oxidation of phenolic compounds (including the p-nitrophenol product of the OPH reaction) and to minimize surface fouling associated with such oxidation processes is the way to the new OPH-CNT amperometric biosensor [140].

11.5 CONCLUSION AND FUTURE OF SENSOR TECHNIQUES FOR OPC DETECTION

Thus, the ideal sensor technology would accomplish three aims, namely detection, identification, and warning sufficiently quickly, accurately, and with high sensitivity. The effective testing of OPCs requires new sensor technology that will be extremely sensitive, reliable, and fast. It should be miniaturized, use few consumables (i.e., have a smaller logistics footprint), require significantly less maintenance than current equipment used to monitor chemical toxic agents, and detect in real time [141]. A sensor should be able to detect chemical agents at threshold concentrations in a minimum of 2–5 seconds. It must have the specificity to distinguish target agent in complex samples [142]. Obviously the incorporation of all the features within a sensor is very complicated and challenging. Solutions to the technological challenge may be realized by the creation of sensor systems based on new sensor technologies such as array-based chips, emerging "lab-on-a-chip" devices, and a "micro total analysis system" [111–113,116]. In recent years, the miniaturization of biochemical and physical processes and their integration onto a single microchip has become a dominant goal of sensor research and development. Recent advances in such areas as nanotechnology, microelectromechanical systems, and microfluidics present new technological possibilities for producing fast, extremely sensitive, and inexpensive "smart" sensing systems for field application. Advances in microfabrication methods of silicon chips make it possible to replicate and produce sensor arrays coated with specific sensing components with a high degree of reliability, and at a low cost [143]. Miniaturization opens the door for placing complete analytical systems in remote environments. Sensors based on digital technology have emerged. Sensors are being developed to incorporate a standard communication interface that enables them to automatically identify themselves and describe their function when they are plugged into a network system. The next generation of field detection devices should be fully automated devices with integrated sample preparation and sensing elements, able to discriminate potential agents in a multianalyte environment, and report results autonomously. Full automation of assays and improved specimen-processing procedures can overcome many of the problems associated with first-generation tests, reducing human error and increasing the accuracy of results.

Rapid and precise sensors capable of detecting pollutants at the molecular level could greatly enhance our ability to protect human health and the environment.

REFERENCES

1. Snyder J. W. and Check W., Bioterrorism threats to our future: The role of the clinical microbiology laboratory in detection, identification, and confirmation of biological agents (Washington, DC: ASM Press, 2001).
2. Okumura T., Suzuki K., Fukuda A., Kohama A., Takasu N., Ishimatsu S., and Hinohara S., The Tokyo subway sarin attack: Disaster management, Part 1: Community emergency response, *Acad. Emerg. Med.*, 5, 613–617, 1998.
3. Cole L., *Enearta: Chemical and Biological Warfare, the Eleventh Plague: the Politics of Biological and Chemical Warfare* (New York, NY: W H Freeman & Company, 1999).

4. Marrs T. C., Maynard R. L., and Sidell F. R., *Chemical Warfare Agents, Toxicology and Treatment* (New York: John Wiley & Sons, 1996).

5. Sidell F. R., Takafuji E. T., and Franz D. R. (eds.), *Textbook of Military Medicine*, Part I: *Warfare, Weaponry, and the Casualty*, Vol. 3.: *Medical Aspects of Chemical and Biological Warfare* (Washington, DC: Borden Institute, Walter Reed Medical Center, 1997).

6. Marrs T. C., Maynard R. L., and Sidell F. R., *Chemical Warfare Agents, Toxicology and Treatment* (New York: John Wiley & Sons, 1996).

7. Smart J. K., History of Chemical and biological warfare: An American perspective," in *Medical Aspects of Chemical and Biological Warfare*, ed. F. R. Sidell, E. T. Takafuji, and D. R. Franz (Washington, DC: Office of the Surgeon General, 1997) 9–86.

8. Dunn M. A., Hackley B. E., Jr., and Sidell F. R., Pretreatment for nerve agent exposure, in *Textbook of Military Medicine: Medical Aspects of Chemical and Biological Warfare*, ed. F. R. Sidell, E. T. Takafuji, and D. R. Franz (Washington, DC: Borden Institute, Walter Reed Medical Center, 1997) 181–196.

9. Somani S. M., *Chemical Warfare Agents* (New York: Academic Press, 1992).

10. www.opcw.gov, *OPCW Press Releases to the UN*, 1997.

11. www.opcw.gov, *OPCW Press Releases to the UN*, 1998.

12. www.opcw.gov, *OPCW Press Releases to the UN*, 1999.

13. www.opcw.gov, *OPCW Press Releases to the UN*, 2000.

14. www.opcw.gov, *OPCW Press Releases to the UN*, 2001.

15. www.opcw.gov, *OPCW Press Releases to the UN*, 2002.

16. www.opcw.gov, *OPCW Press Releases to the UN*, 2003.

17. www.opcw.gov, *OPCW Press Releases to the UN*, 2004.

18. http://www.fas.org/nuke/guide/usa/doctrine/army/mmcch/NervAgnt.htm

19. Cole L. A., *CBW Terrorism Deconstructed*, ed. Jonathan B. Tucker, (Cambridge, MA: MIT Press) 2000.

20. Cole L. A., *The Ultimate Terrorist*, ed. Jessica Stern, (Cambridge MA: Harvard University Press) 1999.

21. Cole L. A., The poison weapons taboo: Biology, culture and policy, *Politics. Life. Sciences.*, 17, 119–132, 1998.

22. Rusyniak D. E. and Nanagas K. A., Organophosphate poisoning, *Semin. Neurol.*, 24(2), 197–204, 2004.

23. Arya N., Pesticides and human health, *Canadian journal of public health, Revue canadienne de santé publique*, 96(2), 89–92, 2005.

24. Morozova V. S., Levashova A. I., and Eremin S. A., Determination of pesticides by enzyme immunoassay, *J. Anal. Chem.*, 60(3), 202–217, 2005.

25. Barlow S. M., Agricultural chemicals and endocrine-mediated chronic toxicity or carcinogenicity, *Scand. J. Work. Environ. Health.*, 31(1), 141–145, 2005.

26. Garry V. F., Pesticides and children, *Toxicol. Appl. Pharmacol.*, 198(2), 152–163, 2004.

27. Ahmed F. E., Analyses of pesticides and their metabolites in foods and drinks, *Trends Anal. Chem.*, 20(11) 649, 2001.

28. Hutson D. H. and Roberts T. R., *Environmental Fate of Pesticide* (Chichester, New York: John Wiley & Sons Ltd, 1990).

29. Jones O. A. H., Voulvoulis N., Lester J. N., Human pharmaceuticals in the aquatic environment—A review, *Environ. Technol.*, 22(12), 1383–1394, 2001.

30. Brandli R. C., Bucheli T. D., Kupper T., Furrer R., Stadelmann F. X., and Tarradellas J., Persistent organic pollutants in source-separated compost and its feedstock materials—A review of field studies, *J. Environ. Qual.*, 34(3), 735–760, 2005.

31. Tadeo J. L., Sanchez-Brunete C., Albero B., and Gonzalez L., Analysis of pesticide residues in juice and beverages, *Crit. Rev. Anal. Chem.*, 34(2), 121–131, 2004.
32. http://www.fas.org/nuke/guide/usa/doctrine/army/mmcch/NervAgnt.htm
33. Roldan-Tapia L., Parron T., and Sanchez-Santed F., Neuropsychological effects of long-term exposure to organophosphate pesticides, *Neurotoxicol. Teratol.*, 27(2), 259–266, 2005.
34. Augerson W., *A Review of the Scientific Literature as it Pertains to Gulf War Illnesses.* *Vol. 5: Chemical and Biological Warfare Agents* (Santa Monica, CA: Rand Corporation, 2000). http://www.rand.org/pubs/monograph_reports/MR1018.5/index.html
35. Somani S. M., *Chemical Warfare Agents* (New York: Academic Press, 1992).
36. Marrs T. C., Maynard R. L., and Sidell F. R., *Chemical Warfare Agents, Toxicology and Treatment* (New York: John Wiley & Sons, 1996).
37. Sidell F. R., Takafuji E. T., and Franz D. R., eds., *Textbook of Military Medicine*, Part I: *Warfare, Weaponry, and the Casualty*, Vol. 3., *Medical Aspects of Chemical and Biological Warfare*, (Washington, DC: Borden Institute, Walter Reed Medical Center, 1997).
38. Smart J. K., History of chemical and biological warfare: An American perspective, in *Medical Aspects of Chemical and Biological Warfare*, ed. F. R. Sidell, E. T. Takafuji, and D. R. Franz, Department of the Army, Borden Institute, 9–86, 1997.
39. Sidell F. R., Urbanetti J. S., Smith W. J., and Hurst C. G., Vesicants, in F. R. Sidell, E. T. Takafuji, and D. R. Franz, 1997.
40. Dunn P., The chemical war: Journey to Iran, *NBC Defense and Technology Int.*, p. 1, 1986.
41. Sidell F. R. and Hurst C. G., Long-term health effects of nerve agents and mustard, in *Medical Aspects of Chemical and Biological Warfare*, ed. F. R. Sidell, E. T. Takafuji, and D. R. Franz, 229–246, 1997.
42. Muir B., Cooper D. B., Carrick W. A., Timperley C. M., Slater B. J., and Quick S., Analysis of chemical warfare agents III. Use of bis-nucleophiles in the trace level determination of phosgene and perfluoroisobutylene, *J. Chromatogr. A.*, 1098 (1–2), 156–165, 2005.
43. Kanu A. B., Haigh P. E., and Hill H. H., Surface detection of chemical warfare agent simulants and degradation products, *Anal. Chim. Acta.*, 553(1–2), 148–159, 2005.
44. Sadik O. A., Wanekaya A. K., and Andreescu S., Advances in analytical technologies for environmental protection and public safety, *J. Environ. Monit.*, 6(6), 513–522, 2004.
45. Hernandez F., Sancho J. V., and Pozo O. J., Critical review of the application of liquid chromatography/mass spectrometry to the determination of pesticide residues in biological samples, *Anal. Bioanal. Chem.*, 382(4), 934–946, 2005.
46. Bigalke H. and Rummel A., Medical aspects of toxin weapons, *Toxicology*, 214(3), 210–220, 2005.
47. Varfolomeyev S., Kurichkin I., Eremenko A., and Efremenko E., Chemical and biological safety. Biosensors and nanotechnological methods for the detection and monitoring of chemical agents, *Pure. Appl. Chem.*, 74, 2311–2316, 2002.
48. Geerdink R. B., Niessen W. M. A., and Brinkman U. A. T., Trace-level determination of pesticides in water by means of liquid and gas chromatography, *J. Chromatogr. A*, 970(1–2), 65–93, 2002.
49. Hernandez F., Sancho J. V., and Pozo O. J., Critical review of the application of liquid chromatography/mass spectrometry to the determination of pesticide residues in biological samples, *Anal. Bioanal. Chem.*, 382(4), 934–946, 2005.
50. Sherma J., Recent advances in the thin-layer chromatography of pesticides: A review, *J. AOAC. Int.*, 86(3), 602–611, 2003.
51. Sherma J., Thin-layer chromatography of pesticides—A review of applications for 2002–2004, *ACTA Chromatographica*, 15, 5–30, 2005.

52. Medana C., Calza P., Baiocchi C., and Pelizzetti E., Liquid chromatography tandem mass spectrometry as a tool to investigate pesticides and their degradation products, *Curr. Org. Chem.*, 9(9), 859–873, 2005.

53. Budde W. L., Analytical mass spectrometry of herbicides, *Mass. Spectrom. Rev.*, 23(1), 1–24, 2004.

54. Malik A. K. and Faubel W., A review of analysis of pesticides using capillary electrophoresis, *Crit. Rev. Anal. Chem.*, 31(3), 223–279, 2001.

55. Aulakh J. S., Malik A. K., Kaur V., and Schmitt-Kopplin P., A review on solid phase micro extraction-high performance liquid chromatography (SPME-HPLC) analysis of pesticides, *Crit. Rev. Anal. Chem.*, 35(1), 71–85, 2005.

56. Ulrich S., Solid-phase microextraction in biomedical analysis, *J. Chromatogr. A*, 902(1), 167–194, 2000.

57. Plaza G., Ulfig K., and Tien A. J., Immunoassays and environmental studies, *Pol. J Environ. Studies*, 9(4), 231–236, 2000.

58. Schobel U., Barzen C., and Gauglitz S., Immunoanalytical techniques for pesticide monitoring based on fluorescence detection, *Fresenius. J. Anal. Chem.*, 366(6–7), 646–658, 2000.

59. Eremin S. A. and Smith D. S., Fluorescence polarization immunoassays for pesticides, *Comb. Chem. High. Throughput. Screen.*, 6(3), 257–266, 2003.

60. Recio R., Ocampo-Gomez G., Moran-Martinez J., Borja-Aburto V., Lopez-Cervantes M., Uribe M., Torres-Sanchez L., and Cebrian M. E., Pesticide exposure alters follicle-stimulating hormone levels in Mexican agricultural workers, *Environ. Health. Perspect.*, 113(9), 1160–1163, 2005.

61. Knopp D., Applications of Immunological methods for the determination of Environmental-Pollutants in Human Biomonitoring—A review, *Anal. Chim. Acta.*, 311(3), 383–392, 1995.

62. Simonian A. L., Flounders A. W., and Wild J. R., FET-based biosensors for the direct detection of organophosphate neurotoxins, *Electroanalysis*, 16(22), 1896–1906, 2004.

63. Sole S., Merkoci A., and Alegret S., Determination of toxic substances based on enzyme inhibition. Part I. Electrochemical biosensors for the determination of pesticides using batch procedures, *Crit. Rev. Anal. Chem.*, 33(2), 89–126, 2003.

64. Sole S., Merkoci A., and Alegret S., Determination of toxic substances based on enzyme inhibition. Part II. Electrochemical biosensors for the determination of pesticides using flow systems, *Crit. Rev. Anal. Chem.*, 33(2), 127–143, 2003.

65. Farre M. and Barcelo D., Toxicity testing of wastewater and sewage sludge by biosensors, bioassays and chemical analysis, *Trends Anal. Chem.*, 22(5), 299–310, 2003.

66. Suri C. R., Raje M., and Varshney G. C., Immunosensors for pesticide analysis: Antibody production and sensor development, *Crit. Rev. Biotechnol.*, 22(1), 15–32, 2002.

67. Ricci F. and Palleschi G., Sensor and biosensor preparation, optimization and applications of Prussian blue modified electrodes, *Biosens. Bioelectron.*, 21(3), 389–407, 2005.

68. Mulchandani A., Chen W., Mulchandani P., Wang J., and Rogers K. R., Biosensors for direct determination of organophosphate pesticides, *Biosens. Bioelectron.*, 16(4–5), 225–230, 2001.

69. Velasco-Garcia M. N. and Mottram T., Biosensor technology addressing agricultural problems, *Biomed. Eng.*, 84(1), 1–12, 2003.

70. Tang L., Zeng G. M., Huang G. H., Shen G. L., Xie G. X., Hong Y. X., and Christine W. C., Toxicity testing in environmental analysis—application of inhibition based enzyme biosensors, *Transactions of Nonferrous Metals Society of China*, 14(1), 14–17, 2004.

71. Hart A. L., Collier W. A., and Janssen D., The response of screen-printed enzyme electrodes containing cholinesterases to organo-phosphates in solution and from commercial formulations, *Biosens. Bioelectron.*, 12(7), 645–654, 1997.

72. Stenersen J., *Chemical Pesticides: Mode of Action and Toxicology* (Boca Raton, FL: CRC Press, 2004).

73. Schulze H., Vorlova S., Villatte F., Bachmann T. T., Schmid R. D., Design of acetylcholinesterases for biosensor applications, *Biosens. Bioelectron.*, 18(2–3), 201–209, 2003.

74. Espinosa M., Atanasov P., and Wilkins E., Development of a disposable organophosphate biosensor, *Electroanalysis*, 11(14), 1055–1062, 1999.

75. Wilkins E., Carter M., Voss J., and Ivnitski D., A quantitative determination of organophosphate pesticides in organic solvents, *Electrochem. Commun.*, 2(11), 786–790, 2000.

76. Shi M. A., Lougarre A., Alies C., Frémaux I., Tang Z. H., Stojan J., and Fournier D., Acetylcholinesterase alterations reveal the fitness cost of mutations conferring insecticide resistance, *BMC. Evol. Biol.*, 6, 45, 2004. See http://www.biomedcentral.com/content/pdf/1471-2148-4-45.pdf

77. Simon S. and Massoulié J., Cloning and expression of acetylcholinesterase from electrophorus. Splicing pattern of the 3' exons in vivo and in transfected mammalian cells, *J. Biol. Chem.*, 272(52), 33045–33055, 1997.

78. De StefanoL., Moretti L., Rendina I., and Rotiroti L., Pesticides detection in water and humic solutions using porous silicon technology, *Sens. Actuators B*, 111, 522–525, 2005.

79. Rotiroti L., De Stefano L., Rendina N., Moretti L., Rossi A. M., and Piccolo A., Optical microsensors for pesticides identification based on porous silicon technology. *Biosens. Bioelectron.*, 20(10), 2136–2139, 2005.

80. Simonian A. L., Good T. A., Wang S. S., and Wild J. R., Nanoparticle-based optical biosensors for the direct detection of organophosphate chemical warfare agents and pesticides, *Anal. Chim. Acta.*, 534(1), 69–77, 2005.

81. Nabok A., Haron S., and Ray A., Registration of heavy metal ions and pesticides with ATR planar waveguide enzyme sensors, *Appl. Surf. Sci.*, 238(1–4), 423–428, 2004.

82. Szekacs A., Trummer N., Adanyi N., Varadi M., and Szendro I., Development of a non-labeled immunosensor for the herbicide trifluralin via optical waveguide lightmode spectroscopic detection, *Anal. Chim. Acta.*, 487(1), 31–42, 2003.

83. Hendrickson O. D., Warnmark-Surugiu I., Sitdikov R., Zherdev A. V., Dzantiev B. B., and Danielsson B., Development of microformat imaging microplate and membrane immunoenzyme assays of the herbicide atrazine, *Int. J. Environ. Anal. Chem.*, 85(12–13), 905–915, 2005.

84. Halamek J., Pribyl J., Makower A., Skladal P., and Scheller F. W., Sensitive detection of organophosphates in river water by means of a piezoelectric biosensor, *Anal. Bioanal. Chem.*, 382(8), 1904–1911, 2005.

85. Halamek J., Makower A., Knosche K., Skladal P., and Scheller F. W., Piezoelectric affinity sensors for cocaine and cholinesterase inhibitors, *Talanta*, 65(2), 337–342, 2005.

86. Lebedev M. Y., Eremin S. A., and Skladal P., Development of the piezoelectric biosensor for acetochlor detection, *Anal. Lett.*, 36(11), 2443–2457, 2003.

87. Makower A., Halamek J., Skladal P., Kernchen F., and Scheller F. W., New principle of direct real-time monitoring of the interaction of cholinesterase and its inhibitors by piezolectric biosensor, *Biosens. Bioelectron.*, 18(11), 1329–1337, 2003.

88. Deo R. P., Wang J., Block I., Mulchandani A., Joshi K. A., Trojanowicz M., Scholz F., Chen W., and Lin Y., Determination of organophosphate pesticides at a carbon nanotube/organophosphorus hydrolase electrochemical biosensor, *Analytica Chimica Acta.*, 530, 185–189, 2005.

89. Mulchandani A., Chen W., Mulchandani P., Wang J., and Rogers K., Biosensors for direct determination of organophosphate pesticides, *Biosens. Bioelec.*, 16, 225–230, 2001.

90. Mulchandani A., Mulchandani P., Chen W., Wang J., and Chen L., Amperometric thick-film strip electrodes for monitoring organophosphate nerve agents based on immobilized organophosphorus hydrolase, *Anal. Chem.*, 71, 2246–2249, 1999.

91. Mulchandani A., Pan S., and Chen W., Fiber-optic enzyme biosensor for direct determination of organophosphate nerve agents, *Biotechnol. Prog.*, 15, 130–134, 1999.

92. Montesinos T., Perez-Munguia S., Valdez F., and Marty J. L., Disposable cholinesterase biosensor for the detection of pesticides in water-miscible organic solvents, *Anal. Chim. Acta.*, 431(2), 231–237, 2001.

93. Ivnitskii D. M. and Rishpon J., A Potentiometric Biosensor for pesticides based on the Thiocholine Hexacynoferrate (III) Reaction, *Biosens. Bioelectron.*, 9(8), 569–576, 1994.

94. Pandey P. C., Upadhyay S., Pathak H. C., Pandey C. M. D., and Tiwari I., Acetylthiocholine/acetylcholine and thiocholine/choline electrochemical biosensors/sensors based on an organically modified sol-gel glass enzyme reactor and graphite paste electrode, *Sens. Actuators B*, 62(2), 109–116, 2000.

95. Stoytcheva M., Sharkova V., and Magnin J. P., Electrochemical approach in studying the inactivation of immobilized acetylcholinesterase by arsenate(III), *Electroanalysis*, 10(14), 994–998, 1998.

96. Pritchard J., Law K., Vakurov A., Millner P., and Higson S. P. J., Sonochemically fabricated enzyme microelectrode arrays for the environmental monitoring of pesticides, *Biosens. Bioelectron.*, 20(4), 765–772, 2004.

97. Rippeth J. J., Gibson T. D., Hart J. P., Hartley I. C., and Nelson G., Flow-injection detector, incorporating a screen-printed disposable amperometric biosensor for monitoring organophosphate pesticides, *Analyst.*, 122, 1425–1429, 1997.

98. Givens R. B., Lin M.-H. Taylor D. J., Mechold U., Berry J. O., and Hernandez V. J., Inducible expression, enzymatic activity, and origin of higher plant homologues of bacterial RelA/SpoT stress proteins in Nicotiana tabacum, *J. Biol. Chem.*, 279(9), 7495–7504, 2004.

99. Vakurov A., Simpson C. E., Daly C. L., Gibson T. D., and Millner P. A., Acetylecholinesterase-based biosensor electrodes for organophosphate pesticide detection II. Immobilization and stabilization of acetylcholinesterase, *Biosens. Bioelectron.*, 20(1), 2324–2329, 2005.

100. Vakurov A., Simpson C. E., Daly C. L., Gibson T. D., and Millner P. A., Acetylcholinesterase-based biosensor electrodes for organophosphate pesticide detection I. Modification of carbon surface for immobilization of acetylcholinesterase, *Biosens. Bioelectron.*, 20(6), 1118–1125, 2004.

101. Nunes G. S., Jeanty G., and Marty J. L., Enzyme immobilization procedures on screen-printed electrodes used for the detection of anticholinesterase pesticides—Comparative study, *Anal. Chim. Acta.*, 523(1), 107–115, 2004.

102. Hermanson G. T. and Smith P. K., Use of the Munnich reaction for immobilization or conjugation of active hydrogen containing compounds-preparation of affinity supports and hapten carrier conjugates, *FASEB J.*, 6(5), A1729–A1729, 1992.

103. Franchina J. G., Lackowski W. M., Dermody D. L., Crooks R. M., Bergbreiter D. E., Sirkar K., Russell R. J., and Pishko M. V., Electrostatic immobilization of glucose oxidase in a weak acid, polyelectrolyte hyperbranched ultrathin film on gold: Fabrication, characterization, and enzymatic activity, *Anal. Chem.*, 71(15), 3133–3139, 1999.

104. Bahulekar R., Ayyangar N. R., and Ponrathnam S., Polyethyleneimine in Immobilization of biocatalysts, *Enzyme. Microb. Technol.*, 13(11), 858–868, 1991.

105. Gamiz-Gracia L., Garcia-Campana A. M., Soto-Chinchilla J. J., Huertas-Perez J. F., and Gonzalez-Casado A., Analysis of pesticides by chemiluminescence detection in the liquid phase, *Trends Anal. Chem.*, 24(11), 927–942, 2005.

106. Mallat E., Barcelo D., Barzen C., Gauglitz G., and Abuknesha R., Immunosensors for pesticide determination in natural waters, *Trends Anal. Chem.*, 20(3), 124–132, 2001.

107. Suri C. R., Raje M., and Varshney G. C., Immunosensors for pesticide analysis: Antibody production and sensor development, *Crit. Rev. Biotechnol.*, 22(1), 15–32, 2002.

108. Plekhanova Y. V., Reshetilov A. N., Yaznina E. V., Zherdev A. V., and Dzantiev B. B., A new assay format for electrochemical immunosensors: Polyelectrolyte-based separation on membrane carriers combined with detection of peroxidase activity by pH-sensitive field-effect transistor, *Biosens. Bioelectron.*, 19(2), 109–114, 2003.

109. Navratilova I. and Skladal P., The immunosensors for measurement of 2,4-dichlorophenoxyacetic acid based on electrochemical impedance spectroscopy, *Bioelectrochemistry*, 62(1), 11–18, 2003.

110. Grennan K., Strachan G., Porter A. J., Killard A. J., and Smyth M. R., Arazine analysis using an amperometric immunosensor based on single-chain antibody fragments and regeneration-free multi-calibrant measurement, *Anal. Chim. Acta.*, 500(1–2), 287–298, 2003.

111. Kricka L. J., Microarrays, biochips and nanochips: Personal laboratories for the 21st Century, *Clin. Chem. Acta.*, 307, 219–223, 2001.

112. Chow A. W., Lab-on-a-chip: Opportunities for chemical engineering, *AIChE. J.*, 48, 1590–1595, 2002.

113. Kopf-Sill A. R., Success and challenges of lab-on-a-chip, *Lab. Chip.*, 2, 42N–47N, 2002.

114. Min J. H. and Baeumner A. J., Characterization and optimization of interdigitated ultramicroelectrode arrays as electrochemical biosensor transducers, *Electroanalysis*, 16(9), 724–729, 2004.

115. Wang J., Chen G., Muck A., Chatrathi M. P., Mulchandani A., and Chen W., Microchip enzymatic assay of organophosphate nerve agents, *Anal. Chim. Acta.*, 505(2), 183–187, 2004.

116. Mile B., Chemistry in court, *Chromatographia*, 62(1–2), 3–9, 2005.

117. Heo J. and Crooks R. M., Microfluidic biosensor based on an array of hydrogel-entrapped enzymes, *Anal. Chem.*, 77(21), 6843–6851, 2005.

118. Yunus K. and Fisher A. C., Voltammetry under microfluidic control, a flow cell approach, *Electroanalysis*, 15(22), 1782–1786, 2003.

119. Wei X. L. and Mo Z. H., The progress of bio-cell sensor and cell chip, *Progress in Biochem. Biophys.*, 31(9), 855–859, 2004.

120. Rossier J., Reymond F., and Michel P. E., Polymer microfluidic chips for electrochemical and biochemical analyses, *Electrophoresis*, 23(6), 858–867, 2002.

121. Qiu H. B., Yin X. B., Yan J. L., Zhao X. C., Yang X. R., and Wang E. K., Simultaneous electrochemical and electrochemiluminescence detection for microchip and conventional capillary electrophoresis, *Electrophoresis*, 26(3), 687–693, 2005.

122. Gao Z. X., Wang Y., Fang Y. J., Zhou H. Y., Wang T., Wang H. Y., Wang S. Q., and Dai S. G., Research on protein chip technologies for the detection of atrazine and papaverine, *Chinese J. Analytical Chemistry*, 33(4), 455–458, 2005.

123. Wang J., Electrochemical detection for microscale analytical systems, *Talanta*, 56, 223–231, 2002.

124. Ahn K. C., Koivunen M. E., Gee S., Kennedy I., and Hammock B. D., Application of europium oxide nanoparticles as a fluorescent reporter for immunoassay, Abstracts of papers of *The American Chemical Society*, 227(1), U65–U65, 2004.

125. Sotiropoulou S., Gavalas V., Vamvakaki V., and Chaniotakis N. A., Novel carbon materials in biosensor systems, *Biosens. Bioelectron.*, 18(2–3), 211–215, 2003.

126. Vetcha S., Abdel-Hamid I., Atanasov P., Ivnitski D., Wilkins E., and Hjelle B., Portable immunosensor for the fast amperometric detection of anti-Hantavirus antibodies, *Electroanalysis*, 12, 1034–1038, 2000.

127. Krishnan R., Ghindilis A. L., Atanasov P., and Wilkins E., Development of an ampero-metric immunoassay based on a highly dispersed immunoelectrode, *Anal. Lett.*, 29, 2615–2631, 1996.

128. Lu L. Y. F. and Wang J., Disposable carbon nanotube modified screen-printed bio-sensor for amperometric detection of organophosphorus pesticides and nerve agents, *Electroanalysis*, 16(1–2), 145–149, 2004.

129. Sotiropoulou S. and Chaniotakis N. A., Carbon nanotube array-based biosensor, *Anal. Bioanal. Chem.*, 375(1), 103–105, 2003.

130. Parikh K., Cattanach K., Rao R., Suh D. S., Wu A. M., and Manohar S. K., Flexible vapour sensors using single walled carbon nanotubes, *Sens. Actuators B*, 113(1), 55–63, 2006.

131. Pritchard J., Law K., Vakurov A., Millner P., and Higson S. P. J., Sonochemically fab-ricated enzyme microelectrode arrays for the environmental monitoring of pesticides, *Biosens. Bioelectron.*, 20(4), 765–772, 2004.

132. Barton A. C., Collyer S. D., Davis F., Gornall D. D., Law K. A., Lawrence E. C. D., Mills D. W., et. al., Sonochemically fabricated microelectrode arrays for biosensors offering widespread applicability: Part I, *Biosens. Bioelectron.*, 20(2), 328–337, 2004.

133. Myler S., Davis F., Collyer S. D., and Higson S. P. J., Sonochemically fabricated microelectrode arrays for biosensors: Part II, Modification with a polysiloxane coating, *Biosens. Bioelectron.*, 20(2), 408–412, 2004.

134. Losito I., De Giglio E., Cioffi N., and Malitesta C., Spectroscopic investigation on polymer films obtained by oxidation of *o*-phenylenediamine on platinum electrodes at different pHs, *J. Math. Chem.*, 11(7), 1812–1817, 2001.

135. Pritchard J., Law K., Vakurov A., Millner P., and Higson S. P. J., Sonochemically fabricated enzyme microelectrode arrays for the environmental monitoring of pesti-cides, *Biotechnol. Bioeng.*, 20, 765–772, 2004.

136. Lin Y. H., Lu F., and Wang J., Disposable carbon nanotube modified screen-printed biosensor for amperometric detection of organophosphorus pesticides and nerve agents, *Electroanalysis*, 16(1–2), 145–149, 2004.

137. Zhao L. Y., Liu H. Y., and Hu N. F., Assembly of layer-by-layer films of heme pro-teins and single-walled carbon nanotubes: Electrochemistry and electrocatalysis, *Anal. Bioanal. Chem.*, 384(2), 414–422, 2006.

138. Wang J., Carbon-nanotube based electrochemical biosensors: A review, *Electro-analysis*, 17(1), 7–14, 2005.

139. Wang J., Nanomaterial-based electrochemical biosensors, *Analyst*, 130(4), 421–426, 2005.

140. Deo R. P., Wang J., Block I., Mulchandanic A., Joshi K., Trojanowicz M., Scholz F., Chen W., and Lin Y., Determination of organophosphate pesticides at a carbon nano-tube/organophosphorus hydrolase electrochemical biosensor, *Anal. Chim. Acta.*, 530, 85–189, 2005.

141. Beaver G. A. and Guiochon L. A., Progress and future of instrumental analytical chem-istry applied to the environment, *Anal. Chim. Acta.*, 524(1–2), 1–14, 2004.

142. Hurst M. R. and Wilkins E., Chemical and biological warfare: Should rapid detec-tion techniques be researched to dissuade usage? A review, *American Journal of Applied Sciences*, 2(4), 796–805, 2005.

143. Radke S. M. and Alocilja E. C., A microfabricated biosensor for detecting foodborne bioterrorism agents, *IEEE. Sens. J.*, 5(4), 744–750, 2005.

12 Reversible Inhibition of Enzymes for Optical Detection of Chemical Analytes

Brandy J. Johnson
Center for Bio/Molecular Science and Engineering
Naval Research Laboratory
Washington, D.C.

Amanda Oliver
Department of Math, Science, and Engineering
Northern Oklahoma College
Stillwater, Oklahoma

H. James Harmon
Physics Department
Oklahoma State University
Stillwater, Oklahoma

CONTENTS

12.1   INTRODUCTION

Proteins perform a wide range of tasks in organisms including structural and mechanical considerations, transport, storage, and catalysis. The proteins responsible for catalysis are enzymes, which reduce the activation energy for various reactions. The active site of a given enzyme is generally a shape and charge complement to the substrate for the reaction catalyzed by that enzyme and provides both selectivity and binding affinity for specific substrates and inhibitors. Though enzymes posses a high degree of selectivity, molecules other than the intended substrate may be bound either at the active site or to other regions of the enzyme. Other molecules, which bind an enzyme and reduce the rate of catalysis, are called inhibitors. Competitive inhibitors bind at the enzyme active site and prevent catalysis while noncompetitive inhibitors bind to other areas of the enzyme and may prevent or hinder catalysis.

Enzymes are used in a range of processes for the rubber, paper, brewing, and dairy industries as well as applications such as protein stain removers in laundry detergents, predigestion of baby foods, and clarification of fruit juices. Enzymes have also been widely described in a variety of sensor applications including those based on optical, thermal, electrochemical, and acoustic transduction mechanisms. The cholinesterases, organophosphorus hydrolase, glucose oxidase, urease, and peroxidases are some of the more frequently described enzymes. Enzymes are a nearly ideal material for use as recognition elements in sensor applications as they possess, to varying degrees, selectivity and sensitivity to their substrates and inhibitors that rival the selectivity and sensitivity of specific antibodies. Extensive literature is available on the purification of different proteins [1–11] as well as methods for either immobilizing or entrapping them [12–14] for use in sensor applications [14–20]. We do not attempt to review these fields, but instead to describe a unique sensor system employing enzymes as recognition elements with porphyrins as optical transducers.

Porphyrins are a family of intensely colored organic molecules sharing a similar basic structure (Figure 12.1). The sensitivity of porphyrins to environmental conditions such as pH, ionic strength, and hydrophobicity can be observed through changes in the spectrophotometric characteristics of the molecules (Figure 12.2). The intense spectral characteristics as well as a strong sensitivity to the immediate environment of the molecule result from the 22 π-electrons of the parent molecule. This sensitivity can be used for discrimination between various analytes as the changes in the spectrophotometric characteristics of the porphyrin are specific to the analyte involved [21–23]. The basic porphyrin structure can be modified by substituent groups around the periphery of the nearly flat parent structure or through incorporation of a metal in the central position via coordination to the four-pyrrole nitrogen atoms. Modification of the porphyrin structure can be used to adapt the specific spectrophotometric characteristics or the binding characteristics of the molecule to suit a particular application. These characteristics of porphyrins have been applied to detection schemes for a variety of different analytes [24].

12.2   DETECTION

Optical detection for the system described here is based on evanescent wave absorbance spectroscopy (EWAS). This method takes advantage of the reflections of light

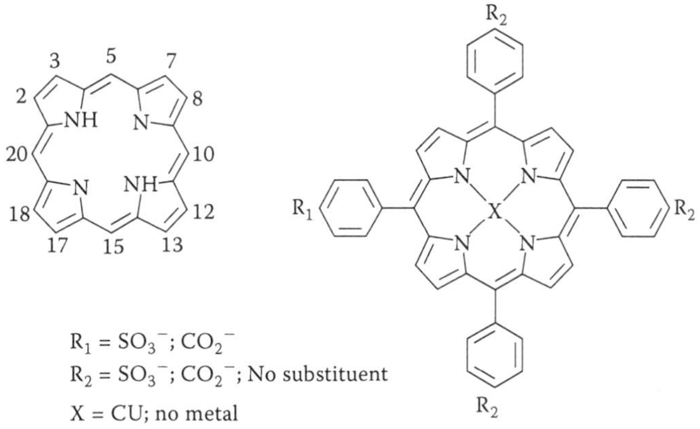

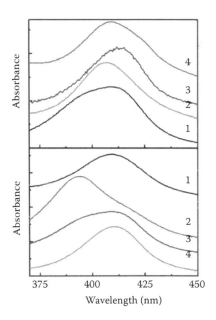

$R_1 = SO_3^-; CO_2^-$
$R_2 = SO_3^-; CO_2^-;$ No substituent
$X = CU;$ no metal

FIGURE 12.1 Porphyrin structure. Modification of substituent groups on the periphery of the porphyrin as well as incorporation of a metal into the center of the macrocycle can be used to obtain porphyrins, which inhibit the activity of various enzymes.

FIGURE 12.2 Porphyrin spectra. The spectrophotometric characteristics of iron-complexed $TPPS_4$, like those of other porphyrins, are sensitive to changes in hydrophobicity, ionic strength, and pH. The first panel shows the absorbance spectrum of Fe-TPPS (0.8 μM) in 50 mM NaPi pH 7 (1) and those of the porphyrin in 50 mM NaPi pH 7 with 50% ethanol (2), 50% methanol (3), or 50% DMSO (4). The second panel shows the impact of ionic strength on the absorbance characteristics comparing spectra collected in 50 mM NaPi pH 7 (3) and 500 mM NaPi pH 7 (1). Also shown are absorbance spectra collected in 50 mM NaPi for pH 4.8 (2) and pH 9.2 (4).

as it travels through a waveguide. When light traveling in one medium strikes the interface between this medium and one of a different refractive index at an angle greater than the critical angle, the evanescent wave extends into the second medium (Figure 12.3). When the second medium is an absorbing material, it is possible to monitor the spectrophotometric characteristics of that material using EWAS. For the detection scheme described here, the enzyme recognition element and the porphyrin transduction element are placed in the second medium. Alternatively, optical spectroscopic measurements can be made with the surface normal (perpendicular) to the light beam path. Evanescent measurement is preferable when transversing the medium by the light beam is not desirable (in pipes, high speed air flows, etc.).

The waveguide used is a simple glass microscope slide. Light is coupled into one side of the microscope slide while the detector is placed on the opposite side (Figure 12.3). The intense spectrophotometric characteristics of the porphyrins eliminate the need for laser light sources; a light emitting diode (LED) is employed. The spectrophotometer used is an Ocean Optics USB2000 (Dunedin, FL). This small portable instrument is capable of measuring 280 to 800 nm at 1 nm resolution with output to a desktop or laptop computer or a pocket PC. The total weight of the system pictured in Figure 12.4 including Compaq Pocket PC, sample holder, Ocean Optics USB2000 with lithium battery pack, fiber-optic cable, and diode light source is less than 1 kg. Data analysis has not yet been automated and requires the use of spectral analysis

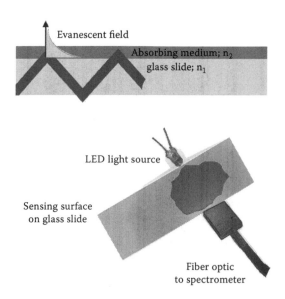

FIGURE 12.3 Evanescent wave absorbance spectroscopy. Reflection of light at the interface between two media of differing refractive indices at angles greater than the critical angle results in the generation of an evanescent wave. This wave decays exponentially with increasing distance from the interface between the media. In the system described here, the light is introduced into a glass slide (n_1) with an analyte sensitive coating on the surface. The light passing through the slide is collected on the side opposite the light emitting diode.

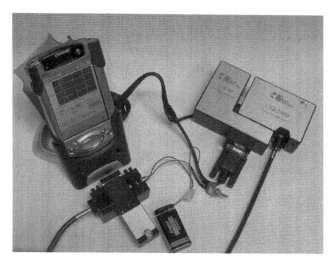

FIGURE 12.4 Prototype system. This is a photograph of the prototype system including Compaq Pocket PC, sample holder, and Ocean Optics USB2000 spectrophotometer with lithium battery pack, fiber-optic cable, and diode light source.

software such as Grams32 (Galactic Industries, Salem, NH), which has been used here.

Because of the design of the sample holder and the simplicity of the optical arrangement, changing the light source can be accomplished by simply choosing a different LED and inserting it into the mount on the sample holder. For experiments with acetylcholinesterase, butyrylcholinesterase, glucose oxidase, or organophosphorus hydrolase as the recognition element, an LED of maximum wavelength 434 nm with a half bandwidth of 83 nm was used [25–31]. The use of carbonic anhydrase as the recognition element requires the use of a pair of diodes in order to illuminate the entire necessary region of the spectrum; maximum wavelength 434 nm (HBW 83 nm) and maximum wavelength 380 nm (HBW 17 nm) [32].

12.3 IMMOBILIZATION

The sensing mechanism used here requires immobilization of the enzymes onto the surface of the glass waveguides. The commonly employed method of amino-functionalized glass uses glutaraldehyde activation to obtain the active surface. Further modifications using various cross-linkers and blockers as well as variation of enzyme density is used to obtain surfaces for optimal analyte response and longevity under storage conditions. The enzymes acetylcholinesterase, butyrylcholinesterase, organophosphorus hydrolase, and glucose oxidase were immobilized using identical protocols with only enzyme concentration being varied [26,28,30,31]. Immobilization of carbonic anhydrase necessitated modification of the basic protocol [32].

All steps in the immobilization process are completed at room temperature and were terminated by rinsing with phosphate-buffered saline (PBS) (50 mM sodium phosphate pH 9 with 0.5 M sodium chloride). ProbeOn™ Plus microscope slides, which are amino functionalized and bear a positive charge, are obtained from Fisher Biotech (Pittsburgh, PA). Glutaraldehyde (GA) (Sigma, St. Louis, MO) activation of the slide surfaces is accomplished by incubation with 0.17 M glutaraldehyde in 50 mM sodium phosphate buffer (NaPi) at pH 8 for 25 minutes followed by rinsing with copious amounts of PBS. The GA activated surface is then incubated with 6.4 mM amino-terminated Starburst® (PAMAM) dendrimer (generation 4; Aldrich, Milwaukee, WI) in 50 mM NaPi pH 8 for 90 minutes followed by rinsing with PBS. TRIS (1 M Tris(hydroxymethyl)aminomethane with HCl pH 9) is used to block unreacted GA on the slide surface during a 20-minute incubation followed by rinsing with PBS. The GA/dendrimer surface is again activated by incubation with glutaraldehyde, and a second layer of dendrimer was applied followed by TRIS blocking. A final GA incubation is used to activate the terminal amino groups of the dendrimer surface in order to facilitate binding of the enzyme. The enzyme is incubated with the surface for 90 minutes at concentrations of 200 nM for acetylcholinesterase and butyrylcholinesterase, 8 µM for organophosphorus hydrolase, and 60 nM for glucose oxidase. Enzyme binding was followed by rinsing with PBS, TRIS blocking, PBS rinsing, and a final rinse with 50 mM NaPi pH 7 to remove any excess salt before storage.

Immobilization of carbonic anhydrase using the dendrimer method described earlier resulted in surfaces with very low enzymatic activity. The dendrimer is used to increase the surface density of amino groups in order to increase the density of enzyme bound. Carbonic anhydrase is a smaller protein than the others mentioned and, like other enzymes, may suffer from steric hindrance if packed too tightly. The following immobilization protocol was used with success to improve the level of enzymatic activity of the waveguide surfaces. ProbeOn™ Plus microscope slides were GA activated as described previously and rinsed with PBS. The activated surfaces were incubated with 0.9 mM LysLysLys (Sigma) in 50 mM NaPi pH 7 for 90 minutes followed by rinsing with PBS and TRIS blocking. The peptide-bearing surface was again GA activated followed by incubation with 12 µM carbonic anhydrase for 90 minutes and TRIS blocking. Before storage, slides were rinsed with PBS and 50 mM NaPi pH 7.

The planar waveguides are stored at room temperature under mild vacuum in three layer food saver bags using a FoodSaver (Vac360) from Tilia (San Francisco, CA). The transduction elements/reporter molecules are applied to the surfaces by incubating a porphyrin solution with the immobilized enzymes at pH 7. Application of the porphyrin to the immobilized enzyme surface has been suggested to adversely affect the slide lifetime [27]; however, recent experiments have shown that AChE-porphyrin slides stored under mild vacuum as described are as sensitive and responsive as freshly made slide surfaces.

12.4   DETECTION MECHANISM

Figure 12.5 illustrates the use of the enzyme and porphyrin together for the detection of substrates and inhibitors of the enzyme. Porphyrins are chosen so that they interact with the enzyme as a competitive inhibitor, that is, they interfere with binding of the substrate through interaction with the enzyme at the active site. Exposure of the porphyrin-enzyme complex to other competitive inhibitors and substrates of the enzyme results in dissociation of the porphyrin from the enzyme [28]. The spectrophotometric characteristics of the porphyrins are highly sensitive to their immediate environment. A particular absorbance spectrum is observed for a porphyrin free in solution (Figure 12.5, Panel A). This spectrum changes upon interaction of the porphyrin with the enzyme giving a particular spectrophotometric signature that can be used to indicate porphyrin-enzyme association; the spectra of different porphyrin-enzyme complexes

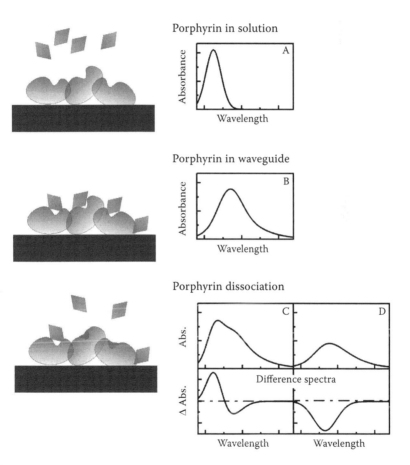

FIGURE 12.5 Detection mechanism. Panel A shows an idealized absorbance spectrum for a porphyrin in solution. When the porphyrin interacts with the immobilized enzyme, the pre-exposure spectrum is observed (Panel B). Upon exposure of the porphyrin-enzyme surface to the target analyte, the porphyrin dissociates from the enzyme resulting in unique postexposure absorbance spectra depending on whether the sample is applied as a vapor (Panel C) or in solution (Panel D). Difference spectra are calculated as the postexposure absorbance spectrum minus the preexposure absorbance spectrum. The rhombus represents porphyrin, the concave shape represents enzyme, and the black rectangle represents immobilization support.

TABLE 12.1
Limits of Detection for Various Analytes

Porphyrin	Enzyme	Interaction Peak (nm)	Analyte	LOD (ppb)
TPPS$_1$	Acetylcholinesterase	446	Sarin (solution)	0.100
			Sarin (vapor)	(250 pg)
			Eserine salicylate	0.037
			Diazinon	0.045
			Galanthamine	0.050
			Scopolamine	0.100
			Tetracaine	0.250
			Triton X-100	83
	Butyrylcholinesterase	421	Eserine salicylate	0.050
			Amitriptyline	0.072
			Drofenine	0.091
			Triton X-100	40
	Carboxypeptidase	438/423	Eserine salicylate	0.010
CuC$_1$TPP	Organophosphorus hydrolase	412	Paraoxon	0.007
			Coumaphos	0.250
			Diazinon	0.800
			Malathion	1
CTPP$_4$	Glucose oxidase	427	Glucose	200 (20 mg/dl)
CuC$_1$TPP	Carbonic anhydrase	404	CO$_2$ (vapor)	10^3 (0.1%)
			Saccharin	10
			1,10-Phenanthroline	100
			Cysteine	1500

are uniquely different as shown in Table 12.1. Dissociation of the porphyrin from the enzyme results in loss of this characteristic spectrophotometric signature. Detection is accomplished by comparing a pre-exposure spectrum of the waveguide with a postexposure spectrum.

There are two possible modes of dissociation of the porphyrin from the enzyme active site. If solution samples are used, the dissociated porphyrin is blotted or rinsed away from the surface with the excess analyte solution. In this case, the absorbance intensity due to the porphyrin-enzyme complex is reduced in the waveguide spectrum by the percentage of complexes impacted (Figure 12.5, Panel D). The difference spectrum calculated as pre-exposure minus postexposure shows a loss in absorbance only at the characteristic peak for that particular porphyrin-enzyme interaction (Figure 12.6). Thus, it is possible to discriminate the binding of analyte(s) to different enzymes co-immobilized on a "mixed bed" surface containing different enzyme-porphyrin complexes.

The application of sample followed by blotting of the excess liquid was used for all solution samples to obtain the results described here. It is, however, not essential

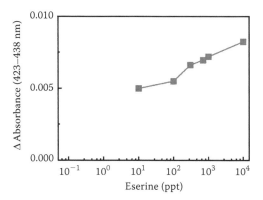

FIGURE 12.6 Change in absorbance of the TPPS$_1$-carboxypeptidase A surface in the presence of various concentrations of eserine (physostigmine).

that the solution be blotted away from the surface. If the waveguides were used in conjunction with a flow chamber that allowed for maintenance of a constant solution thickness above the slide surface and, therefore, a constant path length, pre- and postexposure spectra could be collected with solution in contact with the slide surface. In this case, the absorbance spectrum should reflect not only a reduction in intensity of the spectrophotometric characteristics of the porphyrin-enzyme complex but an increase in intensity of the spectrophotometric characteristics of the porphyrin in solution and/or the porphyrin upon nonspecific interaction with the surface of the waveguide (Figure 12.5, Panel C). The difference spectrum would then also show a loss in absorbance at the characteristic peak for the porphyrin-enzyme interaction and an increase in absorbance of porphyrin either in solution or interacting nonspecifically with the waveguide.

If a vapor sample is applied, the porphyrin dissociates from the enzyme active site; however, there is little solvent present to allow extensive freedom of movement. The result is that the porphyrin, while no longer interfering with the enzyme active site, may continue to interact with the waveguide surface in close proximity to the enzyme and possibly maintain interaction with the enzyme through other areas of the protein surface. The effects on the spectrophotometric characteristics of the waveguide are a decrease in intensity of absorbance at the wavelength of the porphyrin-enzyme complex and an increase in absorbance intensity at the wavelength indicative of nonspecific interaction of the porphyrin with the waveguide.

12.5 SPECIFIC SYSTEMS

12.5.1 ACETYLCHOLINESTERASE

The enzyme acetylcholinesterase (AChE) catalyzes the hydrolysis of acetylcholine to acetate and choline to terminate nerve impulse transmission at cholinergic synapses. This enzyme binds nerve agents such as VX and sarin. Inhibition of AChE by these compounds and the closely related organophosphate pesticides results in constant nerve impulse transmission some symptoms of which are salivation, sweating, nausea, headache, and muscle contraction which can lead to death by suffocation. Enzymatic activity is also inhibited by carbamate pesticides, some anesthetics, and other drugs. The exceptionally high sensitivity of cholinesterases to the presence of these compounds has lead to application of the enzymes in a range of detection applications [19,33–35]. The methods primarily use a comparison of the enzymatic rate under known conditions in the absence/presence of sample to identify the presence of an inhibitor.

The enzymatic activity of AChE from electric eel is inhibited by monosulfonate tetraphenyl porphyrin (TPPS₁) the structure of which can be seen in Figure 12.1 with $R_1 = SO_3$, $R_2 =$ no substituent group, and no metal incorporated [36]. A Lineweaver-Burk plot of enzymatic rates in the absence/presence of TPPS₁ determines the type of inhibition resulting from the presence of the porphyrin. The Lineweaver-Burk plot is the plot of the double-reciprocal form of the initial enzymatic rate versus the substrate concentration. Intersection of the lines generated in the absence and presence of inhibitor at the y-axis shows no change in maximal velocity but a change in the Michaelis constant K_m, an indication of the substrate binding affinity, and indicates competitive inhibition by the porphyrin. Competitive inhibition involves competition of the inhibitor for occupation of the active site of the enzyme.

The immobilized TPPS₁-AChE complex is used to detect sarin, diazinon, eserine salicylate, galanthamine, scopolamine, tetracaine, and Triton X-100 in solution and for the detection of sarin in vapor phase [27–29]. Limits of detection for each target are presented in Table 12.1. Interaction of an AChE inhibitor with the TPPS₁-AChE complex results in dissociation of the porphyrin from the enzyme with a corresponding loss at the 446 nm absorbance peak characteristic for the interaction with changes similar to those shown in Figure 12.5, Panel D. The magnitude of the loss in absorbance at 446 nm is related to the concentration of the target and the association constant between the target and the enzyme when other sample parameters remain constant [27]. If the sample matrix is changed, as in the case of sarin samples prepared in water and isopropanol as opposed to those prepared in 50 mM NaPi pH 7, the dependence on concentration of the change in absorbance varies [29].

12.5.2 BUTYRYLCHOLINESTERASE

Butyrylcholinesterase (BChE), also known as serum cholinesterase, catalyzes the hydrolysis of butyrylcholine at rates similar to those of acetylcholine hydrolysis by AChE. Though the function of BChE is poorly understood, it shows 73% similarity and 53% amino acid sequence identity with AChE from electric eel. There is some evidence that BChE is responsible for the activation or inactivation of compounds such as heroine and cocaine and that it is involved in defense against cholinesterase inhibitors [37]. AChE and BChE are both capable of hydrolysis of acetylcholine or butyrylcholine with the rates of catalysis varying and show differing affinities for different substrates and inhibitors. The level of BChE activity in serum samples has been used to monitor exposure to cholinesterase inhibitors such as organophosphates [38–40].

Monosulfonate tetraphenyl porphyrin (TPPS₁) is also a competitive inhibitor of BChE [26]. The characteristic absorbance peak for the interaction of TPPS₁ with immobilized BChE is observed at 421 nm (as compared to 446 nm for AChE). The BChE-TPPS$_1$ complex is applicable to detection of diazinon, eserine, amitriptyline, drofenine, and Triton X-100 [26]. Limits of detection are presented in Table 12.1. The varying sensitivity of the enzymes AChE and BChE to different inhibitors is indicated by the detection limits for eserine and Triton X-100 [25]. The AChE-TPPS$_1$ complex is slightly more sensitive to eserine than BChE-TPPS$_1$ while the BChE-TPPS$_1$ complex is more sensitive to Triton X-100 than is the AChE-TPPS$_1$. The loss in absorbance intensity at 421 nm upon exposure to competitive inhibitors of the enzyme shows log-linear dependence on concentration. Exposure of the BChE-TPPS$_1$ surface to noncompetitive inhibitors of BChE, which do not bind at the active site, does not result in a change in the absorbance spectrum, substantiating the claim of porphyrin binding at the active site.

12.5.3 ORGANOPHOSPHORUS HYDROLASE

The active site of organophosphorus hydrolase (OPH) contains two metal atoms (zinc in the wild-type enzyme) and catalyzes hydrolysis of numerous organophosphate compounds including pesticides as well as chemical warfare agents such as sarin and soman. Rates of OPH catalyzed hydrolysis of organophosphates exceed those of chemical hydrolysis by NaOH at 4°C by factors of 40 to 2450 [41–43]. The enzyme has been described for use in sensor systems with exceptional detection limits reported for response times on the order of 10 seconds [44–48]. However, the presence of OP/CWA is detected by the inhibition of enzymatic rate determined by comparing rate measurements in the presence and absence of the analyte.

Several porphyrins bind OPH with unique spectrophotometric characteristics resulting; however, in order to form a porphyrin-enzyme complex that is sensitive to the presence of substrates of the enzyme, a copper-complexed porphyrin is necessary [30]. Two candidates, a copper-complexed TPPS$_1$ and copper-complexed TPPC$_1$ (mono(4-carboxy phenyl) porphyrin; $R_1 = CO_2^-$, $R_2 = SO_3^-$) inhibit the activity of OPH in a mixed manner. Mixed inhibition is the inhibition of enzyme activity in a manner such that the maximal enzymatic rate and the concentration needed to achieve half of that rate are both changed. The intersection of the curves in the absence and presence of the inhibitor occurs in the second quadrant of the Lineweaver-Burk plot. Mixed type inhibition involves the interaction of the inhibitor at two or more locations on the enzyme with one of these being the active site. The spectrophotometric characteristics of the porphyrin-enzyme complex are different depending on whether the apo or wild-type enzyme is bound by the copper-complexed porphyrins; however, the spectrophotometric characteristics are identical for the interaction of TPPS$_1$ or TPPC$_1$ with either version of the enzyme. Other porphyrins such as zinc- and iron-complexed TPPS$_1$ as well as the metal-free TPPS$_1$ and TPPC$_1$ do not inhibit the enzymatic activity of OPH.

The spectrophotometric characteristics of the OPH-CuTPPC$_1$ complex are more sensitive to exposure to OPH substrates than those of the OPH-CuTPPS$_1$ complex resulting in a lower limit of detection. The characteristic peak for the interaction of CuTPPC$_1$ with immobilized OPH is at 412 nm. The absorbance intensity of this peak decreases upon exposure of the surface to diazinon, malathion, coumaphos, and paraoxon. The limits of detection presented in Table 12.1 again reflect the association constants between the targets and the enzyme. The change in absorbance at 412 nm upon exposure of the porphyrin-enzyme complex to OPH substrates is log-linearly dependent on the substrate concentration.

The systems described thus far are capable only of discrimination between substrates or inhibitors and those compounds that do not bind the enzyme active site. The hydrolysis of compounds such as coumaphos and paraoxon produces products that absorb in the visible spectrum. If a fixed path length standing drop were used for sample application (described earlier), the absorbance increase at wavelengths indicative of these products together with the decrease in the absorbance of the characteristic peak of the porphyrin-enzyme complex would allow further identification of the analyte detected.

12.5.4   CARBONIC ANHYDRASE

Carbonic anhydrase (CA) is a zinc metalloenzyme involved in mammalian respiration, which catalyzes the hydration of carbon dioxide. Copper-complexed $TPPC_1$ competitively inhibits CA enzymatic activity as does copper-complexed $TPPS_1$ [32]. Experiments comparing the spectrophotometric characteristics of the two porphyrins in the presence of CA and apo-CA indicate that the zinc atoms in the active site of the enzyme are indeed involved in the interaction between the porphyrins and the enzyme. The metal-free porphyrins $TPPS_1$ and $TPPC_1$ do not inhibit the enzymatic activity of CA. Further, the spectrophotometric characteristics of these porphyrins in the presence of apo-CA were identical to those in the presence of wild-type CA, indicating the lack of involvement of the active site-coordinated zinc in the porphyrin-enzyme interaction for metal-free porphyrins.

The characteristic peak for the interaction between immobilized CA and copper-$TPPC_1$ is at 404 nm. Exposure of the porphyrin-enzyme complex to inhibitors of CA such as 1,10-phenanthroline, saccharine, and cysteine results in a loss in the absorbance intensity of the surface at 404 nm with a log-linear relationship between intensity change and concentration. Limits of detection are given in Table 12.1. Detection of gaseous CO_2 in the range between 1,000 (0.1%) and 10,000 (1.0%) ppm is possible using the CA-$TPPC_1$ surface. CO_2 levels are of interest in closed systems such as large office buildings, airplane cabins, shuttle missions, and the International Space Station. Typical outdoor levels of CO_2 are around 300 ppm. When levels reach 2,500 to 5,000 ppm symptoms such as tiredness and headaches may develop with loss of consciousness occurring at around 100,000 ppm.

12.5.5   GLUCOSE OXIDASE

Glucose oxidase (GOD) is widely used in systems for the detection of glucose levels in body fluids (i.e., diabetes) and for the removal of residual sugar in food and agricultural products [17]. Glucose oxidase is highly selective for β-D-glucose catalyzing its oxidation to gluconic acid and showing very low activity for even similar sugars such as 2-deoxy-D-glucose. The enzyme is easily immobilized and enzymatic activity of GOD is preserved over a range of pH values and ionic strengths though it is inhibited by the presence of mercury, copper, and silver. The porphyrin meso-tetra(4-carboxyphenyl)porphine (TPPC$_4$; R$_1$ = R$_2$ = CO$_2^-$ in Figure 12.1) inhibits the enzymatic activity of GOD in a reversible manner [31]. Formation of an immobilized GOD-TPPC$_4$ complex gives a characteristic peak for the interaction of the porphyrin with the active site region of the enzyme at 427 nm. The change in absorbance at this peak upon exposure to glucose can be used to determine glucose concentrations between 20 and 200 mg/dl (1.1 and 11.1 mM). The absorbance change shows linear dependence on glucose concentration up to 200 mg/dl with saturation effects becoming apparent after this point.

12.5.6 CARBOXYPEPTIDASE

Carboxypeptidase A (approx. 35 kDa MW, Sigma Chemical) functions both as a peptidase and an esterase; it is in this latter mode that it can serve as a detector for cholinesterase inhibitors. Unlike the other enzymes such as AChE or BChE, it does not have a serine residue in the active site. TPPS$_1$ forms a complex with the enzyme and, upon challenge with the cholinesterase inhibitor eserine (physostigmine) in water, exhibits a change in the absorbance spectrum with a new peak and a marked increase in absorbance at 423 nm. This suggests TPPS$_1$ may not be completely displaced from the active site. For actual sensor operations, the use of an enzyme such as BChE or carboxypeptidase in place of (or in addition to) AChE will allow for potential identification of the analyte based on different specificities/sensitivities of the enzyme. Enzymes such as OPH, which are not readily available, may be difficult to obtain in large quantities; the supply of AChE is often limited perhaps due to the capture of electric eels, while proteins such as BChE (from horse blood) and carboxypeptidase (pancreas) are more readily available from slaughterhouses.

12.5.7 MULTIENZYME SYSTEM

Each of the surfaces described thus far is capable of discriminating between competitive inhibitors of the immobilized enzyme and other compounds; noncompetitive inhibitors (that bind elsewhere than the active site) do not displace the porphyrin from the active site. They are not useful for discrimination within the class of compounds that competitively inhibit the enzyme and function best to detect a wide spectrum of cholinesterase inhibitors. The cholinesterase-based surfaces, for example, would give the same positive response in the presence of the nerve agent sarin that would be observed in the presence of malathion, which is used as a pesticide. Both of these compounds are toxic and will be detected; but it may be desirable to know which of the two is being detected. Multi-protein arrays have been described to improve the ability of these enzyme-based techniques for identification of specific inhibitors [49–51]. This type of discrimination requires a library of information containing the expected response for various targets across a range of concentrations. Combining the response of OPH and organophosphorus acid anhydrase (OPAA) surfaces with those of surfaces consisting of various cholinesterases possessing differing sensitivities and selectivity should provide a more specific identification of the causative agent.

As a first step in this direction, a surface consisting of electric eel AChE and horse serum BChE was designed. Both of these enzymes are inhibited by $TPPS_1$ and the characteristic absorbance peaks for the porphyrin-enzyme complexes are different (421 vs. 446 nm). This allows for co-immobilization of the two enzymes from a simple mixture of equal concentrations onto the entire slide surface [25]. Exposure to those compounds inhibiting BChE competitively results in a loss in absorbance at 421 nm while compounds inhibiting AChE competitively result in a loss at 446 nm. Compounds inhibiting both enzymes result in a loss at both 421 nm and 446 nm. This combination of two enzymes allows for class discrimination of those compounds, which are inhibitors of BChE, inhibitors of AChE, inhibitors of both enzymes, and inhibitors of neither enzyme. Detection limits are approximately the same for the dual enzyme system as those observed for the single enzyme systems.

12.6 ADDITIONAL COMMENTS

We have described a novel and highly sensitive system based on the reversible inhibition of enzymes by porphyrins. The detection system utilizes only the binding of the analyte/agent to the enzyme without concern for catalysis. As such, the concentration-response curves and LOD are directly related to the K_i value of the enzyme for the inhibitor (or K_m value in enzymes like OPH that use the agent as substrate). Plotting the inverse of the absorbance change versus the inverse of agent concentration yields a plot similar to a Lineweaver-Burk plot to determine kinetic parameters. Here, the graphically determined "K_i" is the concentration at which half the enzyme is bound and corresponds to the LD_{50} value of the agent. As such, the absorbance changes for this sensor should be linearly related to the LD_{50} of the agent; this has been observed with AChE [27]. The surfaces are remarkably stable even when stored under "mild" conditions at room temperature. This stability is imparted by the stability of immobilized enzymes, a widely recognized phenomenon.

All of the surfaces described here employ a glass microscope slide as the support though it should be possible to adapt these techniques to any amino-functionalized surface such as activated nylon or cellulose. The glass microscope slide is used as a planar waveguide allowing the absorbance spectrum of the immobilized porphyrin-enzyme surface to be collected by evanescent wave absorbance spectroscopy. The measurement technique provides advantages over more traditional approaches in that the sample is not applied between the light source and the detector so correction for the absorbance of the sample is not necessary. In addition, the absorbance spectrum is collected over a 2.54 cm path length (1 inch) rather than a path length equal to the thickness of the surface coating, so a very thin layer (basically one enzyme thick) can be used without negatively impacting the signal-to-noise ratio. The thin layer of recognition elements also prevents issues associated with diffusion of analyte through multilayer surfaces. The spectrum collected is an average of the area covered by the 2.54 cm path length as well as the 1 cm wide linear fiber optic. This average gives excellent reproducibility even when porphyrin-enzyme surfaces are produced in small batches. Variations between the slides or even across a single slide become obvious when measurements are taken perpendicular to the surface.

Measurement protocols similar to the one described here are also used in applications such as biosensors based on immunoarray technology. The novel aspect of the detection protocols described is the use of reversible inhibition of an enzyme by a colorimetric agent for selective recognition and optical transduction. The porphyrin interacts with the enzyme at the site where the target (substrate or inhibitor) interacts with the enzyme. When the porphyrin-enzyme complex is exposed to the target, the porphyrin dissociates from the enzyme and the enzyme binds the analyte. This dissociation does not occur in the presence of uncompetitive inhibitors. The porphyrin-enzyme complex possesses a unique set of spectrophotometric characteristics that are different from those of the porphyrin when it is not bound to the enzyme. The changes in the absorbance spectrum of the waveguide surface that occur upon dissociation of the porphyrin from the enzyme are used to indicate the presence of the target. In addition, the magnitude of the intensity change is directly related to the number of porphyrins dissociating from enzymes allowing determination of analyte concentration. Specificity is imparted primarily by the ability of an analyte(s) to bind at the active site just as the specificity of immunosensors is imparted by the ability of specific antigens to bind to their complementary antibodies.

Because a direct event is used to indicate analyte presence and there are no diffusion considerations, this detection technique is real-time suffering only from time requirements imposed by the spectrophotometer and data handling. We have demonstrated that this system is capable of detection of organophosphates in solution at levels below the current safe drinking water standards. We have also demonstrated the utility of the system for the detection of vapor phase analytes such as carbon dioxide and sarin. The enzyme surfaces have also been shown to be insensitive to common interferents including diesel exhaust, cologne, window cleaner, cigarette smoke, and so forth [30]. Reaction of the sample with the surface is complete in less than 1 second. Total time from sample application to result is dependent on subsequent data collecting and processing activities.

REFERENCES

1. Scopes R. K., *Protein Purification: Principles and Practice*, 3rd ed. (New York: Springer, 1993).
2. Cutler P., *Protein Purification Protocols* (Totowa, NJ: Humana Press, 2003).
3. Roe S. (ed.), *Protein Purification Applications*, 2nd ed. (New York: Oxford University Press, 2001).
4. Hilbrig F. and Freitag R., Analytical Technologies in the Biomedical and Life Sciences, *J. Chromatogr. B*, 790, 79–90, 2003.
5. Rito-Palomares M., The practical application of aqueous two-phase processes for the recovery of biological products, *J. Microbiol. Biotechnol.*, 12, 535–543, 2002.
6. Glatz Z., Affinity ultrafiltration BŌLKOVIN, *Chemicke Listy*, 94, 490–493, 2000.
7. Silva M. E. and Franco T. T., Liquid-liquid extraction of biomolecules in downstream processing—A review paper, *Braz. J. Chem. Eng.*, 17, 1–17, 2000.
8. Chiou S. H. and Wu S. H., Evaluation of commonly used electrophoretic methods for the analysis of proteins and peptides and their application to biotechnology, *Anal. Chim. Acta.*, 383, 47–60, 1999.
9. Galaev I. Y., New methods of protein purification. Displacement, *Biochemistry-Moscow Chromatography*, 63, 1258–1265, 1998.
10. Galaev I. Y., New methods of protein purification. Expanded bed chromatography, *Biochemistry-Moscow*, 63, 619–624, 1998.

11. Charcosset C., Purification of proteins by membrane chromatography, *J. Chem. Technol. Biotechnol.*, *71*, 95–110, 1998.

12. Taylor R. F. (ed.), *Protein Immobilization* (New York: Marcel Dekker, 1991).

13. Guisán J. M., *Immobilization of Enzymes and Cells* (Totowa, NJ: Humana Press, 2005).

14. Mulchandani A. and Rogers K. (eds.), *Enzyme and Microbial Biosensors: Techniques and Protocols* (Totowa, NJ: Humana Press, 1998).

15. White B. J. and Harmon H. J., Immobilized Protein and Enzyme Based Sensors, In *Encyclopedia of Sensors*, ed. C. A. Grimes, E. C. Dickey, M. V. Pishko (Valencia, CA: American Scientific Publishers, 2005, 8000).

16. Evtugyn G. A., Budnikov H. C., and Nikolskaya E. B., Biosensors for the determination of environmental inhibitors of enzymes, *Russ. Chem. Rev.*, 68, 1041–1064, 1999.

17. Raba J. and Mottola H. A., Glucose oxidase and an *analytical* reagent, *Crit. Rev. Anal. Chem.*, 25, 1–42, 1995.

18. Krawczyk T. K. V., Analytical applications of inhibition of enzymatic reactions, *Chem. Anal.*, 43, 135–158, 1998.

19. Skladal P., Biosensors based on cholinesterase for detection of pesticides, *Food Technol. Biotechnol.*, 34, 43–49, 1996.

20. Sapsford K. E., Shubin Y. S., Delehanty J. B., Golden J. P., Taitt C. R., Shriver-Lake L. C., and Ligler F. S., Fluorescence-based array biosensors for detection of bio-hazards, *J. Appl. Microbiol.*, 96(1), 47–58, 2004.

21. Mauzerall D., Spectra of molecular complexes of porphyrins in aqueous solution, *Biochemistry*, 4, 1801–1810, 1965.

22. Schneider H.-J. and Wang M., Ligand-porphyrin complexes: Quantitative evaluation of stacking and ionic contributions, *J. Org. Chem.*, 59, 7464–7472, 1994.

23. Shelnutt J. A., Molecular complexes of copper uroporphyrin with aromatic acceptors, *J. Phys. Chem.*, 87, 605–616, 1983.

24. Malinski T., Porphryin-Based Electrochemical Sensors, In *The Porphyrin Handbook*, ed. K. M. Kadish, K. M. Smith, R. Guilard, 1st ed., (New York: Academic Press, 6, 231–256, 2000).

25. White B. J., Legako J. A., and Harmon, H. J., Spectrophotometric detection of cholinesterase inhibitors with an integrated acetyl-/butyrylcholinesterase surface, *Sens. Actuators B*, 89, 107–111, 2003.

26. White B. J., Legako J. A., and Harmon H. J., Rapid reagent-less detection of competitive inhibitors of butyrylcholinesterase, *Sens. Actuators B*, 91, 138–142, 2003.

27. White B. J., Legako J. A., and Harmon H. J., Extended lifetime of reagentless detector for multiple inhibitors of acetylcholinesterase, *Biosens. Bioelectron.*, 18, 729–734, 2003.

28. White B. J., Legako J. A., Harmon H. J., Reagentless detection of a competitive inhibitor of immobilized acetylcholinesterase, *Biosens. Bioelectron.*, 17, 361–366, 2002.

29. White B. J. and Harmon H. J., Effect of propranolol and local anesthetics on myoglobin and cytochrome oxidase, *Sensor Letters*, 3, 1–5, 2005.

30. White B. J. and Harmon H. J., Optical solid-state detection of organophosphates using organophosphorus hydrolase, *Biosens. Bioelectron.*, 20, 1977–1983, 2005.

31. White B. J. and Harmon H. J., Novel optical solid-state glucose sensor using immobilized glucose oxidase, *Biochem. Biophys. Res. Commun.*, 296, 1069–1071, 2002.

32. White B. J. and Harmon H. J., Enzyme-based detection of GB (Sarin) using planar waveguide absorbance spectroscopy, *Sensor Letters*, 3, 36–41, 2005.

33. Sole S., Merkoci A., and Alegret S., Determination of toxic substances based on enzyme inhibition. Part II. Electrochemical biosensors for the determination of pesticides using flow systems, *Crit. Rev. Anal. Chem.*, 33, 127–143, 2003.

34. Everett W. R. and Rechnitz G. A., Enzyme-based electrochemical biosensors for determination of organophosphorous and carbamate pesticides, *Anal. Lett.*, 32, 1–10, 1999.

35. Marco M. P. and Barcelo D., Environmental applications of analytical biosensors, *Meas. Sci. Technol.*, 7, 1547–1562, 1996.

36. White B. J. and Harmon H. J., Interaction of monosulfonate tetraphenyl porphyrin, a competitive inhibitor, with acetylcholinesterase, *Biosens. Bioelectron.*, 17, 463–469, 2002.

37. Çokuğraş A. N., Butyrylcholinesterase: Structure and physiological importance, *Turk. J. Biochem.*, 28, 54–61, 2003.

38. Sanchez-Hernandez J. C., Evaluating reptile exposure to cholinesterase-inhibiting agrochemicals by serum butyrylcholinesterase activity, *Environ. Toxicol. Chem.*, 22, 296–301, 2003.

39. Fossi M. C., Leonzio C., Massi A., Lari L., and Casini S., Serum esterase inhibition in birds: A nondestructive biomarker to assess organophosphorus and carbamate contamination, *Arch. Environ. Contam. Toxicol.*, 23, 99–104, 1992.

40. Chu S. Y., Depression of serum cholinesterase activity as an indicator of insecticide exposure-consideration of the analytical and biological variation, *Clin. Biochem.*, 18, 323–326, 1985.

41. Omburo G. A., Kuo J. M., Mullins L. S., and Raushel F. M., Characterization of the zinc binding site of bacterial phosphotriesterase, *J. Biol. Chem.*, 267, 13278–13283, 1992.

42. Munnecke D. M., Hydrolysis of organophosphate insecticides by an immobilized-enzyme system, *Biotechnol. Bioeng.*, 21, 2247–2261, 1979.

43. Lejeune K. E., Dravis B. C., Yang F. X., Hetro A. D., Doctor B. P., and Russell A. J., Fighting nerve agent chemical weapons with enzyme technology, In *Enzyme Engineering Xiv*, ed. A. I. Laskin, L. Goa-Xiang, Vol. 864, 153–170, 1998.

44. Simonian A. L., diSioudi B. D., and Wild J. R., An enzyme based biosensor for direct determination of diisopropyl fluorophosphate, *Anal. Chim. Acta.*, 389, 189–196, 1999.

45. Russell R. J., Pishko M. V., Simonian A. L., and Wild J. R., Poly(ethylene glycol) hydrogel-encapsulated fluorophore-enzyme conjugates for direct detection of organophosphorus neurotoxins, *Anal. Chim.*, 71, 4909–4912, 1999.

46. Mulchandani A., Pan S. T., and Chen W., Fiber-optic enzyme biosensor for direct determination of organophosphate nerve agents, *Biotechnol. Prog.*, 15, 130–134, 1999.

47. Mulchandani A., Mulchandani P., Chen W., Wang J., and Chen L., Amperometric thick-film strip electrodes for monitoring organophosphate nerve agents based on immobilized organophosphorus hydrolase, *Anal. Chim.*, 71, 2246–2249, 1999.

48. Mulchandani A., Kaneva I., and Chen W., Detoxification of organophosphate nerve agents by immobilized *Escherichia coli* with surface-expressed organophosphorus hydrolase, *Biotechnol. Bioeng.*, 63, 216–223, 1999.

49. Solna R., Dock E., Christenson A., Winther-Nielsen M., Carlsson C., Emneus J., Ruzgas T., and Skladal P., Amperometric screen-printed biosensor arrays with co-immobilised oxidoreductases and cholinesterases, *Anal. Chim. Acta.*, 528, 9–19, 2005.

50. Revzin A. F., Sirkar K., Simonian A., and Pishko M. V., Glucose, lactate, and pyruvate biosensor arrays based on redox polymer/oxidoreductase nanocomposite thin-films deposited on photolithographically patterned gold microelectrodes, *Sens. Actuators B*, 81, 359–368, 2002.

51. Kukla A. L., Kanjuk N. I., Starodub N. F., and Shirshov Y. M., Multienzyme electrochemical sensor array for determination of heavy metal ions, *Sens. Actuators B*, 57, 213–218, 1999.

52. Stoll D., Templin M. F., Bachmann J., and Joos T. O., Protein microarrays: Technologies And Applications, ed. Gerald A. Urban, Vol. 16, BioMEMS Springer US PP245–267, Thursday, February 01, 2007.

13 Sensing of Biowarfare Agents

A. Bogomolova
Fractal Systems Inc.
Belleair Beach, Florida

CONTENTS

13.1 INTRODUCTION

A wide range of microorganisms, of both bacterial and viral origin, as well as purified protein toxins can be turned into bioweapons. Early detection of bioweapons is vital for proper biological defense. Biosensors for biowarfare agents have been under development for more than 20 years. The ideal sensor would analyze unknown samples in minutes, have programmable operation for unattended sample analysis, and be capable of multiple agent analysis for a number of agents. There are two possible approaches: (1) "protein sensing", that is, detection of the protein component of the pathogens and toxins with minimal sample preparation and (2) "genetic sensing", that is, detection of the DNA or RNA, purified from pathogenic bacteria (or toxin-producing bacteria) and viruses. Although the genetic sensing generally possesses better sensitivity, it requires unavoidable nucleic acid purification step and typically utilizes polymerase chain reaction (PCR) amplification of the genetic material, requiring either highly trained personnel or costly and bulky automated equipment setup. Besides, biowarfare protein toxins cannot be detected. Protein sensors use highly specific recognition of the proteins by antibodies, receptors, or, lately, aptamers. Sample preparation is minimal and individual proteins, bacteria, and viruses can be detected in environmental, food, or clinical samples. Comparison of the existing approaches toward individual and multispecific detection of biowarfare agents is the subject of this review.

13.2 BIOWARFARE AGENTS

"Biological warfare is the intentional use of micro-organisms and toxins to produce disease and death in humans, livestock and crops. The attraction of bioweapons in war, and for use in terrorist attacks is attributed to easy access to a wide range of disease-producing biological agents, to their low production costs, to their non-detection by routine security systems, and to their easy transportation from one place to another" [1]. A wide range of microorganisms, of both bacterial and viral origin, can be turned into bioweapons. Only in the United States and the former Soviet Union the potential of the following microorganisms for biowarfare has been researched: anthrax, Argentinian hemorrhagic fever, Bolivian hemorrhagic fever, bubonic plague, brucellosis, Ebola virus, glanders, Lassa fever, Marburg virus, monkey pox, meliodosis, Rift Valley fever, smallpox, tularemia, typhus, Venezuelan equine encephalitis virus, and Q fever [2,3]. Most of them were weaponized, that is, adapted for aerosol delivery. In addition to that, purified toxins, such as ricin, botulinum toxin, cholera toxin, shiga toxin, and staphylococcal enterotoxin B (SEB) can be used as biowarfare agents. Biological weapons are relatively easy to make once the methods are developed. Even primitive aerosol release of a bioweapon by terrorist groups in a highly populated area under the right weather conditions can cause a disaster. There are no vaccines for bacterial infections such as brucellosis, glanders, and meliodosis and for lethal viral infections, such as Ebola and Marburg. Early detection of bioweapons is vital for proper biological defense.

13.3 PROTEIN-BASED SENSING OF BIOWARFARE AGENTS

Protein sensing involves detection of the protein component of the pathogenic bacteria, viruses, and individual toxins. It requires minimal sample preparation. Specificity is achieved by using antibodies, receptors, or aptamers raised or selected to specifically bind surface proteins or whole bacteria, bacterial spores, viral particles, or individual toxins.

13.3.1 IMMUNOSENSORS

Most known biowarfare sensors utilize sandwich-type immunoassays, when specific toxin or bacteria is bound by immobilized "capture" antibody in the first step, and is made "visible" by a "reporter" antibody, carrying fluorescent, luminescent, or other detectable label in the second step. These sensors can utilize the antibodies specific to single pathogen (individual sensors) or antibodies specific to several pathogens for simultaneous detection of multiple agents (multispecific sensors). Multispecific sensors are highly preferential for biodefense and are reviewed below. Immunofluorescent detection, probably the most popular method for sensing of biowarfare agents, was first reported in the late 1960s. It has been also widely used for multispecific sensing, as recently reviewed by Sapsford [4]. Ligler's group at the Naval Research Laboratory first reported a multifunctional immunofluorescent sensor for biowarfare targets in 1998 [5], detecting 5 ng/mL of SEB, 25 ng/mL of ricin, and 15 ng/mL of F1 antigen of *Yersinia pestis* [6], and have been optimizing the approach since then. They adapted the rapid 14-minute immunoassay for testing bacterial and viral analytes (*B. globigii*, 10^5 CFU/mL and MS2 bacteriophage, 10^7 pfu/mL) [7], increased the number of simultaneously detected targets to 9, adding cholera toxin, botulinum toxins, *B. anthracis*, and *F. tularensis*, and lowered the detection limit to 0.5 ng/mL for toxins and ~10^4 CFU for bacterial targets [5,8,9]. The fluorescence detection was performed by CCD camera and the toxins and bacteria were detected in clinical fluids, environmental samples, and foods. Their work resulted in the creation of The Multi-Analyte Array Biosensor (MAAB) instrument with fluidics module, limiting the number of manipulations to only the initial sample loading [10]. Another antibody-based microarray has been reported for the multiplexed detection of cholera toxin, diphtheria toxin, anthrax lethal factor and protective antigen, SEB, and tetanus toxin C fragment in spiked samples, using both direct and competitive formats [11]. Quantitative multispecific sensing with fluorescent or chemiluminiscent detection has also been developed for ricin, viscumin, SEB, tetanus, and diphtheria toxins, and lethal factor of anthrax [12].

Nanogram sensitivity with excellent detection time of 15 minutes was reported for cholera toxin (8 ng/mL), SEB (4 ng/mL), ricin (10 ng/mL), and *Bacillus globigii* (6.2 x 10^4 CFU/mL) by using flow immunofluorescent assay with a scanning confocal microscope equipped with a 635-nm laser as a detector [13]. Fluorescent labels have been tested for multispecific sensing. Thus, using fluorescence resonance energy transfer (FRET) microspheres, conjugated to antibodies as the fluorescent reporters, a close detection limit was reported for two simulants: 1 ng/mL for ricin (A chain) and 5 ng/mL for crude spore preparation of *Bacillus globigii* [14]. Using lanthanide (Eu^{3+})-labeled detector antibody, which generated a fluorescent signal through dissociation of the Europium from the antibody, immunoassay on commercially available DELFIA time-resolved fluorometry system (Perkin-Elmer) with enhanced sensitivity can be completed in 2 hours for *F. tularensis*, Botulinum toxins A&B, and SEB with a detection range for the toxins between 4 and 20 pg/mL in buffer [15].

Using quantum dots, highly luminescent semiconductor nanocrystals conjugated to antibodies, four toxins (cholera toxin, ricin, shiga-like toxin 1, and SEB) were detected and quantified simultaneously in a multiplexed sandwich immunoassay [16]. Various optical and electrochemical approaches, or their combination, were incorporated into the multispecific sensing, resulting in improved sensitivity and speed. Thus, as early as in 1995, an immunomagnetic electrochemiluminescence approach, utilizing magnetic beads-conjugated capture antibodies and [Ru(bpy)$_3$$^{2+}$]-labelled reporter antibodies, has been reported [17]. The 40-minute sensing protocol, registering ectrochemiluminescence, evoked from the [Ru(bpy)$_3$$^{2+}$]-tagged reporter antibodies by application of an electrical potential, showed femtogram sensitivity for botulinum toxinA, cholera beta subunit, ricin, and SEB and was able to detect 100 spores of *B. anthracis*.

A bidiffractive grating immuno biosensor was reported for four separate antigens: SEB, ricin, botulinum toxin, and *F. tularensis* [18]. An optical flow-cell multichannel immunosensor, based on solid phase enzyme-linked immunosorbent assay (ELISA) with peroxidase-labeled antibodies-mediated detection, was developed for sensing toxins (SEB), model bacteria, and virus particles. The sensing and signal-transducing component of the sensor consisted of a light-emitting diode and a photodetector [19]. An electric-field-driven fluoroimmunoassay was developed on an active electronic microchip, where 1-minute application of an electric field was used to accelerate detection of SEB and cholera toxin B [20].

A silicon-based light-addressable potentiometric biowarfare sensor utilizing three independent approaches, sandwich immunoassays, nucleic acid hybridization assays, and enzyme inhibition assays, was developed [21]. Immunoassay sensitivity was 5 pg/mL for protein (SEB), 2 ng/mL for virus (Newcastle disease virus), and 20 ng/mL for bacteria (Brucella melitensis); while in gene probe assay format sensitivity of 0.30 fmol (1.8×10^8 copies per 60-µl) of single-stranded target DNA was achieved. Excellent sensitivity was achieved in redox enzyme-amplified electrochemical sandwich-type immunoassay using active individually addressable microelectrodes on the chip, CombiMatrix's VLSI array [22]. Capture antibodies were tagged with oligonucleotides and chaperoned to individual electrodes with immobilized complementary oligonucleotides by the self-assembly process, reporter antibodies were HRP-labeled. Detection limits were 300 pg/mL (50 µL sample, 300 amoL) for ricin, single spores for *B. globigii*, 10^6 pfu for M13 bacteriophage.

A sophisticated two-step approach, with surface plasmon resonance detecting the binding of the toxin(s) to antibodies immobilized on a surface of a sensor chip, followed by identification of the bound toxin(s) by matrix-assisted laser desorption/ionization time-of-flight mass spectrometry has been developed for food analysis [44]. SEB was readily detected in milk and mushroom samples on multiaffinity sensor chip surfaces at 1 ng/mL simultaneously with toxic-shock syndrome toxin-1. Specificity of the immunochemical sensing is defined by the specificity of the used antibody and can be quite high. Sensitivity of immunochemical detection typically lay in the picogramm/femtomole range, and individual sensors generally have lower detection limits compared to multispecific sensors. Contrary to the pathogens of bacterial and viral nature, which can be detected with much higher sensitivity with genetic sensing, discussed in the next section, toxins cannot be detected genetically. Detection limits for toxins are summarized in Table 13.1.

13.3.2 APTAMER-BASED SENSORS

Aptamers are short DNA or RNA oligonucleotides, specifically selected from oligonucleotide library for binding of a selected protein. The sequence of the aptamer determines its secondary structure, which is responsible for highly specific *binding with the selected protein* due to hydrophobic interactions and hydrogen bonding.

Aptamers can substitute antibodies or can be used together with antibodies in sandwich-type protein-sensing assays. Aptamers offer several advantages compared to antibodies, such as their small size, resulting in higher density of immobilized binding sites, better stability and heat resistance, lower cost, and, most importantly, ease of renewal and multiple reusability. Since aptamer-based sensing is a relatively new area, their general properties and capabilities will be briefly overviewed before proceeding to biodefense-specific aptamers.

Aptamer-based protein sensors have been developed since 1998. Thus, 0.7 amol of thrombin in 140-pL sample (0.5 µM concentration) was detected in a single-aptamer sensor through binding to a fluorescently labeled DNA aptamer by evanescent-wave-induced fluorescence anisotropy in less than 10 minutes [45]. Performance of DNA

TABLE 13.1
Individual Toxin Sensing Detection Limits

Toxin	Detection Limit	Method and Detection Time	References
Ricin	100 pg/mL	Immunochromatography, colloidal gold, silver enhancement; 10 minutes	[23]
		Evanescent wave fiber-optic immunosensor	[24]
		Colorimetric and chemiluminescence ELISA	[25]
BoNT/B	50 pg/mL	Immunochromatography, colloidal gold, silver enhancement; 10 minutes	[26]
	15 pg/mL	Sandwich hybrid receptor-immunoassay, (with trisialoganglioside GT1b-liposomes); 20 minutes	[27]
	8 nM/L	Micromechanosensor, cleavage of synaptobrevin 2 substrate; 15 minutes	[28]
Cholera toxin	8 fmol	Heterogeneous immunoassay on supported bilayer membranes in a microfluidic device; 25 minutes	[29]
	0.5 µg/mL	Lipid monolayers on quartz crystal microbalance	[30]
	1.0×10^{-8} M/L	Differential pulse voltammetry on redox-functionalized diacetylene lipids containing cell surface ganglioside GM1	[31]
	10 fg/ml (8 zmol in 70 µL)	Sandwich immunoassay with GM1-liposomes; 20 minutes	[32]
	1.0×10^{-13} M	Microgravimetric quartz crystal microbalance transduction	[33]
	200 ng/mL	Fluororescently labeled toxin captured and immobilized on the surface of the optical waveguide GM1; 20 minutes	[34]
SEB	0.5 ng/mL	Magnetoelastic immunosensor	[35]
	2.5 µg/mL	Piezoelectric immunosensor	[36]
	5 ng/mL	Wavelength modulation-based SPR biosensor	[37]
	ng/mL	Fiber optic surface plasmon resonance; 10 minutes	[38]
	femtomol	Surface plasmon resonance with amplification	[39]
	0.4 ng/mL	Impedance electrochemical immunosensor	[40]
	100 pg	Fluorescent magneto-immunoassay; 30 minutes	[41]
SEA, SEB SEC2, SED, SEE	0.5 ng/g 1 ng/g	Automated VIDAS SET2 bioMerieux	[42]
Shiga toxin	femtomol	Flow cytometry with immunoassay; 3 hours	[43]

aptamers was found equivalent to that of antibodies in a quartz crystal biosensor, used to detect free unlabeled proteins in real time in various complex protein mixes. Aptamers were equivalent to antibodies in terms of specificity and sensitivity (both receptor types selectively detected 0.5 nmol/L of IgE, used as analyte) and exceeding performance of antibodies in regeneration and overall stability [46]. Using sandwich approaches with 2 different aptamers, immobilized on the electrode capture aptamer and labeled with pyrroquinoline quinone glucose dehydrogenase reporter aptamer, thrombin was measured electrochemically at 10 nM concentration [47].

Aptamers possess a number of additional useful properties. Thus, an ability to switch between binding a complementary oligonucleotide and a protein target was used in a real-time sensing approach: fluorophore-labeled DNA aptamer formed a duplex with the oligonucleotide, modified with a quenching moiety. Upon target introduction, the aptamer bound the target complex with a resulting fluorescent signal [48].

Aptamers undergo significant conformational transition from unfolded to a folded structure upon binding to an analyte. The conformational change of the negatively charged oligonucleotide DNA aptamer was detected by adding a water-soluble, cationic polythiophene derivative, which transduced the new complex formation into an optical (colorimetric or fluorometric) signal without any labeling of the probe or of the target. This simple and rapid methodology has enabled the detection of human thrombin in the femtomole range [49].

Aptamers can be designed to carry multiple functionalities. For example, in a modular aptamer approach, the protein recognition event is transduced into fluorescence changes through allosteric regulation of noncovalent interactions with a fluorophore. Such modular RNA aptamer includes a reporting domain, which signals the binding event of an analyte through binding to a fluorophore; a recognition domain, which binds the analyte; and a communication module, which serves as a conduit between recognition and signaling domains. The approach was tested on several analytes (adenosine triphosphate (ATP), flavin mononucleotide (FMN), and theophylline) [50].

Aptamers have a strong potential for *multiplex sensing* of proteins. Thus, a chip-based biosensor was developed for specific detection and quantification of cancer-associated proteins in complex biological mixtures using immobilized fluorescently labeled DNA and RNA aptamers. Fluorescence polarization anisotropy was used for solid- and solution-phase measurements of target protein binding [51].

Aptamers, specifically binding to biowarfare agents, are being actively researched. Thus, there are already known and tested DNA aptamers to anthrax spores [52], to cholera toxin and SEB [53], to tularemia bacteria [54], and to Shiga toxin [55]. These aptamers have been already tested for detection of relevant bacteria and toxins, and their performance was found to be equal to or exceeding that of antibodies in similar detection formats. Thus, DNA aptamers specific to nonpathogenic Sterne strain *B. anthracis* spores have been used for detection of spore components with a dynamic range from 10 to 6×10^6 anthrax spores [52]. Biotinylated aptamers specific to cholera whole toxin and SEB were bound to streptavidin-coated magnetic beads and used for the detection of ruthenium trisbypyridine [$Ru(bpy)_3^{2+}$]-labeled cholera toxin and SEB by electrochemiluminescence in the low nanogram to low picogram ranges [53]. DNA aptamers specific to *F. tularensis* are tested in ELISA-like and Western dot-blot format. Cocktail of aptamers was used both to capture tularemia bacteria and to detect it through the complex of biotin-labeled aptamers with streptavidin conjugated horseradish peroxidase (HRP) in ELISA format and through streptavidin conjugated alkaline phosphatase (AP) in dot-blot format. The performance of the aptamers was equal to or exceeding the performance of antibodies in similar detection formats. A linear detection range for aptamer-ELISA was 250 ng–2 µg [54]. Performance of Shiga toxin-specific aptamer, conjugated to fluorescent quantum dots, was found to be equal to or exceeding that of commercially available fluorescent antibody [55]. DNA aptamers specific to the spores of a different strain of anthrax, vaccine strain A. 16R *B. anthracis*, have been selected. Hybrid sandwich assay was developed for detecting anthrax spores by using biotinylated aptamers and antispore antibodies [56].

RNA aptamer specifically binding and inhibiting the activity of the catalytic ricin A-chain (RTA) has been selected [57]. Initially 80 nucleotides long, it has been shortened to 31-nucleotide aptamer that contained all sequences and structures necessary for interacting with RTA. Inhibition properties of this aptamer exceeded known ricin inhibitors. This aptamer along with others was used to develop a multispecific sensor. Aptamers were immobilized on beads, introduced into micromachined chips on the electronic tongue sensor array, and used for the detection and quantification of proteins. The lowest detected ricin concentration using this sensor was 1 pM (320 ng/mL) [58].

Aptamer-based multispecific biowarfare sensors are under development in a number of laboratories, but nothing has been published yet. An aptamer-based sensor array has been reported using chip-based microsphere array format ("electronic tongue") using lysozyme- and ricin-specific aptamers [58]. Overall, aptamer-based sensing approaches have good perspectives and one day they might substitute or become as popular as immunochemical sensing.

Protein-based sensing is fast and relatively simple; however, it is not as sensitive as genetic sensing which is discussed in Section 13.4.

13.4 GENETIC SENSING

Genetic (DNA or RNA) sensing methods work through registration of a hybridization event, which is a duplex formation between immobilized oligonucleotides of the sensor and a complementary target DNA of pathogenic bacteria or virus in the sample. Genetic sensing technologies became sensitive, automated, and miniaturized with the advancement of genomics work. Obviously, this method of detection requires preparation of the genetic material out of the sample with possible pathogen, which unavoidably lengthens and complicates the sensing procedure. DNA/RNA sensors have a number of advantages compared to immunosensors for detection of multiple biowarfare agents. Oligonucleotide probes (which will specifically recognize DNA or RNA in any given bacteria or virus) are cheap and easy to make, and they are stable in a wider range of conditions, compared to antibodies (used to recognize protein components of bacteria or virus in immunosensing). Detailed comparison of immunosensing to nucleic acid-based detection was performed by Iqbal et al. [59], concluding that genetic sensing is more specific and sensitive than immunological-based detection, while the latter is faster and more robust.

Specificity of genetic sensing is determined by the uniqueness of the selected genetic target sequence. Cross-reactivity between closely related species might become a problem if the target sequence is not selected carefully or if the stringency of hybridization reaction is not selected right. However, with the right conditions point mutations are readily determined and the specificity of genetic sensing is excellent. Sensitivity of this method is very high, down to 1 CFU (colony-forming unit) for bacteria or few viral particles. The major drawback of genetic sensing (in addition to the need for genetic material purification) is the need for amplification of the genetic material, which is typically done by PCR amplification of DNA or RNA. PCR amplification protocols are getting faster; thus, in real-time PCR, the products of PCR reactions are registered by fluorescence, using labeled probes that hybridize to the amplicons as they are formed. PCR equipment is getting more portable (HANAA, Handheld Advanced Nucleic Acid Analyzer), developed by Livermore National Laboratory in 1999, battery-powered, handheld PCR instrument, with theoretically a possible detection time as short as 7 minutes [3]. Some specific examples of application of advanced DNA-sensing technology to biodefense are discussed next.

13.4.1 EXISTING GENETIC SENSING METHODS FOR BIOWARFARE AGENTS

(a) *Smallpox* DNA can be detected in 3 hours, by LightCycler PCR, with high sensitivity of 5–10 copies per sample. (b) A 5' nuclease PCR assay targeting the plasminogen activator gene (*pla*) of *plague* was showing excellent sensitivity of 2.1×10^5 copies of the *pla* target or 1.6 pg of total cell DNA. The assay detected *Y. pestis* in experimentally infected fleas and monkey blood and oropharyngeal swabs. The signal is detected and interpreted by the ABI 7700 Sequence Detector (Applied Biosystems, Foster City, Calif.), a combination of thermal cycler, laser, and detection per software system [60]. (c) PCR-based assay, with FRET registration, was used to detect ciprofloxacin-resistant (Cp^r) mutants of plague. It was possible to distinguish the wild type from antibiotic-resistant type, and detect approximately 10 pg of DNA (or 4 CFU, colony forming units) of wild-type *Y. pestis* KIM 5 or Cp^r mutants in crude lysates [61].

Bacillus anthracis causes disease through the means of two different toxins and antiphagocytic capsule which are encoded on two virulence plasmids, pXO1 and pXO2, and their genes are most often used to detect anthrax. (d) Using real-time PCR with primers specific to pag and cap A genes on pXO1 and pXO2 plasmids, along with sap gene as a chromosomal marker, anthrax spores were detected from soil samples with a detection limit of 100 ng of DNA (or 10 spores) per ml [62]. (e) It is possible to detect 1 spore of anthrax in air samples with Light Cycler [63]. (f) with nested real-time PCR, 1 spore in soil samples [64]. (g) In another approach, RNA was extracted from lysed cells, amplified using nucleic acid sequence-based amplification (NASBA), and rapidly identified through the detection of transcription of atxA gene, which is essential for the bacteria to produce the toxin proteins and therefore to cause disease. The process takes 12 hours for the detection of one viable *B. anthracis* spore or 1.5 fmol of target mRNA. This approach is based on an oligo-nucleotide sandwich hybridization assay format and uses a membrane flow-through system. Signal amplification is provided when the target sequence hybridizes to a second oligonucleotide probe that has been coupled to dye-encapsulating liposomes. The dye in the liposomes provides a signal that can be read visually or quantified with a handheld reflectometer [65]. (h) A number of B. anthracis mutants with high-level resistance to resistance to ciprofloxacin have been obtained and genetically characterized. All of them had at least one mutation in gyrA, which was enough to confer resistance [66]. Additional mutations in parC and gyrB, present simul-taneously with gyrA mutations, were found in higher level ciprofloxacin-resistant isolates. A recent review [67] provides the comparative analysis of existing anthrax-detecting methods up to date.

(i) PCR-based detection of *tularemia* is based on identification of the outer membrane protein (Fop A) and *tul4* genes [68,69]. (j) The best detection limit was achieved by using TaqMan 5' hydrolysis fluorogenic PCR and handheld BioSeeq thermocycler, 50 fg of DNA (approximately 25 genome equivalents) can be detected in infected tissues in 4 hours [69].

Another advantage of genetic sensing is the ease of multispecific sensing, using array approaches. This is very beneficial, since one can use multiple probes for each single pathogen, for example, a genetic probe, plus a probe for pathogenecity (if carried by a plasmid, like in case of anthrax), plus probes for antibiotic-resistance-conferring mutations, since a number of biowarfare pathogens are already known to carry some antibiotic resistance and can be further genetically engineered in this direction.

Examples of multispecific biowarfare sensors include MAGIchip (microarray of gel-immobilized compounds), used to detect DNA of variola major virus in 6 hours and distinguish it from other members of the family of orthopox viruses. Sensitivity in the range 100–1000 DNA copies/sample was achieved and analysis was com-pleted in 6 hours (the method includes PCR amplification/labeling of the target DNA sequence and fluorescence detection) [70]; BARC (Bead Array Counter), which uses DNA hybridization, magnetic microbeads, and giant magnetoresistive sensor chips and was reported to detect DNA of anthrax, plague, tularemia, cholera, botulinum, brucella, campylobacter, and vaccinia [71].

Automated PCR-based methods have been developed for a number of possible biowarfare pathogens. Thus, an array of PCR microchips for rapid, parallel testing of samples for pathogenic microbes was developed (Advanced Nucleic Acid Analyzer, ANAA), utilizing 10 silicon reaction chambers with thin-film resistive heaters and solid-state optics was used to detect several bacterial pathogens, in as little as 16 minutes, with detection limits of 10^5–10^7 organisms/L (10^2–10^4 organisms/mL) [72]. In addition to that, DNA-based sensing adds a possibility to conduct a more extensive genetic analysis of a given pathogen, following a real-time PCR detection, similar to what was done for *B. anthracis, Y. pestis,* and *F. tularensis* [72]. Recent reviews examine in detail the applications of DNA array methodology and a variety of PCR-based techniques to biodefense [73–75].

However, high cost along with the need for trained personnel and clean environment limits the use of PCR-based genetic sensing in field conditions. Avoiding PCR amplification could immensely simplify field-testing and will make DNA biosensors more applicable to field use. PCR-free genosensing approaches are getting increasing attention lately both for clinical and biodefense applications. Many DNA detection assays have been developed using radioactive labels, molecular fluorophores, and chemiluminescence schemes. Recent application of nanostructure-based labels for biodefense applications has been reviewed [76].

13.4.2 Perspectives on Genetic Sensing: Electrochemical Detection

Electrochemical methods of hybridization detection present an interesting alternative to radioactive, chemi- and fluorescent hybridization detection and are described here in more detail since they have not been reviewed in the literature. They have a high potential for automation and miniaturization and are simple to use and require only basic electrochemical equipment. Electrochemical methods can be separated into *indirect* (using electrochemical labels and hybridization markers) and *direct* (using the ability of DNA to contribute to conductivity of a conductive polymer, which changes upon hybridization). It is worth noting that most of the reported work so far was performed with oligonucleotide synthetic targets, and some of them were related to biowarfare pathogens. However, further intensive studies are required for the described approaches to be used for real-life sensing of long genomic DNA or RNA of biowarfare pathogens. Indirect methods of electrochemical hybridization detection generally have better detection limits but take longer and require more steps, compared to direct methods.

They include the following:

(a) Using electroactive enzyme as a label such as HRP or AP: Detection of electrical current, generated by HRP was adapted to DNA sensing with a detection limit as low as 10 fM [77]. Using an HRP-labeled detector probe, it was possible to detect 3000 copies of 38-b.p. DNA oligonucleotide in 10 μl droplet (0.5 fM) [78]. It is possible to substitute HRP by bilirubin oxidase (which catalyzes the reduction of ambient O_2 to water and obviates the need for adding H_2O_2 without sensitivity reduction) [79]. When AP-conjugated detector probe was used to precipitate an insoluble insulating product, the reported detection limit was 1.2 pM [80]. Using AP-labeled detector probe to generate the electroactive label, it was possible to measure nM concentrations of target [81]; or to achieve a 20-fM detection limit after ferrocenyl enhancement [82]. Enzyme dendritic structures can be used for sensitivity enhancement [83], and different enzymes for multispecific measurements. Thus, using AP and β-galactosidase, it was possible to differentiate the signals of 2 DNA targets in connection to chronopotentiometric measurements of their electroactive phenol and alpha-naphthol products with fM detection limits [84].

(b) Using electroactive labels for oligonucleotides, modified with *transition metal complexes* (ruthenium, osmium, iron, rhodium, and copper), one can detect DNA electrochemically with cyclic voltammetry. Popular use of ferrocene-modified oligonucleotides as detector probes resulted in femtomole detection levels [85]. Utilizing stem-loop structure of ferrocene-tagged capture probe, which undergoes conformational change upon hybridization and converts from loop to stem, distancing the ferrocene tag from the electrode, it was possible to achieve a 5 fM detection limit [86]. Using dipyridophenazine and phenantroline-dione complexes of osmium, it was possible to achieve a detection limit of few picomoles [87]. By electrocatalytic detection, with $Ru(NH_3)_6^{3+}$ as an electroactive redox DNA label and $Fe(CN)_6^{3-}$ as an oxidizer, it was possible to monitor DNA hybridization at nanomolar levels [88]. With a combination approach, labeling target DNA with modified nucleobases (7,8-dihydro-8-oxoguanine or 5-aminouridine), and registering their oxidation in the presence of $Os(bpy)_3^{2+}$, it was possible to register 40 aM of the desired target [89].

(c) Nanoparticles as electroactive labels are becoming increasingly popular. By labeling with silver nanoparticles, 1 pM detection limit was achieved through determination of solubilized silver ions by anodic stripping voltammetry [90]. Using a detector probe, labeled with Cu-Au alloy nanoparticle, with release of the copper metal atoms by oxidative metal dissolution and determination of the solubilized Cu^{2+} ions, the detection limit of 5 pM/L was achieved [91]. Using cadmium sulfide (CdS) nanoclusters as the oligonucleotide labeling tag, with further dissolution of the CdS nanoclusters and determination of the dissolved cadmium ions, it was possible to measure 0.2 pM/L of target DNA [92]. An elegant approach, using three different nanoparticles (zinc sulfide, cadmium sulfide, and lead sulfide) have been used to differentiate the signals of three DNA targets by stripping voltammetry of heavy metals at femtomolar amounts of DNA [93]. Gold nanoparticles, conjugated to DNA, allowed down to 0.5 pM limit of detection [94]. Detector probes labeled with gold nanoparticles, facilitating further silver deposition, lead to measurable conductivity changes with target DNA concentration (anthrax lethal factor sequence) of 500 femtomoles [95]. The same group (Mirkin's group) achieved even better results with the bio-bar-code amplification (BCA) approach, which utilizes oligonucleotide-modified nanoparticles for signal amplification. They were able to detect 10 copies in the 30 μL sample (5×10^{-19} M sensitivity) using scattered light measurement [96].

Another way to detect a hybridization event electrochemically is to use a hybridization marker; a compound, which binds selectively, double-stranded (ds) or single-stranded (ss) DNA. Intercalation agents selectively bind dsDNA with high affinity ($K > 10^6\,M^{-1}$) by intercalating its aromatic heterocyclic groups between the base pairs of dsDNA. Using transition metal complexes and differential pulse voltammetry for detection, it was possible to achieve a pM detection range with Os(II)-dipyridophenazine, with linear calibration 1–100 pM [87]; nanomolar range with tris(1,10-phenanthroline)cobalt [97]; micromolar range with ferrocenyl naphthalene diimide [98]; and with hemin, iron complex of porphyrin [99]. The use of electrocatalytic amperometry was associated with further sensitivity increase (down to 600 fM) when registering DNA hybridization through electrocatalytic intercalator including Os complex [100]. ssDNA markers, such as OsO_4, can also be incorporated into DNA-sensing [101].

Direct electrochemical detection of DNA hybridization allows omitting the step of pre- or posthybridization labeling or posthybridization incubation with hybridization markers. Several electrochemical properties of DNA can be used for DNA sensing. For example, DNA hybridization can be monitored through the oxidation signal of guanine in the target DNA, when electro-inactive inosine-substituted capture probes are used [102,103]. Using dsDNA's ability to transport charge along nucleotide stacking, the perturbation of the double helix π-stack, introduced by a mismatched nucleotide, can be detected through reduction electron flow by measuring the attenuation of the charge transfer [104].

Direct registration of DNA hybridization advanced a lot with the use of electroactive films of conductive polymers as electrodes. Conductive polymers, which consist of conjugated backbones that are easily oxidized or reduced (doped) with a concomitant increase or decrease in conductivity, with each polymer having its own redox characteristics.

DNA is electroactive and can serve as a dopant due to the negative charge of phosphates, contributing to the polymer conductivity. Oligonucleotide probes can be incorporated into growing films of a conducting polymer [105–109], or the surface of the polymer electrode can be derivatized with covalent attachment of oligonucleotides [110–112]. Hybridization with a complementary DNA target causes changes in conductivity of the electrode, which can be measured by amperometry [105,107], impedance spectroscopy [110], cyclic voltammetry [111], differential pulse voltammetry [112] or photocurrent spectroscopy [106]. The detection limits of most direct DNA-sensing approaches are in the micromolar range, however, the latest publications on direct DNA sensing report lower detection limits: Using the polythiophene derivative poly (thiophen-3-yl-acetic acid 1,3-dioxo-1,3-dihydro-isoindol-2-yl ester), chemically modified with an oligonucleotide probe, it was possible to measure target DNA with 1 nM detection limit by monitoring changes upon DNA hybridization (carried out for 3 hours) by cyclic voltammetry [111]. Besides, it was possible to monitor DNA hybridization in real time through capacitance measurement, with a detection limit of 0.5 nM and rapid response (seconds) upon addition of the complementary strand, using an oligonucleotide-modified silicon chip [113].

While electrochemical genosensing might become very useful for biowarfare detection in the future, currently available and most reliable systems combine both protein- and DNA-sensing approaches, and have been reported to be used for biodefense sensing. Thus, Nanogen developed stacked "lab-on-a-chip" microarray for performing automated electric-field-driven immunoassays and DNA hybridization assays [114]. A fully automated, autonomous pathogen detection system (APDS) was created for continuous monitoring the air for biological threat agents (bacteria, viruses, and toxins). The APDS performs continuous aerosol collection, sample preparation, and detection using multiplexed immunoassay followed by confirmatory PCR using real-time TaqMan assays for aerosolized *B. anthracis*, *Y. pestis* and botulinum toxin [115]. By coupling highly selective antibody- and DNA-based assays, the probability of a false positive is extremely low.

REFERENCES

1. DaSilva E. J., Biological warfare, bioterrorism, biodefence and the biological and toxin weapons convention Electronic, *Journal of Biotechnology*, 2(3), 99–129, 1999.
2. Alibek K. and Handelman S., *Biohazard* (New York: Dell Publishing, 2000).
3. Miller J., Engelberg S., and Broad W., *Germs: Biological Weapons and America's Secret War* (New York: Simon & Shuster, 2001).
4. Sapsford K. E., Shubin Y. S., Delehanty J. B., Golden J. P., Taitt C. R., Shriver-Lake L. C., and Ligler F. S., Fluorescence-based array biosensors for detection of biohazards, *J. Appl. Microbiol.*, 96(1), 47–58, 2004.

5. Ligler F. S., Taitt C. R., Shriver-Lake L. C., Sapsford K. E., Shubin Y., and Golden J. P., Array biosensor for detection of toxins, *Anal. Bioanal. Chem.*, 377(3), 469–477, 2003.

6. Wadkins R. M., Golden J. P., Pritsiolas L. M., and Ligler F. S., Detection of multiple toxic agents using a planar array immunosensor, *Biosens. Bioelectron.*, 13(3–4), 407–415, 1998.

7. Rowe C. A., Tender L. M., Feldstein M. J., Golden J. P., Scruggs S. B., MacCraith B. D., Cras J. J., and Ligler F. S., Array biosensor for simultaneous identification of bacterial, viral, and protein analytes, *Anal Chem.*, 71(17), 3846–3852, 1999.

8. Rowe-Taitt C. A., Cras J. J., Patterson C. H., Golden J. P., and Ligler F. S., A ganglioside-based assay for cholera toxin using an array biosensor, *Anal. Biochem.*, 281(1), 123–133, 2000.

9. Taitt C. R., Anderson G. P., Lingerfelt B. M., Feldstein M. J., and Ligler F. S., Nine-analyte detection using an array-based biosensor, *Anal. Chem.*, 74(23), 6114–6120, 2002.

10. Taitt C. R., Golden J. P., Shubin Y. S., Shriver-Lake L. C., Sapsford K. E., Rasooly A., and Ligler F. S., A portable array biosensor for detecting multiple analytes in complex samples, *Microb. Ecol.*, 47(2), 175–185, 2004.

11. Rucker V. C., Havenstrite K. L., Herr A. E., Antibody microarrays for native toxin detection, *Anal. Biochem.*, 339(2), 262–270, 2005.

12. Rubina A. Y., Dyukova V. I., Dementieva E. I., Stomakhin A. A., Nesmeyanov V. A., Grishin E. V., and Zasedatelev A. S., Quantitative immunoassay of biotoxins on hydrogel-based protein microchips, *Anal. Biochem.*, 340(2), 317–329, 2005.

13. Delehanty J. B. and Ligler F. S., A microarray immunoassay for simultaneous detection of proteins and bacteria, *Anal. Chem.*, 74(21), 5681–5687, 2002.

14. Wang L., Cole K. D., Gaigalas A. K., and Zhang Y. Z., Fluorescent nanometer microspheres as a reporter for sensitive detection of simulants of biological threats using multiplexed suspension arrays, *Bioconjug. Chem.*, 16(1), 194–199, 2005.

15. Peruski A. H., Johnson L. H. 3rd, and Peruski L. F. Jr., Rapid and sensitive detection of biological warfare agents using time-resolved fluorescence assays, *J. Immunol. Methods.*, 263(1–2), 35–41, 2002.

16. Goldman E. R., Clapp A. R., Anderson G. P., Uyeda H. T., Mauro J. M., Medintz I. L., and Mattoussi H., Multiplexed toxin analysis using four colors of quantum dot fluororeagents, *Anal. Chem.*, 76(3), 684–688, 2004.

17. Gatto-Menking D. L., Yu H., Bruno J. G., Goode M. T., Miller M., and Zulich A. W., Sensitive detection of biotoxoids and bacterial spores using an immunomagnetic electrochemiluminescence sensor, *Biosens. Bioelectron.*, 10(6–7), 501–507, 1995.

18. O'Brien T., Johnson L. H. 3rd, Aldrich J. L., Allen S. G., Liang L. T., Plummer A. L., Krak S. J., and Boiarski A. A., The development of immunoassays to four biological threat agents in a bidiffractive grating biosensor, *Biosens. Bioelectron.*, 14(10–11), 815–828, 2000.

19. Koch S., Wolf H., Danapel C., and Feller K. A., Optical flow-cell multichannel immunosensor for the detection of biological warfare agents, *Biosens. Bioelectron.*, 14(10–11), 779–784, 2000.

20. Ewalt K. L., Haigis R. W., Rooney R., Ackley D., and Krihak M., Detection of biological toxins on an active electronic microchip, *Anal. Biochem.*, 289(2), 162–172, 2001.

21. Lee T. and Shim Y., Direct DNA hybridization detection based on the oligonucleotide-functionalized conductive polymer, *Anal. Chem.*, 73, 5629–5632, 2001.

22. Dill K., Montgomery D. D., Ghindilis A. L., Schwarzkopf K. R., Ragsdale S. R., and Oleinikov A. V., Immunoassays based on electrochemical detection using microelectrode arrays, *Biosens. Bioelectron.*, 20(4), 736–742, 2004.

23. Shyu R. H., Shyu H. F., Liu H. W., and Tang S. S., Colloidal gold-based immunochromatographic assay for detection of ricin, *Toxicon.*, 40(3), 255–258, 2005.
24. Narang U., Anderson G. P., Ligler F. S., and Burans J., Fiber optic-based biosensor for ricin, *Biosens. Bioelectron.*, 12(9–10), 937–945, 1997.
25. Poli M. A., Rivera V. R., Hewetson J. F., and Merrill G. A., Detection of ricin by colorimetric and chemiluminescence ELISA, *Toxicon.*, 32(11), 1371–1377, 1994.
26. Chiao D. J., Shyu R. H., Hu C. S., Chiang H. Y., and Tang S. S., Colloidal goldbased immunochromatographic assay for detection of botulinum neurotoxin type B, *J. Chromatogr. B. Analyt. Technol. Biomed. Life. Sci.*, 809(1), 37–41, 2004.
27. Ahn-Yoon S., DeCory T. R., Baeumner A. J., and Durst R. A., Ganglioside-liposome immunoassay for the ultrasensitive detection of cholera toxin, *Anal. Chem.*, 75(10), 2256–2261, 2003.
28. Liu W., Montana V., Chapman E. R., Mohideen U., and Parpura V., Botulinum toxin type B micromechanosensor, *Proc Natl Acad Sci U S A.*, 100(23), 13621–13625, 2003.
29. Phillips K. S. and Cheng Q., Microfluidic immunoassay for bacterial toxins with supported phospholipid bilayer membranes on poly(dimethylsiloxane), *Anal. Chem.*, 77(1), 327–334, 2005.
30. Stine R., Pishko M. V., and Schengrund C. L., Heat-stabilized glycosphingolipid films for biosensing applications, *Langmuir.*, 20(15), 6501–6506, 2004.
31. Cheng Q., Zhu S., Song J., and Zhang N., Functional lipid microstructures immobilized on a gold electrode for voltammetric biosensing of cholera toxin, *Analyst.*, 129(4), 309–314, 2004.
32. Ahn-Yoon S., DeCory T. R., and Durst R. A., Ganglioside-liposome immunoassay for the detection of botulinum toxin, *Anal. Bioanal. Chem.*, 378(1), 68–75, 2004.
33. Alfonta L., Willner I., Throckmorton D. J., and Singh A. K., Electrochemical and quartz crystal microbalance detection of the cholera toxin employing horseradish peroxidase and GM1-functionalized liposomes, *Anal. Chem.*, 73(21), 5287–5295, 2001.
34. Rowe-Taitt C. A., Hazzard J. W., Hoffman K. E., Cras J. J., Golden J. P., Ligler F. S., Simultaneous detection of six biohazardous agents using a planar waveguide array biosensor, *Biosens. Bioelectron.*, 15(11–12), 579–589, 2000.
35. Ruan C., Zeng K., Varghese O. K., and Grimes C. A., A staphylococcal enterotoxin B magnetoelastic immunosensor, *Biosens. Bioelectron.*, 20(3), 585–591, 2004.
36. Lin H. C. and Tsai W. C., Piezoelectric crystal immunosensor for the detection of staphylococcal enterotoxin B, *Biosens. Bioelectron.*, 18(12), 1479–1483, 2003.
37. Homola J., Dostalek J., Chen S., Rasooly A., Jiang S., and Yee S. S., Spectral surface plasmon resonance biosensor for detection of staphylococcal enterotoxin B in milk, *Int. J. Food. Microbiol.*, 75(1–2), 61–69, 2002.
38. Slavik R., Homola J., and Brynda E., A miniature fiber optic surface plasmon resonance sensor for fast detection of Staphylococcal enterotoxin B, *Biosens. Bioelectron.*, 17(6–7), 591–595, 2002.
39. Naimushin A. N., Soelberg S. D., Nguyen D. K., Dunlap L., Bartholomew D., Elkind J., Melendez J., and Furlong C. E., Detection of Staphylococcus aureus enterotoxin B at femtomolar levels with a miniature integrated two-channel surface plasmon resonance (SPR) sensor, *Biosens. Bioelectron.*, 17(6–7), 573–584, 2002.
40. DeSilva M. S., Zhang Y., Hesketh P. J., Maclay G. J., Gendel S. M., and Stetter J. R., Impedance based sensing of the specific binding reaction between staphylococcus enterotoxin B and its antibody on an ultra-thin platinum film, *Biosens. Bioelectron.*, 10, 675–682, 1995.

41. Alefantis T., Grewal P., Ashton J., Khan A. S., Valdes J. J., and Del Vecchio V. G., A rapid and sensitive magnetic bead-based immunoassay for the detection of staphylococcal enterotoxin B for high through put screening, *Mol. Cell. Probes.*, 18(6), 379–382, 2004.

42. Vernozy-Rozand C., Mazuy-Cruchaudet C., Bavai C., and Richard Y., Comparison of three immunological methods for detecting staphylococcal enterotoxins from food, *Lett. Appl. Microbiol.*, 39(6), 490–494, 2004.

43. Tazzari P. L., Ricci F., Carnicelli D., Caprioli A., Tozzi A. E., Rizzoni G., Conte R., and Brigotti M., Flow cytometry detection of Shiga toxins in the blood from children with hemolytic uremic syndrome, *Cytometry. B. Clin. Cytom.*, 61B(1), 40–44, 2004.

44. Nedelkov D., Rasooly A., and Nelson R. W., Multitoxin biosensor-mass spectrometry analysis, a new approach for rapid, real-time, sensitive analysis of staphylococcal toxins in food, *Int. J. Food. Microbiol.*, 60(1), 1–13, 2000.

45. Potyrailo R. A., Conrad R. C., Ellington A. D., and Hieftje G. M., Adapting selected nucleic acid ligands (aptamers) to biosensors, *Anal. Chem.*, 70(16), 3419–3425, 1998.

46. Liss M., Petersen B., Wolf H., and Prohaska E., An aptamer-based quartz crystal protein biosensor, *Anal. Chem.*, 74(17), 4488–4495, 2002.

47. Ikebukuro K., Kiyohara C., and Sode K., Novel electrochemical sensor system for protein using the aptamers in sandwich manner, *Biosens. Bioelectron.*, 20(10), 2168–2172, 2005.

48. Nutiu R. and Li Y., Structure-switching signaling aptamers, *J. Am. Chem. Soc.*, 23, 125(16), 4771–4778, 2003.

49. Ho H. A. and Leclerc M., Optical sensors based on hybrid aptamer/conjugated polymer complexes, *J. Am. Chem. Soc.*, 126(5), 1384–1387, 2004.

50. Stojanovic M. N. and Kolpashchikov D. M., Modular aptameric sensors, *J. Am. Chem. Soc.*, Aug, 126(30), 9266–9270, 2004.

51. McCauley T. G., Hamaguchi N., and Stanton M., Aptamer-based biosensor arrays for detection and quantification of biological macromolecules, *Anal. Biochem.*, 319(2), 244–250, 2003.

52. Bruno J. G. and Kiel J. L., In vitro selection of DNA aptamers to anthrax spores with electrochemiluminescence detection, *Biosens. Bioelectron.*, 14(5), 457–464, 1999.

53. Bruno J. G. and Kiel J. L., Use of magnetic beads in selection and detection of biotoxin aptamers by electrochemiluminescence and enzymatic methods, *Biotechniques*, 2002 Jan; 32(1), 178–180, 2002 and 32(1), 182–183, 2002.

54. Kiel J. L., Holwitt E. A., Vivekananda J., and Franz V., Elisa-like format for comparing DNA capture elements (aptamers) to antibody in diagnostic efficacy, *Scientific Conference on Chemical & Biological Defense Research 15–18 November*, Hunt Valley, Maryland, 2004.

55. Kiel J. L., Holwitt E. A., Parker J. E., Vivekananda J., and Franz V., Nanoparticle-labeled DNA Capture Elements for Detection and Identification of Biological Agents, in "Optically Based Biological and Chemical Sensing for Defence", ed. J. C. Carrano, A. Zukauskas, *Proceedings of SPIE*, Vol. 5617, SPIE, Bellingha, WA, 2004.

56. Zhen B., Song Y. J., Guo Z. B., Wang J., Zhang M. L., Yu S. Y., and Yang R. F., In vitro selection and affinity function of the aptamers to Bacillus anthracis spores by SELEX appears in Sheng Wu Hua Xue, Yu Sheng Wu Wu Li Xue Bao, *Acta. Biochim. Biophys. Sin. (Shanghai).*, 34(5), 635–642, 2002.

57. Hesselberth J. R., Miller D., Robertus J., and Ellington A. D., In vitro selection of RNA molecules that inhibit the activity of ricin A-chain, *J. Biol. Chem.*, 275(7), 4937–4942, 2000.

58. Kirby R., Cho E. J., Gehrke B., Bayer T., Park Y. S., Neikirk D. P., McDevitt J. T., and Ellington A. D., Aptamer-based sensor arrays for the detection and quantitation of proteins, *Anal. Chem.*, 76(14), 4066–4075, 2004.

59. Iqbal S. S., Mayo M. W., Bruno J. G., Bronk B. V., Batt C. A., and Chambers J. P., A review of molecular recognition technologies for detection of biological threat agents, *Biosens. Bioelectron.*, 15(11–12), 549–578, 2000.

60. Higgins J. A., Ezzell J., Hinnebusch B. J., Shipley M., Henchal E. A., and M. Sofi Ibrahim M. S., 5' Nuclease PCR Assay to Detect *Yersinia pestis*, *J. Clin. Microbiol.*, 36(8), 2284–2288, 1998.

61. Lindler L. E., Fan W., and Jahan N., Detection of ciprofloxacin-resistant *Yersinia pestis* by fluorogenic PCR using the light cycler, *J. Clin. Microbiol.*, 39(10), 3649–3655, 2001.

62. Ryu C., Lee K., Yoo C., Seong W. K., and Oh H. -B., Sensitive and rapid quantitative detection of Anthrax spores isolated from soil samples by real-time PCR, *Microbiol. Immunol.*, 47(10), 693–699, 2003.

63. Makino S. and Cheun H. I., Application of the real-time PCR for the detection of airborne microbial pathogens in reference to the anthrax spores, *J. Microbiol. Methods.*, 53(2), 141–147, 2003.

64. Cheun H. I., Makino S. I., Watarai M., Erdenebaatar J., Kawamoto K., and Uchida I., Rapid and effective detection of anthrax spores in soil by PCR, *J. Appl. Microbiol.*, 95(4), 728–733, 2003.

65. Hartley H. A. and Baeumner A. J., Biosensor for the specific detection of a single viable *B. anthracis* spore, *Anal. Bioanal. Chem.*, 376(3), 319–327, 2003.

66. Price L. B., Vogler A., Pearson T., Busch J. D., Schupp J. M., and Keim P., In vitro selection and characterization of Bacillus anthracis mutants with high-level resistance to ciprofloxacin, *Antimicrob. Agents. Chemother.*, 47(7), 2362–2365, 2003.

67. Edwards K. A., Clancy H. A., and Baeumner A. J., Bacillus anthracis, toxicology, epidemiology and current rapid-detection methods, *Anal. Bioanal. Chem.*, 11, 1–12, 2005.

68. Higgins J. A., Hubalek Z., Halouzka J., Elkins K. L., Sjostedt A., Michelle Shipley M., and Sofi Ibrahim M. S., Detection of francisella tularensis in infected mammals and Vectors using a probe-based polymerase chain reaction, *Am. J. Trop. Med. Hyg.*, 62(2), 310–318, 2000.

69. Emanuel P. A., Bell R., Dang J. L., McClanahan R., David J. C., Burgess R. J., Thompson J., Collins L., and Hadfield T., Detection of Francisella tularensis within infected mouse tissues by using a hand-held PCR thermocycler, *J. Clin. Microbiol.*, 41(2), 689–693, 2003.

70. Lapa S., Mikheev M., Shchelkunov S., Mikhailovich V., Sobolev A., Blinov V., Babkin I., et al., Species-level identification of orthopoxviruses with an oligonucleotide microchip, *J. Clin. Microbiol.*, 40(3), 753–757, 2002.

71. Edelstein R. L., Tamanaha C. R., Sheehan P. E., Miller M. M., Baselt D. R., Whitman L. J., and Colton R. J., The BARC biosensor applied to the detection of biological warfare agents, *Biosens. Bioelectron.*, 14, 805–813, 2000.

72. Belgrader P., Benett W., Hadley D., Long G., Mariella Jr. R., Milanovich F., Nasarabadi S., Nelson W., Richards J., and Stratton P., Rapid pathogen detection using a microchip PCR array instrument, *Clin. Chem.*, 44, 2191–2194, 1998.

73. Ivnitski D., O'Neil D. J., Gattuso A., Schlicht R., Calidonna M., and Fisher R., Nucleic acid approaches for detection and identification of biological warfare and infectious disease agents, *Biotechniques.*, 35(4), 862–869, 2003.

74. Hashsham S. A., Wick L. M., Rouillard J. M., Gulari E., and Tiedje J. M., Potential of DNA microarrays for developing parallel detection tools (PDTs) for microorganisms

relevant to biodefense and related research needs, *Biosens. Bioelectron.*, 20(4), 668–683, 2004.

75. Draghici S., Chen D., and Reifman J., Applications and challenges of DNA microarray technology in military medical research, *Mil. Med.*, 169(8), 654–659, 2004.

76. Mirkin C., Thaxton S., and Rosi N. L., Nanostructures in biodefense and molecular diagnostics, *Expert. Rev. Mol. Diagn.*, 4(6), 749–751, 2004.

77. Jun Huang T., Liu M., Knight L. D., Grody W. W., Miller J. F., and Ho C., An electrochemical detection scheme for identification of single nucleotide polymorphisms using hairpin-forming probes, *Nucleic. Acids. Res.*, 30(12), e55, 2002.

78. Zhang Y., Kim H. H., Heller A., Enzyme-amplified amperometric detection of 3000 copies of DNA in a 10-microL droplet at 0.5 fM concentration, *Anal. Chem.*, 75(13), 3267–3269, 2003.

79. Zhang Y., Pothukuchy A., Shin W., Kim Y., and Heller A., Detection of approximately 10(3) copies of DNA by an electrochemical enzyme-amplified sandwich assay with ambient O(2) as the substrate. *Anal Chem.*, 76(14), 4093–4097, 2004.

80. Lucarelli F., Marrazza G., and Mascini M., Enzyme-based impedimetric detection of PCR products using oligonucleotide-modified screen-printed gold electrodes, *Biosens. Bioelectron.*, 20(10), 2001–2009, 2005.

81. Carpini G., Lucarelli F., Marrazza G., and Mascini M., Oligonucleotide-modified screen-printed gold electrodes for enzyme-amplified sensing of nucleic acids, *Biosens. Bioelectron.*, 20(2), 167–175, 2004.

82. Kim E., Kim K., Yang H., Kim Y. T., and Kwak J., Enzyme-amplified electrochemical detection of DNA using electrocatalysis of ferrocenyl-tethered dendrimer, *Anal. Chem.*, 75(21), 5665–5672, 2003.

83. Dominguez E., Rincon O., and Narvaez A., Electrochemical DNA sensors based on enzyme dendritic architectures, an approach for enhanced sensitivity, *Anal. Chem.*, 76(11), 3132–3138, 2004.

84. Wang J., Kawde A. N., Musameh M., and Rivas G., Dual enzyme electrochemical coding for detecting DNA hybridization, *Analyst.*, 127(10), 1279–1282, 2002.

85. Wlasoff W. A. and King G. C., Ferrocene conjugates of dUTP for enzymatic redox labelling of DNA, *Nucleic. Acids. Res.*, 30(12), e58, 2002.

86. Fan C., Plaxco K. W., and Heeger A. J., Electrochemical interrogation of conformational changes as a reagentless method for the sequence-specific detection of DNA, *Proc Natl Acad Sci USA.*, 100(16), 9134–9137, 2003.

87. Maruyama K., Mishima Y., Minagawa K., and Motonaka J., DNA sensor with a dipyridophenazine complex of osmium (II) as an electrochemical probe, *Anal. Chem.*, 74(15), 3698–3703, 2002.

88. Lapierre M. A., O'Keefe M., Taft B. J., and Kelley S. O., Electrocatalytic detection of pathogenic DNA sequences and antibiotic resistance markers, *Anal. Chem.*, 75(22), 6327–6333, 2003.

89. Gore M. R., Szalai V. A., Ropp P. A., Yang I. V., Silverman J. S., and Thorp H. H., Detection of attomole quantities of DNA targets on gold microelectrodes by electrocatalytic nucleobase oxidation, *Anal. Chem.*, 75(23), 6586–6592, 2003.

90. Cai H., Xu Y., Zhu N., He P., and Fang Y., An electrochemical DNA hybridization detection assay based on silver nanoparticle label, *Analyst.*, 127, 803–808, 2002.

91. Cai H., Zhu N., Jiang Y., He P., and Fang Y., Cu & Au alloy nanoparticle as oligonucleotides labels for electrochemical stripping detection of DNA hybridization, *Biosens. Bioelectron.*, 18(11), 1311–1319, 2003.

92. Zhu N., Zhang A., He P., and Fang Y., Cadmium sulfide nanocluster-based electrochemical stripping detection of DNA hybridization, *Analyst.*, 128(3), 260–264, 2003.

93. Wang J., Liu G., and Merkoci A., Electrochemical coding technology for simultaneous detection of multiple DNA targets, *J. Am. Chem. Soc.*, 125(11), 3214–3215, 2003.

94. Zhang Z. L., Pang D. W., Yuan H., Cai R. X., and Abruna H. D., Electrochemical DNA sensing based on gold nanoparticle amplification, *Anal. Bioanal. Chem.*, 381(4), 833–838, 2005.

95. Park S.-J., Taton T. A., and Mirkin C. A., Array-based electrical detection of DNA with nanoparticle probes, *Science*, 295, 1503–1506, 2002.

96. Nam J. M. and Stoeva S. I., Mirkin C. A., Bio-bar-code-based DNA detection with PCR-like sensitivity, *J. Am. Chem. Soc.*, 126(19), 5932–5933, 2004.

97. Wang J., Cai X., Rivas G., Shiraishi H., Farias P. A. M., and Dontha N., DNA electrochemical biosensor for the detection of short DNA sequences related to the Human Immunodeficiency Virus, *Anal. Biochem.*, 68, 2629–2634, 1996.

98. Yamashita K., Takagi M., Kondo H., and Takenaka S., Electrochemical detection of nucleic base mismatches with ferrocenyl naphthalene diimide, *Anal. Biochem.*, 306, 188–196, 2002.

99. Kara P., Ozkan D., Kerman K., Meric B., Erdem A., and Ozsoz M., DNA sensing on glassy carbon electrodes by using hemin as the electrochemical hybridization label, *Anal. Bioanal. Chem.*, 373, 710–716, 2002.

100. Tansil N. C., Xie H., Xie F., and Gao Z., Direct detection of DNA with an electrocatalytic threading intercalator, *Anal. Chem.*, 77(1), 126–134, 2005.

101. Hassmann J., Misch A., Schulein J., Krause J., Graβl B., Muller P., and Bertling W. M., Development of a molecular diagnosis assay based on electrohybridization at plastic electrodes and subsequent PCR, *Biosens. Bioelectron.*, 16(9–12) 857–863, 2001.

102. Wang J., Kawde A. N., Erdem A., and Salazar M., Magnetic bead-based label-free electrochemical detection of DNA hybridization, *Analyst.*, 126(11), 2020–2024, 2001.

103. Kara P., Ozkan D., Erdem A., Kerman K., Pehlivan S., Ozkinay F., Unuvar D., Itirli G., and Ozsoz M., Detection of achondroplasia G380R mutation from PCR amplicons by using inosine modified carbon electrodes based on electrochemical DNA chip technology, *Clin. Chim. Acta.*, 336(1–2), 57–64, 2003.

104. Marques L. P., Cavaco I., Pinheiro J. P., Ribeiro V., Ferreira G. N., Electrochemical DNA sensor for detection of single nucleotide polymorphisms, *Clin. Chem. Lab. Med.*, 41(4), 475–481, 2003.

105. Wang J., Jiang M., Fortes A., and Mukherjee B., New label-free DNA recognition based on doping nucleic-acid probes within conducting polymer films, *Anal. Chim. Acta.*, 402, 7–12, 1999.

106. Lassalle N., Mailley P., Vieil E., Livache T., Roget A., Correia J. P., and Abrantes L. M., Electronically conductive polymer grafted with oligonucleotides as electrosensors of DNA. Preliminary studies of real time monitoring by in situ techniques, *J. Electroanal. Chem.*, 509, 48–57, 2001.

107. Komarova E., Aldissi M., and Bogomolova A., Direct electrochemical sensor for fast reagent-free DNA detection, *Biosens. Bioelectron.*, 21(1), 182–189, 2005.

108. Peng H., Soeller C., Cannell M. B., Bowmaker G. A., Cooney R. P., and Travas-Sejdic J., Electrochemical detection of DNA hybridization amplified by nanoparticles, *Biosens. Bioelectron.*, 21(9), 1727–1736, 2005.

109. Li C. M., Sun C. Q., Song S., Choong V. E., Maracas G., and Zhang X. J., Impedance label less detection-based polypyrrole DNA biosensor, *Front Biosci.*, 10, 180–186, 2005.

110. Lee W. E., Thompson H. G., Hall J. G., and Bader D. E., Rapid detection and identification of biological and chemical agents by immunoassay, gene probe assay and enzyme inhibition using a silicon-based biosensor, *Biosens. Bioelectron.*, 14(10–11), 795–804, 2000.

111. Cha J., Han J. I., Choi Y., Yoon D. S., Oh K. W., and Lim G., DNA hybridization electrochemical sensor using conducting polymer, *Biosens. Bioelectron.*, 18(10), 1241–1247, 2003.

112. Pham M. C., Piro B., Tran L. D., Ledoan T., and Dao L. H., Direct electrochemical detection of oligonucleotide hybridization on poly(5-hydroxy-1,4-naphthoquinone-co-5-hydroxy-3-thioacetic acid-1,4-naphthoquinone) film, *Anal Chem.*, 75(23), 6748–6752, 2003.

113. Wei F., Sun B., Guo Y., and Zhao X. S., Monitoring DNA hybridization on alkyl modified silicon surface through capacitance measurement, *Biosens. Bioelectron.*, 18(9), 1157–1163, 2003.

114. Yang J. M., Bell J., Huang Y., Tirado M., Thomas D., Forster A. H., Haigis R. W., et al., An integrated, stacked microlaboratory for biological agent detection with DNA and immunoassays, *Biosens. Bioelectron.*, 17(6–7), 605–618, 2002.

115. Hindson B. J., McBride M. T., Makarewicz A. J., Henderer B. D., Setlur U. S., Smith S. M., Gutierrez D. M., et al., Autonomous detection of aerosolized biological agents by multiplexed immunoassay with polymerase chain reaction confirmation, *Anal Chem.*, 77(1), 284–289, 2005.

Appendix A

TABLE A.1
The Commercial Chemical Sensor Company/National Laboratory/Organization

Sr. No.	Company	Specialization	Sensor	References
1	AAI-Abtech (Yardley, PA, U.S.)	Chem-bio sensor technologies for the bioanalytical laboratory, monitoring and biomedical diagnostics markets	BioSenSys™ Multianalyte Workstation for biosensor-based immunodiagnostic assays	www.abtechsci.com
2	ABB Inc. Instrumentation Div.	Analytical instruments for the process, environmental, steam and power industries	NH_3, Cl_2, dissolved organics monitors, dissolved O_2, NH_3–NH_2 fluoride, hydrogen purity, oxygen, combustibles, etc.	www.abb.com/instrumentation
3	Abbott Laboratories	Continuum of care, from nutritional products and laboratory diagnostics through medical devices and pharmaceutical therapies	Handheld point of care diagnostics and blood glucose monitoring	www.abbott.com
4	Alpha MOS SA	Electronic digitalization of human senses, such as smell and taste	Olfaction products include PROMETHEUS, FOX, GEMINI, and KRONOS	www.alpha-mos.com
5	Alphasense Limited	Oxygen gas sensors and toxic gas sensor technologies	Oxygen and toxic gas electrochemical sensors, infrared detectors for CO_2, NDIR methane/HC etc.	www.alphasense.com
6	AMETEK Inc.	Control analyzers, industrial process control instrument, and measurement instruments	UV, Vis, and IR process analyzers, residual gas analyzers, gas chromatographs etc.	www.ametek.com
7	Animas	Medical devices and equipments	Insulin pump and glucose sensor	www.animascorp.com
8	AppliedSensor AB	Chemical sensor solutions for air quality, safety, and control	Hydrogen module, CH_4 sensor, CO, NO_2 sensor, flammable vapor, etc.	www.appliedsensor.com

TABLE A.1
Continued

Sr. No	Company	Specialization	Sensor	References
9	Applied Enzyme Technology Ltd.	Biosensors is a highly specialized area of protein stabilization sensors are being developed for the water and agrifood markets	NH3 sensor for water, pyruvate sensor for sweetness in onions, Organophosphate sensors, glucose sensors, etc.	www.aetltd.com
10	Argonne National Laboratory Environmental Science Division (EVS) (Argonne, IL)	Conducts applied environmental research, assessment, and technology development	Organic and metal film arrays on chemiresist transducers for gases, volatile organics, etc.	www.evs.anl.gov/index.cfm
11	Arthur D Little, Inc. (Cambridge, MA) Applied Biotechnology Laboratory	Detection of the Foodborne pathogens *E. coli* O157:H7 and Salmonella spp, warfare	Environmental endocrine disrupters, Foodborne pathogen Listeria monocytogenies, and quantitative assessment of biological warfare agents	Field Analytical Chemistry & Technology, 2(6), 371–377
12	Autoteam GmbH (Berlin, Germany)	Biosensor	BOD microbial electrode	www.auto-team.de
13	Biacore International AB	Scientific instruments that employ affinity-based biosensor technology	A range of sensor chips for detecting proteins, nucleic acids, carbohydrates, viruses or intact cells, etc.	www.biacore.com
14	Biosensors International Group, Ltd.	Manufactures and commercializes innovative medical devices used in interventional cardiology and critical care procedures	Biotrans II Transducer Kits, Thermodilution and Pulmonary Artery Catheters etc.	www.biosensorsintl.com
15	Bio Sensor Technology GmbH	Glucose and lactate membrane type and thick film type biosensors for clinical analyzers	Thick film type biosensors, membrane type biosensors, Glukometer, LactatProfi	www.bst-biosensor.de
16	Biostar Inc. (Boulder, CO) Sekos Inc.	Diagnostic immunoassays, focuses on producing products for the health care and scientific communities	Integrated physiologic sensor system, structural stability sensor	www.biostargroup.com
17	Bayer Health care (Ascensia contour)	Diabetes sensor	Bayer diabetes	www.ascensia.ch/pub/de/home.asp

No.	Company	Description	Website
18	Bio-Technical Resources (Manitowoc, WI)	Development of microbial fermentation and biocatalysis-based processes	Biotechnology of ascorbic acid manufacture, production of glucosamine etc.	www.biotechresources.com
19	BioSensor, LLC, (Honolulu, HI)	Biometric Verification Security System™	Card finger print matching system	www.biosensorhawaii.com
20	Cambridge Consultants Ltd (Cambridge, U.K.)	Healthcare, industrial and consumer products, telecom, informatics, media, and electronics	Biosensor for ATP (bacterial growth), antibody-analyte reactions	www.cambridge consultants.com
21	Central Kagaku Corp. (Tokyo, Japan)	Biosensor Redox, microbial-based electrode	Analyzer and apparatus for water, air, and soil: BOD, COD, VOC, DO, TOC	www.aqua-ckc.jp
22	Cholestech Corp. (Hayward, CA) (Inverness Medical Innovations Incorporated)	Diagnostic tools and immediate risk assessment and therapeutic monitoring of heart disease, diabetes, and other chronic diseases	Cholestech LDX® and GDX™ Systems for testing cholesterol and related lipids, blood glucose and glycemic control, inflammation, and liver enzymes	www.cholestech.com and www.inverness-medical.com
23	City Technology Ltd. (Hampshire, U.K.)	Gas sensors: over 300 sensors, 28 gases, 10 product ranges	Electrochemical sensors, oxygen sensors, toxic sensors, pellistors, infrared sensors	www.citytech.com
24	Cranfield University	Health, environment and water, manufacturing, aerospace, automotive, defense, engineering, and management	Biomedical sensors and biosensors has been built up over twenty-five years etc.	www.cranfield.ac.uk/health
25	Controle Analytique	Trace gas analyzers offering gas measurements by integrating a complete solution	Argon gas, HC and N_2 and chromatographs for impurities in noncorrosive bulk gases etc.	www.cai-ca.com
26	Delphi Ventures	Help build healthcare companies	A provide capital, contacts and leadership to help entrepreneurs realize their vision	www.delphiventures.com
27	Delphian Corporation (Northvale, NJ)	Gas monitoring systems	Catalytic bead sensors (combustible gases), electrochemical sensors (toxic gases & oxygen), etc.	www.delphian.com
28	DENSO Corporation	Advanced automotive technology, systems, and components for all the world's major automakers	NO_x sensor, oxygen sensor, air/fuel ratio sensor	www.globaldenso.com
29	Detcon Inc.	A wide range of industrial grade gas detection sensors, control systems, and process analyzers	Solid state MOS H_2S sensors, electrochemical sensors for a long list of toxic gases, catalytic and infrared sensors for the detection of combustible gas etc.	www.detcon.com

TABLE A.1
Continued

Sr. No	Company	Specialization	Sensor	References
30	DuPont Diagnostic Systems (Glasgow, DE)	DuPont Biosensor Materials are designed for use in medical monitoring, diagnostics, drug delivery, food and beverage testing, environmental sensors	Biosensors, ion selective sensors, medical electrodes, PTF sensors, etc.	www.dupont.com
31	Dräger Safety AG & Co. KgaA (Draegerwerk AG & Company KGaA)	Breathing and protection equipment as well as gas detection and analysis systems	Mobile and fixed gas detection systems equipped with highly sensitive sensors, which detect a wide range of hazardous substances at trace levels	www.draeger.com
32	DropSens, S.L.	Commercializes electrochemical biosensors based on thick-film hybrid technology (screen-printed electrodes)	Point Of Care biosensors, enzymatic sensors, immunosensors and DNA sensors	www.dropsens.com
33	Emerson Electric Company	Sensor (humidity sensor, pH)	Dew point/humidity sensor, pH sensor, etc.	www.emerson.com
34	EnviteC-Wismar (Honeywell)	Sensors and high-quality monitoring equipment for use in healthcare and the manufacturing industry, as well as in the fields of environmental and safety technology	Oxygen sensor, NO_x sensor	www.envitec.com
35	e2v technologies Limited	Sensing products for applications including fire, rescue and security thermal imaging, X-ray spectroscopy, and military surveillance, targeting, and guidance	Biosensors, gas sensors	www.e2v.com
36	Ercon Incorporated	Manufacturer and preferred source of high-performance coating materials designed for diagnostic biomedical electrodes and electrochemical sensors etc.	Sensors for the detection, quantification, and monitoring of bodily fluid and environmental constituents and in other healthcare diagnostics and therapeutics	www.erconinc.com

#	Company	Description	Application/Products	Website
37	Figaro Engineering Inc.	The Figaro Gas Sensor is a gas sensitive semiconductor	Methane, propane, CO, hydrogen, etc., volatile organic vapors (alcohol, ketones, esters, benzols, etc.), and many others	www.figarosensor.com
38	First Sensor Technology	Focus on producing pressure sensors with an operating temperature of up to 225°C	Piezoresistive pressure sensors	www.first-sensor.com
39	FIS Incorporated	Semiconductor gas sensors aiming to realize a gas detection system, which is equivalent to or superior to a human sense of smell	Semiconductor gas sensors for methane, propane, butane, ammonia, alcohol, hydrogen, CO, VOCs, etc.	www.fisinc.com
40	GE Sensing (General Electric Company)	Sensing elements, devices, instruments, and systems that enable our customers to monitor, protect, control, and validate their critical processes and applications	Sensors for gases, moisture, humidity, etc.	www.gesensing.com
41	Palintest Ltd (Halma plc)	Hazard detection and life protection	Chemical sensor for heavy metals	www.palintest.com www.halma.com
42	Genzyme Diagnostics	Different areas; critical raw materials, finished reagent kits for use on clinical chemistry analyzers, and point of care rapid tests	Manufacturers and suppliers of enzymes and chemical substrates to the diagnostics industry	www.genzymediagnostics.com
43	Gwent Electronic Materials Ltd (GEM)	Diagnostic sensor business	Manufacturing materials for the electronics and associated industries, biosensor materials etc.	www.g-e-m.com
44	Home Diagnostics Incorporated	Blood glucose monitoring systems and disposable supplies for people with diabetes worldwide	Blood glucose meters	www.homediagnosticsinc.com
45	Honeywell Sensing and Control	Test and measurement sensors, pressure transducers, torque sensors, instrumentation, accelerometers, and displacement sensors	Humidity sensor, oxygen sensor, pH sensor, etc.	sensing.honeywell.com
46	Horiba Limited	An extensive array of instruments and systems for applications ranging from automotive R&D, process and environmental monitoring, in vitro medical diagnostics	pH meter, ion meter, gas analyzer devices, blood analyzers, and food analyzers	www.horiba.com
47	ICx Technologies	Surface plasmon resonance biosensors for the determination of the affinity and kinetics of biomolecular interactions	Protein kinetics analysis as well as studies of protein-small molecule interactions	www.discoversensiq.com

TABLE A.1
Continued

Sr. No	Company	Specialization	Sensor	References
48	i-STAT Corp. (Princeton, NJ)	Medical diagnostic products for blood analysis, providing critical diagnostic information to health care professionals accurately and immediately at the point of patient care	Blood electrolytes, glucose, urea, nitrogen, hematocrit polymeric and enzyme membranes on ISE-based silicon arrays	www.i-stat.com
49	International Sensor Technology	Fixed and portable instruments for the detection of over 150 different toxic and combustible gases	Solid-state, electrochemical, infrared, photoionization, and catalytic bead sensors	www.intlsensor.com
50	Life Scan (Milpitas, CA) (Johnson & Johnson)	Blood glucose monitoring with the introduction of OneTouch® Technology	Enzyme (glucose oxidase/peroxidase) sensor	www.lifescan.com www.jnj.com
51	Manning Systems (Honeywell)	Sensor technologies for gas specific sensing and "stand-alone" gas safety-monitoring readouts	Various gas sensors	www.manningsystems.com
52	MediSense Inc. (Waltham, MA)	Self-testing blood glucose monitoring systems that enabled people with diabetes to manage their disease effectively	Blood glucose electrode	www.abbottdiagnostics.com
53	Metrika (Bayer)	Portable A1C testing systems for managing diabetes and tracking blood glucose levels over time	A1CNow+ diabetes monitoring system	www.a1cnow.com
54	Metrohm	Ion analysis technologies under one roof	Solid-state working electrodes, pH, ion, and conductivity meters	www.metrohm.com/com
55	Microbics Corp. (Carlsbad, CA)	Light-emitting microorganisms, such as toxins, mutagens (screening)	Fluorescence probe for luminescent bacteria	www.usmicrobics.com www.bugsatwork.com
56	Microsensor Systems Inc.	Surface acoustic wave (SAW) based chemical sensing	SAW sensors for organophosphates, volatile organics, ammonia	www.microsensorsystems.com

#	Company	Description	Application	URL
57	Mine Safety Appliances Company	Sophisticated safety products designed to protect people throughout the world	Sensors for CO, CO_2, NO_2, chlorine, SO_2, and other toxic gases	www.msanet.com
58	Mitsubishi Electric Corp.	Applications in a wide range of fields—from home electronics products, to automobile fuel jet devices	pH-sensitive FET sensor for glucose, triglycerides	www.mitsubishi.com
59	Molecular Devices Corp.	Microarray scanners	BioMicro Systems—Microarray hybridization, Phalanx Biotech—Human and mouse gene expression microarrays etc.	www.moleculardevices.com
60	MST Technology (Honeywell)	Detection capabilities for monitoring toxic gas and chemicals	Sensors for a wide range of toxic, corrosive, and combustible gases	www.mst-technology.com
61	Naval Research Laboratory	Surface acoustic wave (SAW) sensor as noses for gas detection and identification	SAW sensor systems are currently being used to monitor hazardous chemical vapors, chemical warfare agents, potential fires, and environmental pollutants	www.nrl.navy.mil
62	Neotronics Scientific	Electronic nose system	Neotronics NOSE for odors, aromas	www.neotronics.com
63	Nemoto & Company Limited	Safety, security, and health for the benefit of the society	Sensors for methane, butane, hydrogen, CO, CO_2, alcohol, ammonia, H_2S, NO_2, organic solvent	www.nemoto.co.jp
64	NGK Spark Plug Company Limited	Original equipment oxygen sensors	NTK oxygen sensors	www.ngksparkplugs.com
65	Nova Biomedical Corporation	Fast whole blood analyzers to support the care of critically ill patients	Blood testing analyzers, chemistry analyzers for biotechnology, self testing diabetes products	www.novabiomedical.com
66	Novatech Controls Pty Limited	Oxygen probes, analyzers, and sensors	Oxygen sensing probes, oxygen analyzers, water vapor analyzers, portable O_2/CO_2 analyzers, oxygen sensor	www.novatech.com.au
67	Oak Ridge National Laboratory (Oak Ridge, TN)	Chemical and radiological sensors	Metal oxide film arrays on a chemiresist transducer for gases, volatile organics	www.csm.ornl.gov/PR/NS01-10-03.html
68	Omron Electronics LLC	Measuring and testing sensor technology, as well as vision inspection systems	Surface plasmon resonance biosensor chip	www.omron247.com and www.omron.com
69	Oriental Electric Co. Ltd. (Tokyo, Japan)	Enzyme-based biosensor system for monitoring the freshness of fish	Freshness meter Clark oxygen electrode	United States Patent 5288613

TABLE A.1
Continued

Sr. No	Company	Specialization	Sensor	References
70	Osmetech Molecular Diagnostics: Osmetech plc	Electronic odor sensor ("e-nose") technology	eSensor® XT-8 system for nucleic acids, eSensor® CFCD for PCR amplification and mutation detection	www.osmetech.com
71	Pegasus Biotechnology (Agincourt, Ontario, Canada)	Amperometric enzyme electrode to indicate fish and meat freshness	Microfresh analyzer	Journal of Food Science, 2006, 57(1), 77–81
72	PalmSens, The Netherlands	Electrochemical (bio-) sensors, BVT technologies	Platinum sensor immobilized with acetyl-cholinesterase, platinum sensor immobilized with GOD (glucose oxidase)	www.palmsens.com
73	Pharmacia Biosensor AB	Label-free surface plasmon resonance (SPR) based technology for studying biomolecular interactions in real time.	Antibody-coated SPR-based instrument	www.biacore.com
74	Physical Sciences Incorporated	Technologies for aerospace, energy, environmental, manufacturing, and medical markets	Ascent ambient gas sensor, MWGS-1 multi-wavelength gas sensor; pharmaceutical process etc.	www.psicorp.com
75	Polymer Technology Systems Incorporated	Diagnostic tests for chronic diseases such as heart disease, diabetes, and other related medical conditions	Cartouche® cholesterol analyzer, cholesterol strips, HDL strips, triglycerides strips	www.cardiochek.com
76	Prufgerate-Werk Medigen GmbH	Enzyme electrode for blood glucose, lactate	BOD biosensor, glucose analyzer ESAT 6660-2	
77	Q-Sense	Quartz crystal microbalance with dissipation monitoring (QCM-D) technology	Q-Sense E4, Q-Sense Modules, etc.	www.q-sense.com
78	RAE Systems Incorporated	Multisensor chemical and radiation detection monitors and networks for industrial applications and homeland security	Various gas and radiation detection products for personal and area monitoring	www.raesystems.com
79	Roche Holding Limited	DNA tests, diabetes monitoring supplies, and point-of-care diagnostics used in a variety of health care settings	Blood gas analysis, cholesterol and diabetes monitoring, etc.	www.roche.com

80	Rosemount Analytical Inc.	Combustion analysis, process analysis, and emissions monitoring analyzers and solutions	NO/NO$_x$, O$_2$, hydrocarbon analyzers	www.processanalytic.com www.emersonprocess.com/raihome/gas
81	Sandia National laboratories,	Microchemical sensors for in situ monitoring and characterization of volatile contaminants	Develops microchemical sensors for in-situ monitoring of subsurface contaminants	www.sandia.gov/sensor
82	Seiko Instruments (Chiba, Japan)	Membrane-based PZT sensor	Lipid membrane coated piezoelectric crystals: odorants (amyl acetate, citral)	www.sii.co.jp United States Patent 4789804
83	Sensata Technologies BV	Sensors with ceramic capacitive and hermetic technologies for a full range of automotive and heavy vehicle off-road systems	Spreeta SPR sensor, refractive index sensor	www.sensata.com
84	SenseAir® AB	Infrared gas sensors and controllers for both stand-alone operations and integration in other systems	SenseAir® CO$_2$ sensors	www.senseair.com
85	Sensidyne	Detection of toxic and combustible gases, and oxygen enrichment/deficiency in a wide variety of industries	Fixed gas detection systems, gas detector tubes	www.sensidyne.com
86	Sensorex	Quality pH electrodes, ORP electrodes, conductivity sensors, dissolved oxygen probes, chlorine dioxide sensors	pH and ORP electrodes, conductivity sensors, dissolved oxygen sensor line, and sanitizer sensors	www.sensorex.com
87	Servomex (Spectris plc)	World expert in gas analysis	Gas sensors for O$_2$, CO, CO$_2$, methane	www.servomex.com www.spectris.com
88	Sierra Monitor Corp.	Hazardous gas detection systems for combustibles, oxygen deficiency, and toxic gases with digital link to central control system for plant-wide monitoring	Catalytic bead combustible gas; infrared sensors electrochemical sensors for oxygen, and toxic gas	www.sierramonitor.com
89	Smiths Group plc	Sensors to detect and identify explosives, narcotics, weapons, chemical agents, biohazards, and contraband	Automated hardware/software systems to detect biological and chemical agents, explosives/narcotics, etc.	www.smiths-group.com www.smithsdetection.com

TABLE A.1
Continued

Sr. No	Company	Specialization	Sensor	References
90	Teledyne Analytical Instruments	Gas and liquid analyzers for industrial, OEM, and medical applications	Oxygen sensors	www.teledyne-ai.com www.teledyne.com
91	Texas Instruments, Analytical Sensors	Innovative analog and DSP technologies, along with our other semiconductor products	Blood pressure, glucose levels, pulse, tidal carbon dioxide, and various other biometric values	www.ti.com
92	ThermoMetric AB (TA Instruments–Waters LLC)	Detection of changes in temperature due to immobilized enzyme-analyte reactions	Organic acids, urea, sugars, antibiotics, heavy metals, etc.	www.tainstruments. com
93	Thermo Electron Corp.	Thermo Scientific help solve your toughest analytical challenges with our world-class portfolio of products and technologies	Drugs-of-abuse testing and drug monitoring, including immunosuppressant drug testing; antiepileptic drug monitoring; organ transplantation drug monitoring; thyroid hormone testing etc.	www.thermo.com
94	TOA Electronics Ltd.	Analytical and electronic instrumentation	Water quality analyzers, ambient air analyzers	www.dkktoa.co.uk
95	Tokuyama Corporation			www.tokuyama.co.jp/ eng
96	Transducer Research Inc.	Polymeric films on Ion-selective electrodes	CO, CO_2, O_3, H_2S, SO_2, NO_2, NO, Cl_2	www.universalsensors. co.uk
97	Universal Sensors Ltd.	Instrument suitable for analytical/diagnostic laboratories, hospitals, etc.	Polymer coated potentiometric biosensors	
98	Universal Sensors Inc.	Biosensors, electrochemistry, and general analytical support	Amperometric electrode instruments, non-enzyme electrodes, amperometric enzyme electrodes, potentiometric enzyme electrodes	intel.ucc.ie/sensors/ universal
99	VTI-Valtronics Incorporated	Gas-measuring sensors utilizing its core technology of nondispersive infrared	Various CO_2 gas sensors	www.val-tronics.com
100	YSI Life Sciences (YSI Incorporated)	Scientific instruments, sensors, and systems that serve a variety of scientific, environmental, and industrial markets worldwide	YSI enzyme electrode for xylose and glucose	www.ysilifesciences. com www.ysi.com

Appendix B

LIST OF ACRONYMS

CHAPTER 1

ANN	Artificial neural network models
APCVD	Atmospheric-pressure chemical vapor deposition
CeO_2	Cesium oxide
CH_4	Methane
CNTs	Carbon nanotubes
CO	Carbon monoxide
FET	Field effect transistor
GaAs	Gallium arsenide
H2	Hydrogen
H_2S	Hydrogen sulfide
ICs	Integrated circuits
In_2O_3	Indium oxide
$In_2O_3-SnO_2$	Indium oxide-tin oxide
LEL	Lower exposure limits
LPFD	Low-pressure flame deposition
LPG	Liquefied petroleum gas
MOD	Metal organic deposition
MOS	Metal-Oxide-semiconductor
MOSFET	Metal-Oxide-Field Effect Transistor
MWNTs	Multi-walled nanotubes
$Na_2Zr_2Si_3PO_{12-}$	NASICON
Nb_2O_5	Niobium oxide
$NO_X:$	Nitrogen oxides
NO2	Nitrogen dioxide
NO	Nitric oxide
O2	Oxygen
PECVD	Plasma enhanced chemical vapor deposition
Ppb	Parts-per-billion
PVD	Physical vapor deposition
r.f. PECVD	Radio frequency plasma enhanced chemical vapor deposition
SnO_2	Tin oxide
SO_2	Sulfur dioxide
$SrTiO_3$	Strontium Titanate ($SrTiO_3$)
SWNT	Single-walled nanotubes
TiO_2	Titanium oxide

VOC	Volatile organic compound
WO_3	Tungsten oxide
YSZ	Yttria-stabilized zirconia
ZnO	Zinc oxide

CHAPTER 2

Ag	Silver
CdO	Cadmium oxide
C_2H_5OH	Ethanol
CH_3OH	Methanol
CNFs	Carbon nanofibers
CO	Carbon monoxide
Cu	Copper
DMMP	Dimethyl methylphosphonate
EPA	Environmental Protection Agency
EU	European Union
Fe_2O_3	Iron oxide
FTIR	Fourier Transform Infrared spectroscopy
GC-AED	Gas chromatography-atomic emission detection
GC-IR-MS	Gas chromatography-infrared detection-mass spectral detection
HCl	Hydrochloric acid
H_2SO_4	Sulphuric acid
In_2O_3	Indium oxide
LaF_3	Lanthanum fluoride
$LaFeO_3$	Lanthanum iron oxide
LB	Langmuir-Blodgett
LBL	Layer-by-layer
LC-MS	Liquid chromatography-mass spectrometry
LS	Langmuir-Schaefer
MWNTs	Multiwalled carbon nanotubes
NH_3	Ammonia
$(NH_4)_2S_2O_8$	Ammonium persulfate
Ni	Nickel
NMR	Nuclear magnetic resonance
NO	Nitric oxide
O3	Ozone
OAQPS	Office of Air Quality Planning and Standards
OP	Organophosphate
PAn	Polyacrylonitrile
PANI	Polyaniline
PbPc LuPc	Phthalocyanines
PCl_3	Phosphorous trichloride
PC	Polycarbonate
Pd	Palladium
PDDA	Poly(diallyldimethylammonium chloride)

PEOA	Poly(o-ethoxyaniline)
PHTh	Polythiophenes
PM	Particulate Matter
PMMA	Poly(methyl methacrylate)
POAS	Poly(ortho-anisidine)
PPY	Polypyrrole
PS	Polystyrene
PSS	Polystyrene sulphonates
Pt	Platinum
PTAA	Poly(thiophene acetic acid)
Rh	Rhodium
Ru	Ruthenium
SAM	Self-assemble monolayers
SnO_2	Tin oxide
SO_2	Sulphur dioxide
SPAn	Sulfonated polyaniline
Ta_2O_5	Tantalum oxide
TiO_2	Titanium oxide
VDP	Vapor deposition polymerization
WO_3	Tungsten oxide
ZrO_2	Zirconium oxide

CHAPTER 3

CR-Pc_2Lu	Octa-(15-crown-5)-lutetium bisphthalocyanine
CT	Charge-transfer
CuPc A_2	Tris-(2,4-di-t-amylpheoxy)-(8-quinolinoxy) CuPc
CuPaz(t-Bu)$_4$	Copper (II) tetra-(tert-butyl)-5,10,15,20-tetraazaporphyrin
Cu_2Pc_2	Binuclear phthalocyanine
CuPaz	Porphyrazine
DNFMB	2,4-dinitrotrifluoromethoxybenzene
FET	Field-effect transistor
LB	Langmuir-Blodgett
$LuPc_2$	Lutetium bisphthalocyanine
MPc	Metallophthalocyanines
MPcR$_4$	Metal phthalocyanines
MPcR$_8$	Mesogenic octa-substituted phthalocyanine derivatives
MPP	Metalloporphyrins
M-PPIX	Metal complexes of protoporphyrin IX dimethyl ester
OEP	Octaethylporphyrins
PCA	Principal component analysis
PCs	Substituted phthalocyanines
QCM	Quartz crystal microbalance
SAW	Surface acoustic wave
SPR	Surface plasmon resonance
TFBAR	Thin-film bulk acoustic wave resonators

TPP	Tetraphenylporphyrins
TSMRs	Shear mode resonators
VOCs	Volatile organic compounds
ZnPc	Tetra-4-(2,4-di-t-amylphenoxy)

CHAPTER 4

AlN	Aluminum nitride
BVP	Boundary value problem
FEA	Finite element analysis
FEM	Finite Element Models
FPW	Flexural plate wave
GaAs	Gallium arsenide
IDT	Inter-digital transducer
LGS	Langasite
$LiNbO_3$	Lithium niobate
$LiTaO_3$	Lithium tantalate
PVF	Polyvinylidene fluoride
PZT	Lead zirconium titanate
SAW	Surface acoustic wave
SH	Shear-Horizontal
SiC	Silicon carbide
TSM	Thickness shear mode
ZnO	Zinc oxide

CHAPTER 5

AChH	Acetylcholinesterase
Ag-Ab	Antigen–antibody
ANN	Artificial neural networks
BOD	Biochemical oxygen demand
BuChE	Acetylcholine and butyrylcolinesterase
BuTCH	Butyrylthiocholine
ChE	Cholinesterase
ChO	Choline oxidase
CoPC	Cobalt phtalocyanine
EW	Evanescent wave
FIA	Flow injection analysis
GMP	Good manufacturing practice
HACCP	Hazard Analysis Critical Control Point
HRP	Horseradish peroxidise
ISFET	Ion selective field-effect transducers
LAPS	Light-addressable potentiometric sensors
MIP	Molecularly imprinted polymers
PAHs	Polycyclic aromatic hydrocarbons
QA/QC	Quality assurance and quality control

SPR	Surface plasmon resonance
TCNQ	Tetracyano quinodimethane
VOCs	Toxic volatile organic compounds

CHAPTER 6

ANN	Artificial neural network
ArPU	Aromatic polyurethane
BAW	Piezoelectric bulk acoustic wave
CA	Cluster analysis
DFA	Discriminant factorial analysis
ENFET	Enzyme FET
FMS	Fingerprint mass spectrometer
IMS	Ion-mobility spectrometry
ISE	Ion-selective electrodes
ISFET	Ion-selective field affects transistors
LAPV	Large amplitude pulse voltammetry
LDA	Linear discriminate analysis
MO	Metal oxide
MLAPS	Multiple light-addressable potentiometric sensor
PCA	Principal components analysis
PR	Pattern recognition
SAPV	Small amplitude pulse voltammetry
SAS	Sensor array system
SAW	Surface acoustic wave sensors
SIMCA	Soft-independent modeling of class analogy
SPR	Surface plasmon resonance
SPME	Solid phase microextraction
TSM	Thickness shear mode
VCT	Vacuum cook-in-bag/tray technology
VSC	Volatile sulfur compound

CHAPTER 7

AFM	Atomic force microscopy
CM	Carboxymethylated
DDI	DNA-directed immobilization
ELISA	Enzyme-linked immunosorbent assay
Fc-PDA	N-(10,12-pentacosadiynoic)-acetylferrocene
GST	Glutathione S-transferase
HACCP	Hazard Analysis at Critical Control Point
Ha-tag	N-terminal haemagglutinin
His	Histidine
IgG	Immunoglobulin G
LC	High performance liquid chromatography
LODs	Limits of detection

MBP	Maltose binding protein
MS	Gas chromatography coupled with mass spectrometry
PCR	Polymerase chain reaction
PSA	Prostate-specific antigen
RIA	Radioimmunoassay
RID	Radial immunodiffusion
ScFv-cbd$_2$	Two chitin-binding domains.
SPR	Surface Plasmon Resonance
TIR	Total internal reflection
QCM	Quartz Crystal Microbalance

CHAPTER 8

BCA	Breath Collecting Apparatus
GC	Gas chromatography
GC–MS	GC linked with mass spectrometry
PLS	Partial Least Squares
PLS-DA	Partial Least Squares-Discriminant Analysis
VOCs	Volatile organic compounds

CHAPTER 9

ACNA	Associated National Chemical Companies
ANOVA	A one-way analysis of variance
carboplatinum	Diammine(1,1-cyclobutanedicarboxylate)platinum(II) $[Pt(NH_3)_2C_4H_6C_2O_4]$
cisplatin	*cis*-diaminedichloroplatinum(II) *cis*-$[Pt(Cl)_2(NH_3)_2]$
dsDNA	Double-stranded
FF	Fluorescence
LSD	Fisher's least-significant difference
NAMI-A	Imidazolium *trans*-imidazoledimethylsulfoxidetetrachloro-ruthenate $[Ru(III)Cl_4(DMSO)(Im)][ImH]$
NARA	National Animal Research Authority
ssDNA	Single-stranded DNA
SPE	Screen printed electrodes
titanocene	Titanocene dichloride $[(\eta^5\text{-}C_5H_5)_2TiCl_2]$

CHAPTER 10

2ADNT	2-amino-4, 6-dinitrotoluene
4ADNT	4-amino-2, 6-dinitrotoluene
Ag/AgCl	Silver/silver chloride
BSA	Bovine serum albumin
DANT	Diamino isomer, 2,4-diamino-6-nitrotoluene
FMN	Flavine-adenine mononucleotide

NAD	Nicotinamide adenine dinucleotide
NADH	Nicotinamide adenine dinucleotide
NADH·H+	Nicotinamide adenine dinucleotide
NADPH·H+	Nicotinamide adenine dinucleotide phosphate
RDX	Cyclo-1, 3,5-trimethylene- 2,4,6-trinitramine
TAT	Triaminotoluene
TNT	2,4,6-trinitrotoluene
UXO	Unexploded ordinance

CHAPTER 11

2,4-DCPAA	2,4-dichlorophenoxyacetic acid
AchE	Acetylcholinesterase
ATPh	4-aminothiophenol
CCD	Charge-Coupled Device
CHO	Choline oxidase
CNTs	Carbon nanotubes
CWA	Chemical warfare agent
DTSP	3,3'-dithiopropionic acid di-(N-succinimidyl ester)
EIS	Electrochemical impedance spectroscopy
HRP	Horseradish peroxidase
IDA	Interdigitated array
MUA	11-mercaptoundecanoic acid
OPCs	Organophosphorous compounds
OPH	Organophosphate hydrolase
OVA	Ovalbumin
OWLS	Optical waveguide light mode spectroscopy
PANI	Polyaniline
PGA	Polygalacturonic acid
PVSA	Poly(vinylsulphonic acid)
scAb	Single-chain antibody

CHAPTER 12

AchE	Acetylcholinesterase
CA	Carbonic anhydrase
DMSO	Dimethyl sulfoxide
EWAS	Evanescent wave absorbance spectroscopy
GA	Glutaraldehyde
GOD	Glucose oxidase
LED	Light emitting diode
OPAA	Organophosphorus acid anhydrase
OPH	Organophosphorus hydrolase
TPPC$_4$	Meso-tetra(4-carboxyphenyl)porphine
TPPS$_1$	Monosulfonate tetraphenyl porphyrin

CHAPTER **13**

APDS	Autonomous pathogen detection system
ELISA	Enzyme-linked immunosorbent assay
FRET	Fluorescence resonance energy transfer
MAAB	Multi-Analyte Array Biosensor
PCR	Polymerase chain reaction)
RTA	Ricin A-chain
Ru(bpy)$_3^{2+}$]	Ruthenium trisbypyridine
SEB	Staphylococcal enterotoxin B

Index

Note: Page numbers in *italics* and **bold** refer to figures and tables, respectively.